Shidlovskiy

PRINCIPLES OF PYROTECHNICS

Third Edition (revised and enlarged)

Approved by the Ministry of Higher and Intermediate

Special Education, RSFSR, as a textbook for

technological curricula at colleges of chemical technology

1964

NOTICE TO READERS
Concerning any offer found in this publication to sell or transfer products or information that are subject to governmental regulation, such sales or transfers of the product or information will be made in accordance with Federal, State and local laws applicable to the buyer or transferee, and further, explosives transfers are prohibited to felons, fugitives, juveniles and some other persons, as determined by such regulation.

American Fireworks News

Editorial Notes

This document has been prepared from two translations of the original text. No editorial changes have been made to the author's work, except those accidentally occurring during the translations and subsequent rekeying.

The original Russian textbook, *Osnovy Pirotekhniki (1964)* was translated into English by two U.S. military offices: 1. Translation Division, Foreign Technology Division, United States Air Force, Wright Patterson AFB, 23 Oct 64, Publication FTD-HC-23-1704-74; and 2. U.S. Joint Publications Research Service, United States Army, Feltman Research Laboratories, Picatinny Arsenal, May, 1965, Technical Memorandum 16154.

Both documents may be obtained from the National Technical Information Service as their publications AD-A-001 859 and AD-462 474.

When preparing this version, the staff of American Fireworks News used both translations, comparing each paragraph and selecting the more easily understood text. Considerable care was taken to discern the author's intentions when ambiguities were discovered in the translations. Readers finding such problems are urged to obtain the translation documents mentioned above and form their own conclusions.

Among the improvements made for this reprinting was the addition of an extensive subject index. The publishers gratefully acknowledge the assistance of Dr. Ken Kosanke (PyroLabs, Inc.) in making these significant improvements to the text.

The Staff
American Fireworks News
June, 1997

© 1997, Rex E. & S.P., Inc.

ISBN 0-929931-13-0

Printed in U.S.A.

PRINCIPLES OF PYROTECHNICS

Table of Contents

Chapter	Page
Editorial Notes	2
Preface	7
Forward	8

Part One

1. GENERAL CONCEPT OF PYROTECHNIC AGENTS AND PYROTECHNIC COMPOSITIONS
  1. Classification of Pyrotechnic Devices and Compositions ... 9
  2. Combustion of Compositions ... 10
  3. Requirements Placed upon Pyrotechnic Agents and Compositions ... 12
  4. Functions of the Components of a Composition ... 13
  5. Possible Highly Exothermic Reactions ... 14
  6. Combustibility of Various Substances & Mixtures ... 16

2. OXIDIZERS
  1. Choice of Oxidizers and Their Classification ... 18
  2. Physico-Chemical Properties of Oxidizers ... 20
  3. Hygroscopicity of Oxidizers ... 25
  4. Technical Requirements Imposed on Oxidizers ... 27
  5. Production of Oxidizers ... 27

3. FUEL SUBSTANCES
  1. Selection and Classification of Fuels ... 30
  2. High-Energy Fuels ... 32
  3. Technical Requirements Placed on Metal Powders ... 38
  4. Production of Metal Powders ... 39
  5. Inorganic Fuels of Medium Calorific Value ... 40
  6. Organic Fuels ... 41

4. METHODS OF COMPACTION OF COMPOSITIONS AND TESTING OF THE STRENGTH OF PELLETS. ORGANIC POLYMER BINDERS
  1. Methods of Compaction of Compositions; Uses of Binders ... 44
  2. Strength Testing of Pellets ... 45
  3. Factors Affecting the Strength of Pressed Compositions ... 46
  4. Classification of Binders and Their Physicochemical Properties ... 47
  5. Amount of Binder Introduced into the Compound ... 50

5. PRINCIPLES OF FORMULATION AND CALCULATION OF PYROTECHNIC COMPOSITIONS
  1. Calculation of Binary Mixtures ... 52
  2. Formulation and Calculation of Ternary and Multicomponent Mixtures ... 54
  3. Calculation of Compositions with a Negative Oxygen Balance ... 55
  4. Calculation of Metal Chloride Compositions ... 58
  5. Calculation of Compositions with a Fluorine Balance ... 60

6. HEAT OF COMBUSTION OF PYROTECHNIC COMPOSITIONS
1. Calculation of the Heat of Combustion 61
2. Experimental Determination of the Heat of Combustion 65
3. Relationship between the Use of Compositions and Their Heat of Combustion 66

7. GASEOUS COMBUSTION PRODUCTS OF PYROTECHNIC COMPOSITIONS 67

8. COMBUSTION TEMPERATURE OF PYROTECHNIC COMPOSITIONS
1. Calculation of Combustion Temperature of Compositions.................................... 71
2. Experimental Determination of the Combustion Temperature 76
3. Relationship between the Use of a Composition and the Maximum Combustion Temperature.. 82

9. SENSITIVITY OF PYROTECHNIC COMPOSITIONS
1. Determination of the Sensitivity of Compositions to Thermal Effects............... 83
2. Determination of the Sensitivity of Compositions to Mechanical Effects.......... 87
3. Factors Affecting the Sensitivity of Compositions to Initial Impulses................ 91

10. MECHANISM OF COMBUSTION AND FACTORS AFFECTING THE RATE OF COMBUSTION OF PYROTECHNIC COMPOSITIONS
1. Mechanism of Combustion of Compositions... 95
2. Factors Affecting Combustion Rate of Compositions 98

11. EXPLOSIVE PROPERTIES OF PYROTECHNIC COMPOSITIONS............................... 107

12. PHYSICAL AND CHEMICAL STABILITY OF PYROTECHNIC COMPOSITIONS
1. Changes Occurring in Compositions in Storage 114
2. Determination of the Hygroscopicity and Chemical Stability of Compositions............ 119
3. Decreases in the Efficiency of Compositions during Storage and Their Permissible Storage Lives.. 122

Part Two

13. ILLUMINATION COMPOSITIONS
1. Special Requirements Imposed on Illumination Compositions 123
2. Thermal and Luminescent Radiation... 125
3. Luminous Characteristics of Compositions... 127
4. Formulation of Binary Mixtures.. 129
5. Multicomponent Illumination Compositions 134
6. Principles of Flame Photometry and Radiometry of Pyrotechnic Compositions 137
7. Effect of Combustion Conditions on the Radiant Intensity of a Flame and Devices for Stationary Testing 142

14. PHOTOMIXTURES
1. Use of Photomixtures. Nighttime Aerial Photography. Photographic Materials. 146
2. Photoflash Bombs 148
3. Methods of Determining the Characteristics of Photoflashes............................ 149
4. Photomixtures. Requirements Placed on Them. Factors Affecting the Optical Characteristics of Flashes and Properties of Photomixtures...................... 155

15. TRACER COMPOSITIONS
1. Brief Information on the Design of Tracers. Illumination Calculations 156
2. Requirements Placed on Tracer Compositions 159
3. Formulas of Tracer Compositions. Testing Methods 160

16. SIGNAL-FLARE COMPOSITIONS
1. Signaling Systems. Requirements Place on Compositions.. 162
2. Character of the Flame Radiation 162
3. Principles of Formulation of Compositions and Basic Requirements Placed on Their Components 163
4. Yellow Flare Compositions 165
5. Red Flare Compositions 166
6. Green Flare Compositions 168
7. Blue and White Flare Compositions 170
8. Testing Methods 170

17. INCENDIARY COMPOSITIONS
1. Ignitable Materials 173
2. Classification of Incendiary Compositions and Requirements Placed on Them 174
3. Thermites 175
4. Thermite Incendiary Compositions 178
5. Incendiary Compositions with Salt Oxidizers 179
6. Liquid Petroleum Products & Solidified Combustibles. Flame-Thrower Mixtures 180
7. The "Electron" Alloy 183
8. Phosphorus and Its Compounds 183
9. Halogen Compounds of Fluorine 185
10. Other Incendiary Substances and Mixtures 186
11. Methods of Testing Incendiary Compositions. Extinction of Incendiary Compositions 187

18. COMPOSITIONS OF SMOKE-SCREEN
1. General Data on Aerosols 191
2. Optical Properties of Aerosols 193
3. Methods of Preparation of Aerosols 194
4. Compositions of Smoke-Screen and Requirements Imposed on Them 196

19. COMPOSITIONS OF COLORED SMOKES
1. Signaling by Means of Colored Clouds and Methods of Their Preparation 201
2. Dyes Used in Signal Smoke Compositions 203
3. Colored Smoke Compositions 205
4. Methods of Testing Colored Smoke Compositions 207

20. SOLID PYROTECHNIC FUELS
1. Classification and Energy Characteristics 209
2. Service Requirements 217
3. Oxidizers 218
4. Organic and Metallic Fuels 222

21. IGNITION COMPOSITIONS. OTHER TYPES OF PYROTECHNIC COMPOSITIONS
1. Ignition Compositions and Requirements Placed on Them 222
2. Ignition Composition for Rocket Engines 224
3. Other Types of Pyrotechnic Compositions: Gasless, Simulating, and Hissing Compositions 225

22. USE OF PYROTECHNIC COMPOSITIONS IN THE NATIONAL ECONOMY. MATCH COMPOSITIONS 228

23. PRINCIPLES OF TECHNOLOGY OF PYROTECHNIC PRODUCTION
  1. Preparation of Ingredients 235
  2. Preparation of Compositions 240
  3. Compacting (Molding) of Compositions 241
  4. Loading of Articles 243

Brief Sketch of the History of Development of Pyrotechnics in Russia 245

Appendices
  1. Heat of Formation of Oxides, Fluorides, Chlorides, and Sulfides of Certain
     Elements 247
  2. Formulas of Delay Compositions After Ellern 249
  3. Average Values of Pressure and Temperature at Different Heights at Moderate
     Latitudes 249
  4. Heat of Formation of Oxides as a Function of the Atomic Number of the Element 250

Bibliography 251

Index 265

Preface

The Third Edition of the book has been substantially revised and contains new material reflecting the advances made by pyrotechnics in recent years.

The book presents the general theoretical principles of pyrotechnics and modern methods of preparation and calculation of pyrotechnic compositions, and gives information on the properties of various types of pyrotechnic compositions.

It provides detailed descriptions of the properties of the components, fuels, and oxidizers; considerable attention is given to the physical nature of the combustion processes discussed. The properties of various types of pyrotechnic compositions (illumination and ignition compositions, solid rocket fuel compositions, etc.) are discussed separately.

This is a textbook for students in technological colleges. It may also be of interest to industrial scientific personnel, and engineers working in the area of pyrotechnics and related areas (explosives, powder, rocket design, etc.)

Foreword

Since the publication of the second edition (1954), major changes brought about by new advances in defense technology have taken place in pyrotechnics.

A number of new theoretical interpretations have been developed, and new areas of application of pyrotechnic compositions have appeared, particularly in the national economy and in rocket engineering.

In writing this book, the author made use of published works of Soviet as well as foreign experts.

The course presented in this book, entitled "Principles of Pyrotechnics," consists of two parts.
The first part gives a classification of pyrotechnic compositions and data on the principles of their development, and discusses the physicochemical properties common to all types of compositions. The second part deals with individual types of pyrotechnic compositions, their specific properties, and the requirements placed on them.

The description of the arrangement and action of pyrotechnic agents is presented in general outline only, since this is necessary to explain the requirements placed on the compositions.

The principles of technology involved are discussed and a brief sketch of the history of development of pyrotechnics in Russia is given. In addition, the book provides an extensive list of modern literature on pyrotechnics and related areas of knowledge: physics and chemistry of combustion, thermochemistry, illumination, chemistry, etc., with which pyrotechnic specialists are also concerned.

The author expresses his gratitude for their assistance in the preparation of this textbook to Cand. Techn. Sci. I. I. Vernidub, who wrote Chapter 14, Cand. Techn. Sci. Ye. S. Shakhidzhanov, who wrote Chapter 20, and to engineer A. V. Smetan, who wrote Chapter 23, paragraphs 6 and 7 of Chapter 13, and partially paragraph 2 of Chapter 8.

The author thanks Doct. Tech. Sci. Prof. V. G. Pavlyshin, Doct. Tech. Sci. Prof. K. A. Andreyev (deceased), Doct. Tech. Sci. N. F. Zhirov, Doct. Tech. Sci. N. A. Silin, Cand. Tech. Sci. N. N. Ivanova, Cand. Tech. Sci. N. A. Bil'dyukevich, Cand. Tech. Sci. I. M. Bocharskiy, Cand. Tech. Sci. I. V. Bystrov (deceased), Cand. Tech. Sci. N. P. Gadakchyan, Cand. Tech. Sci. A. G. Sokolov, Cand. Tech. Sci. I. A. Chelnokov, and Eng. I. V. Obez'yayev for much useful advice and valuable comments.

The author will be very grateful to readers for sending any comments or pointing out specific deficiencies of this book.

PRINCIPLES OF PYROTECHNICS

Part 1

Chapter 1

GENERAL CONCEPT OF PYROTECHNIC AGENTS AND PYROTECHNIC COMPOSITIONS

The word "pyrotechnics" is derived from the Greek words pyr (fire) and techne (art, skill). Pyrotechnics is the science dealing with the methods of production and properties of pyrotechnic compositions and hardware. When they burn (or explode), pyrotechnic compositions produce light, heat, smoke, sound, or reaction effects, which are utilized in military technology or in the national economy (industry, transportation, agriculture, filmmaking, etc.). In addition, pyrotechnic compositions are used in making firecrackers and fireworks.

There are a number of handbooks on the preparations of fireworks: those of V. M. Solodovnikov [6], P. S. Tsytovich [92], Izzo, [12], Weingart [15]; our book will not discuss products of firework pyrotechnics.

Chapter 22 deals with the application of pyrotechnic compositions in the national economy.

1. CLASSIFICATION OF PYROTECHNIC DEVICES AND COMPOSITIONS

The following items of military importance are equipped with pyrotechnic compositions:

1) illumination devices (cartridges, shells, bombs, etc.) used for illuminating at night;
2) photoillumination devices (photobombs), used in nighttime aerial photography;
3) tracer devices, designed to make the flight trajectory of bullets and shells (and other moving objects) visible and thus to facilitate more accurate aiming at rapidly moving aerial and ground targets;
4) nighttime signaling devices (cartridges, grenades, etc.) used for signals over distances;
5) daytime signaling devices (cartridges, grenades, etc.) used for the same purpose, but in the daytime;
6) ignition devices (bullets, bombs, shells, etc.) designed to destroy the adversary's physical facilities and his means of attack and defense; special incendiary anti-personnel devices are also used;
7) masking devices (smoke projectiles, etc.) used to produce smoke screens;
8) rockets of various applications and flight range; their motors are equipped with rocket compositions (solid pyrotechnic fuel);
9) training and simulation devices used in maneuvers and exercises as well as in battle situations: they simulate the action of various types ammunition (atomic bombs, high-explosive shells and bombs, chemical warfare devices, etc.) or various phenomena taking place on the field of battle (sound or flash of gunfire, fire, etc.) and can thus disorient the adversary's observations service;
10) target designation devices (shells, bombs, etc.), which indicate the adversary's objectives to be destroyed by aviation or artillery.

Pyrotechnic compositions are divided into the following categories:

1) illumination;
2) photoillumination (photomixtures);
3) tracers;
4) compositions for nighttime signal lights;
5) compositions for colored signal smokes;
6) incendiary compositions;
7) compositions of white, gray and black masking smokes;
8) solid pyrotechnic fuel;
9) ignition compositions present in small amounts in all pyrotechnic devices;
10) others (simulation, hissing, etc.).

Many categories of compositions are used in the most varied types of devices. For example, illumination compositions are frequently used in tracer devices; white and black smoke compositions can be used not only in masking devices, but also in tracers, simulation and training devices, etc.

Pyrotechnic compositions can be classified in the following manner according to the nature of the processes involved in their combustion.

Flame Compositions

- White flame compositions. Used in rocket, illumination, photoillumination, training simulation, target designation and incendiary devices.
- Colored flame compositions. Used in tracers, nighttime signaling, training simulation and target designation devices.

Thermite Compositions

- Thermite incendiary compositions. Used in incendiary devices.
- Gasless (low gas) compositions. Used in time fuses and detonators.

Smoke Compositions

- Compositions for white, gray, and black smoke. Used in masking, tracer, simulation training, and target designation devices.
- Substances and Mixtures Burning in Atmospheric Oxygen
- Metals and metal alloys. Used in incendiary and photoillumination devices.
- Phosphorus, its solutions and alloys. Used in incendiary devices.
- Petroleum mixtures. Used in incendiary devices and flame throwers.
- Substances and mixtures igniting on contact with water or air. Used in incendiary and special signaling devices.
-

2. COMBUSTION OF THE COMPOSITIONS

Highly exothermic chemical reactions may take place in the form of combustion. The formation of a flame or generation of light is not an indispensable indication of combustion; for example, in the combustions of pyrotechnic smoke compositions, there is usually no flame, and no emission of light is observed.

The chief indicators setting the process of combustion apart from other forms of chemical reactions are:

1) presence of a moving reaction zone having a high temperature and separating the still unreacted substances from the reaction products; this differentiates the combustion processes from chemical reactions where the temperature is the same or nearly the same at all points of the reacting system;

2) absence of a pressure differential in the reaction zone (in the flame); this clearly differentiates combustion reactions from explosive processes.

The combustion of pyrotechnic compositions is an oxidation-reduction reaction in which the oxidation of some components of the mixture, i.e., the fuels, proceeds simultaneously with the reduction of the other components, i.e., the oxidizers, of the same mixture. Several types of combustions are distinguished according to the degree of homogeneity of the initial system.

A solid or liquid fuel burning in atmospheric oxygen constitutes a heterogeneous system (before the start of combustion). Processes of combustion of homogeneous systems, i.e., explosive gaseous mixtures or individual explosives, are also known.

Pyrotechnic compositions, which are mechanical mixtures of several components, mostly solid and finely divided, are intermediate in degree of homogeneity between condensed fuel and pure substances (or homogeneous mixtures). The degree of homogeneity determines many of the properties of pyrotechnic compositions.

Combustion of pyrotechnic compositions takes place as a result of heat transfer from the reaction zone to the layers in which preparation for the combustion process takes place. The ignition of pyrotechnic compositions is also based on the principle of heat transfer. For combustion to take place, it is necessary to create a local temperature rise in the pyrotechnic composition; this is usually accomplished by direct action of combustibly powder gases on the composition, or by using special ignition compositions.

When a pyrotechnic compositions is set off by a fire impulse, and its combustion takes place in an open space, its rate of combustion is slow, and is measured in millimeters per second in most cases. If, however, the combustion of a pyrotechnic composition takes place in a closed space or if an impact or a detonator capsule is used as the initial impulse, the combustion may change to an explosion whose rate is measured in hundreds and in some cases thousands of meters per second.

Sometimes an acceleration of combustion is also observed when a large number of pyrotechnic compositions burn simultaneously in an open space.

The basic operations must be carried out when preparing a pyrotechnic composition and loading it into a product or device:

1) preparation of the components (grinding and drying);
2) preparation of powdered composition (mixing of the components, granulation and drying of the composition);
3) compaction and molding of the compositions (by pressing or otherwise).

Normal operation of a pyrotechnic composition makes it necessary that the components be thoroughly ground up and uniformly mixed with one another. In a carefully prepared composition, the particles of the individual components, with the exception of thermite, are usually no longer distinguishable with the unaided eye.

Compaction of the composition ensures a deceleration of the combustion process and a decrease of the volume occupied by the composition in the product, and imparts a high mechanical strength to the composition. In most pyrotechnic products, the compositions are found in compacted form. An exception are photoillumination compositions, present in photobombs in the form of powders, and granulated smoke compositions, in certain cases.

The preliminary preparation of the components is safe, since both the fuels and the oxidizers currently used for the preparation of pyrotechnic compositions, when taken separately, are in most cases insensitive to mechanical disturbances (impact, friction) and do not possess explosive properties.

This does not apply to various types of fuels scattered in air in the form of dust. For example, cases of powerful explosions of aluminum dust are known. Another exception are oxidizers such as ammonium perchlorate, ammonium nitrate and metal chlorates, which even in pure form without fuel mixtures may explode in the presence of a strong initial impulse. In some cases, impact or friction may cause local heating and ignition of a component; for example, ignition of red phosphorus being rubbed through metal screens has been observed.

Mixtures of oxidizers and fuels are sensitive to mechanical impulses, and may ignite during impact or friction. Impact or friction may also occasionally constitute a sufficient impulse to produce an explosions in pyrotechnic compositions. For this reason, the operations of preparation and compaction of pyrotechnic compositions in the course of which the latter must be subjected to mechanical actions are usually dangerous.

3. REQUIREMENTS PLACED ON PYROTECHNIC AGENTS AND COMPOSITIONS

The chief requirement placed on pyrotechnics is that their action produce a maximum special effect. For different types of pyrotechnics, the special effect is determined by different factors. This question is treated in detail in the chapters dealing with the description of the properties of individual categories of compositions. Here, only a few examples are given as illustrations.

For tracers, the special effect is determined by a good visibility of the flight trajectory of the bullet or shell. The visibility is determined by the light intensity of the tracer composition and also substantially depends on the flame color.

A desirable special effect of incendiary weapons, i.e., the reliability of their kindling of suitable fuel materials, is determined (in the presence of proper design of the ammunition) by the creation of a sufficient duration of combustion of the compositions and amount and properties of the cinders produced in the course of combustion.

For masking smoke devices, the special effect depends on the speed of formation of the thickest and stablest smoke curtain possible.

The quality of rockets is determined by the range and accuracy of their flight such that the indicated target is reached. The flight range will be greater, the larger the unit impulse I (kg sec/kg) of the rocket composition, and the greater the ratio of the weight of the composition to the total weight of the rocket.

All the pyrotechnics should be safe in handling and storage. The effect produced by their action must not decrease after an extended storage period.

The materials used in the manufacture of pyrotechnics should be as readily available as possible. The manufacturing process should be simple, safe, and should permit mechanization and automation of mass production.

Pyrotechnic compositions should have the following qualities:

1) produce the maximum special effect with a minimum consumption of the composition. [Note: The effectiveness of the action is determined not only by the composition formula, but also by the design of the product, technology of manufacture of the composition, and external conditions (pressure, temperature) under which its combustion takes place.]
2) burn uniformly at a fixed rate;
3) possess chemical and physical stability in extended storage;
4) have the lowest possible sensitivity to mechanical impulses;
5) not be too sensitive to thermal effects (not ignite in the presence of a moderate temperature elevation, etc.);
6) have minimum explosive properties; rare cases in which the presence of explosive properties of the compositions is necessary will be discussed below;
7) the combustion products must not include any poisoning substances, which might interfere with the extinction of fires in the plant;
8) the technological process of manufacture should be relatively simple;
9) must not contain scarce components;
10) must not contain components having toxic effects on the human organism.

Articles made of pyrotechnic compositions should have a sufficient mechanical strength to meet the service requirements.

In preparing pyrotechnic compositions satisfying the above requirements, it is necessary in each individual case to plan the selection of the fuel and oxidizer thoroughly and to calculate their relative amounts. It is also necessary to consider the physicochemical characteristic of the fuel and oxidizer by using available literature data on the properties of pyrotechnic compositions and their individual components.

The development of a composition and the calculation of the relative amount of its components is also substantially complicated by the fact that a binary (oxidizer-fuel) mixture is frequently insufficient to meet all the requirements, so that additional components fulfilling a given special function must be added to the composition.

4. FUNCTIONS OF THE COMPONENTS OF A COMPOSITION

Pyrotechnic composition substances can be divided into the following categories:

a) fuels;
b) oxidizers;
c) binders, i.e., organic polymers giving mechanical strength to the compacted compositions;
d) combustion accelerators and inhibitors;
e) deterrents, i.e., additives decreasing the sensitivity of the compositions to friction;
f) substances which improve processing (lubricant additives, solvents for organic polymers introduced into the composition, etc.).

In addition, the compositions of signal lights include substances imparting a color to the flame, and smoke compositions include smoke-forming substances. In some cases, the same component may have several different functions in the composition, and sometimes as combustion retardants. In signal compositions, strontium nitrate is always the oxidizer, and at the same time imparts a red color to the flame.

Concerning the questions discussed in Sec.1-4 of this chapter, see also the following books: M. A. Budnikov, I. A. Levkovich, I. V. Bystrov, et al. [1], I. V. Bystrov [2], A. G. Gorst [3], N. F. Zhirov [5], Shilling [156], and Ellern [9].

5. POSSIBLE HIGHLY EXOTHERMIC REACTIONS

All chemical reactions involve rupture of the bonds between atoms and formation of different new bonds. If the energy evolved by the formation of the new bonds is higher than that expended in breaking the old bonds, we are dealing with an exothermic reaction. Obviously, the largest amount of heat will be evolved when the bonds being broken are weak and the new ones are very strong.

An examination of the energetics of bonds leads to the conclusion that the strongest bonds are formed by the combination of elements with opposite properties, i.e., typical metals with typical nonmetals, such as fluorine or oxygen. Their reactions form ionic compounds. When nonmetals combine with one another (elements of groups V-VII of the Periodic System), they form polar compounds and mostly energetically unstable bonds.

Metals such as magnesium, aluminum, etc., are frequently used as fuels in pyrotechnics; the fact that fluorine and oxygen are gases under ordinary conditions substantially complicates their use in pyrotechnic products. For this reason, attempts are made to use as oxidizers compounds in which oxygen or fluorine is linked by weak bonds with other nonmetals. However, the majority of such compounds are either gases or low-boiling liquids.

Example: ClF_3, OF_2, NF_3, SF_6, Cl_2O_7, ClO_2, NO_2, N_2O_4, etc.

This fact and the toxicity and excessive chemical activity of these substances prevent their practical application.

These properties largely disappear when (to the detriment of energetics) nonmetal oxides (anhydrides) are combined with metal oxides, for example:

$$Cl_2O_7 + K_2O = 2KClO_4 \quad \text{or} \quad N_2O_5 + K_2O = 2KNO_3, \text{ etc.}$$

These reactions form solid substances, i.e., salts, whose properties are highly suited for use in pyrotechnic compositions. They are less active chemically, their powders can be mixed with standard fuels at ordinary temperature, and these mixtures have an adequate chemical stability.

The following oxidizers are usually employed in pyrotechnics: perchlorates, chlorates and nitrates of alkali (or alkaline earth) metals, such as

$$KClO_4, KNO_3, Ba(NO_3)_2, \text{ etc.}$$

They are convenient to use, but obviously are much less energetically favored than elemental oxygen.

Another type of oxidizers used in pyrotechnics are oxides of low-activity metals. Obviously, the reactions of their displacement by more active metals will take place with the liberation of a significant amount of heat. Such for example is the combustion reaction of iron-aluminum thermite:

$$Fe_2O_3 + 2Al = Al_2O_3 + 2Fe + O.$$

Similarly, one could expect the use of mixtures in which the more active metal would displace the less active one from its fluorine compounds, for example:

$$FeF_3 + Al = AlF_3 + Fe + 70 \text{kcal},$$

which corresponds to 0.5kcal/g of mixture. However, the fluorine content of the earth's crust is low (0.03%); fluorine occurs most often in the form of a compound extremely stable thermally, i.e., fluorspar, whose processing yields most of the other fluorine compounds.

In addition to the economic considerations, the use of fluorides in pyrotechnics as oxidizers is also prevented by the fact that fluorine (an extremely active element) forms very few solid compounds containing weak bonds; hence, in the case of fluorides, it would be difficult to create mixtures of high calorific value. The character of combustion of the mixture FeF_3+Al will be completely different from that of the mixture Fe_2O_3+Al; this is due to the fact that AlF_3 sublimes at 1260°C.

In principle, it would be possible to use higher oxides of certain metals, for instance Mn_2O_7 CrO_3, etc., as oxidizers, but the physicochemical properties of these compounds are not very favorable to their use: Mn_2O_7 is a chemically unstable liquid, and chromic anhydride CrO_3 is strongly hygroscopic. The reaction of these compounds with alkali metal or alkaline earth oxides leads to the formation of compounds suitable for practical use as oxidizers in pyrotechnics, for example:

$$CrO_3+BaO = BaCrO_4, \text{ or } Mn_2O_7+ K_2O = 2KMnO_4$$

Metal peroxides are advantageous because of their higher oxygen content in comparison with the oxides of the same metals, but many are fairly unstable on heating (for example, CaO_2) or toward water (for example, Na_2O_2). In practice, among the peroxides, only barium peroxide BaO_2, and in rare cases, strontium peroxide SrO_2 [9] have been used in pyrotechnics thus far.

Let us now return to the discussion of fuels. In addition to metals, strong bonds with fluorine and oxygen are produced by hydrogen; fairly strong bonds with oxygen are also formed by boron, carbon, silicon, a phosphorus. This means that these elements as well as some of their compounds (boron hydrides, hydrocarbons, etc.) can be used as fuels in pyrotechnics. However, the following questions automatically arises: since many nonmetals (nitrogen, chlorine) form weak bonds with oxygen, and also with hydrogen and carbon, why not combine the atoms of these elements into a molecule in which nitrogen (or chlorine) acts as a buffer in this molecule by separating C and H from oxygen:

$$C \text{ and } H \mid N \mid O$$
(buffer)

When such a molecule is subjected to an energetic external action, the buffer will be knocked out, and the combination of C and H with oxygen with the formation of CO_2 and H_2O will evolve a large amount of heat. Hence, the reactions of intramolecular combustion can also be highly exothermic. This idea is entirely correct and has been applied for a long time.

Substances whose molecule contains a buffer such as nitrogen or chlorine between C and H on the one hand and O on the other hand have been known for a long time: they are nitro compounds or esters of nitric (or perchloric) acid. These substances are capable of intramolecular combustion. However, these substances have a very important disadvantage from the standpoint of pyrotechnics. The absence of heterogeneity in the system, and the small distance between the atoms in the molecule account for the fact that when the molecule is subjected to an energetic external action, it may break up at a high rate, in the form of an explosion. In other words, the substances described are explosive.

The use of the intramolecular combustion of explosives in pyrotechnics is possible, but frequently involves a considerable risk: when the combustion conditions are disturbed (see Chapter 11), the combustion may change to an explosion.

An intermediate stage is possible between the two extremes of heterogeneous systems, or individual explosives, on the other. If the molecule of a given substance containing a nitrogen buffer has few atoms of fuel elements and an excess (obviously, up to a certain limit) of oxygen atoms, such a substance will be capable of an intramolecular combustion reaction, but the amount of heat produced by such a reaction will be fairly low, the combustion temperature will be moderate, and the probability of the combustion-to-explosion transition will be considerably lower. Such substances are ammonium salts of certain oxygen-containing acids, for example, ammonium nitrate or perchlorate. The combustion of these substances evolves free oxygen, and the heat and combustion temperature are comparatively low:

$$2NH_4ClO_4 = 4 H_2O + N_2 + Cl_2 + 2O_2, \quad 2NH_4NO_3 = 4 H_2O + (1-x) N_2 + (1-x)O_2 + 2xNO$$

However, the energetics of these substances can be easily corrected by adding a certain (calculated) amount of fuel (an organic binder can be used) so as to make full use of the excess oxygen of these substances. If the amount of fuel added is significant and causes a disturbance in the homogeneity of the system, the danger of the combustion-to-explosion transition will be reduced.

In addition, another variant can be employed. There are substances with a nitrogen buffer which contain few oxygen atoms and an excess of C and H atoms. Such substances include nitroguanidine $CN_4H_4O_2$, dinitrotoluene $C_7H_6N_2O_4$, polynitrourethanes, polyvinylnitrate, etc. The energetics of these substances can also be easily corrected by adding a certain amount of oxidizer (among substances incapable of an exothermic decomposition process). This case involves a disturbance of homogeneity and thus a considerable decrease of the likelihood of an explosion.

All of the above can be illustrated by the following Table.

Table 1.1 DEGREE OF HETEROGENEITY & EXPLOSIVE PROPERTIES OF PYROTECHNIC COMPOSITIONS

NO.	SYSTEM	DESCRIPTION OF HOMOGENEITY OF THE SYSTEM	EXPLOSIVE PROPERTIES
1	1. Fuel 2. Oxidizer	Heterogeneous	Slight
2	1. Fuel: contains a little nitrogen (buffer) and a little oxygen 2. Added Oxidizer	Heterogeneous	Medium
3	1. Oxidizer: contains a little nitrogen (buffer) and a little H & C 2. Added fuel	Heterogeneous	Medium
4	1. Explosive: contains nitrogen (buffer) oxygen and C+H in stoichiometric proportions	Homogeneous	High

6. COMBUSTIBILITY OF VARIOUS SUBSTANCES & MIXTURES

Pyrotechnics is a young science undergoing rapid development. New problems arise, new requirements are formulated and as a result, new compositions are designed. The pyrotechnist should be able to evaluate the possibility of occurrence of a given chemical reaction in the form of combustion.

In accordance with Berthelot's principle (which is unquestionably valid for highly exothermic reactions taking place at room temperature), when the ambient conditions are properly selected, any chemical system in which an exothermic reaction can take place should prove suitable for the propagation of a combustion reaction therein.

On this basis, the author of the book studied and showed experimentally the combustibility of the system ($H_2O + Mg$) [39], a series of inorganic ammonium salts [237], hydrazine salts, and hydroxylamine sulfite.

Combustion reactions of mixtures of magnesium and many oxygen-containing organic substances such as alcohols, aldehydes, carbohydrates, etc., are certainly possible; such processes are most easily carried out with solid substances, but with greater difficulty when liquid substances are used.

The combustion reaction in the system $3 H_2O + 2Al$ is entirely possible. We will also give a calculation of the exothermic effect of the reactions in the systems ($3 H_2O + 2B$), ($2 H_2O + Si$) and ($4 H_2O + P$) [119, p.540]. For the latter system, it is our view that the reaction may take place according to the equation $8 H_2O + 2P_6(white) = 2H_3PO_4 + 5H_2 + 68 kcal$ which corresponds to 0.33 kcal/g of mixture of the reacting substances.

In some cases, the estimate of the exothermic effect of a reaction has a different meaning, i.e., the estimate is made in order to evaluate the fire hazard (or explosion hazard) of a system.

The explosion hazard of silver perchlorate, recently established experimentally in an unfortunate case [44], could have been predicted on the basis of the thermochemical calculation:

$$AgClO_4 = AgCl + 2O_2 + 22 kcal$$

Substances or mixtures which have practically no tendency to burn (or explode) at room temperature because their decomposition evolves too little heat, acquire this tendency when their stored energy increases, i.e., at elevated temperatures.

A case in point was the giant explosion of potassium chlorate melt in Liverpool in 1899 [19]. The addition of small amounts of fuel (1% of phenol-formaldehyde resin + 5% of MnO_2 catalyst) to potassium chlorate makes it fuel at room temperature.

Extensive material on the question of sudden fires and explosions is given by Langhans [159] and also in the paper of Tomlinson and Audrieth [122].

Chapter 2

OXIDIZERS

A mixture of fuel and oxidizer forms the basis of every pyrotechnic composition.

In order to obtain the amount of heat necessary to produce a required special effect, it would seem sufficient to burn up the fuel simply by using atmospheric oxygen. However, the combustion of fuel substances under such conditions usually takes place more slowly than their combustion involving the oxygen of the oxidizer, [Note 1: The only exception is the case in which there is a large contact surface between the fuel and air, i.e., when a suspension of finely dispersed powder of fuel is present in air.] and therefore, mixtures containing no oxidizer (see Chapter 17, Sec. 6-8) are used for pyrotechnic purposes less often than compositions with oxidizers.

Most frequently, pyrotechnic compositions contain oxygen compounds as oxidizers, but oxidizers containing no oxygen are sometimes encountered.

The following example of an oxygen-free combustion reaction can be given:

$$C_2Cl_6 + 3Mg = 2C + 3MgCl_2. \quad (2.1)$$

In this reaction, hexachloroethane is the oxidizer.

When combining with fluorine, many active metals (magnesium, aluminum) evolve even more heat than when combining with oxygen. Compositions can thus be visualized in which the oxidizers will be fluorinated organic compounds.

For example, a fuel mixture will be one of magnesium and polytetrafluoroethylene (teflon): (The combustibility of the mixture $(C_2F_4)n + Na$ is indicated in the handbook [233, vol. I, p.592].)

$$(C_2F_4)n + 2nMg = 2nC + 2nMgF_2. \quad (2.2)$$

The heat of combustion of this mixture, found by calculation, is about 2.3kcal/g.

The present chapter examines oxidizers in a narrow sense of this term only, i.e., those whose oxidizing action is due to the oxygen they contain, which is consumed by the oxidation of the fuel during the combustion of the mixture.

1. CHOICE OF OXIDIZERS AND THEIR CLASSIFICATION

The oxidizer should be a solid with a melting point not below 50-60C and have the following properties:

1) contain the maximum quantity of oxygen;
2) readily give up oxygen during combustion of the pyrotechnic composition;
3) be a chemical compound stable in the range from -60 to +60C and inert to the action of water;
4) be nonhygroscopic;
5) be readily available and
6) have no toxic effect on the human organism.

However, oxidizers which do not possess all of the above properties are still being used in pyrotechnic compositions: for example, $NaNO_3$ or $NaClO_4$ are hygroscopic, and lead compounds are toxic.

In many cases, it is necessary to tolerate certain negative properties of oxidizers and take them into account in the technological process of manufacture of the compositions and products (for example, the product must be sealed).

Particular care must be taken so that compositions prepared by using the selected oxidizer are not too sensitive to mechanical impulses and do not possess significant explosive properties.

In addition, in selecting the oxidizer for pyrotechnic flame compositions, it is also necessary in many cases to consider the radiation intensity of the decomposition products of the oxidizer in various regions of the spectrum. Thus, compositions of signal lights must not include oxidizers whose introduction alters the flame color (for example, sodium salts must not be introduced into the compositions of red, green and blue lights).

It is also extremely important that the oxidizer provide the required combustion rate of the composition.

The following compounds are used most frequently as oxidizers in pyrotechnic compositions:

SALTS

a) nitrates - $NaNO_3$, KNO_3, $Sr(NO_3)_2$, $Ba(NO_3)_2$;
b) chlorate - $KClO_3$;
c) perchlorates - $KClO_4$, $NaClO_4$. $NaNO_3$

PEROXIDES

Barium peroxide BaO_2.

OXIDES

a) of iron - Fe_3O_4 and Fe_2O_3;
b) of manganese - MnO_2;
c) of lead - Pb_3O_4, PbO_2.

In addition to the above oxidizers, we should note the possibility of using such oxidizers as chromates and bichromates - $BaCrO_4$, $PbCrO_4$, $K_2Cr_2O_7$, potassium permanganate - $KMnO_4$, sodium and barium chlorates - $NaClO_3$ and $Ba(ClO_3)_2$, lead nitrate - $Pb(NO_3)_2$, sulfates $CaSO_4$, $SrSO_4$, $BaSO_4$, as well as certain oxides and peroxides - CuO, SiO_2, strontium peroxide - SrO_2, etc.

Potassium bichromate, used in limited amounts as an additional oxidizer in match compositions, has practically no other uses, since the amount of oxygen it evolves during the decomposition amounts to only 16% of the weight of the oxidizer; it decomposes with comparative ease, since the initial stage of its decomposition involves the absorption of a small amount of heat. It should also be considered that potassium bichromate or permanganate powders have a strong corrosive effect on the mucous membranes of the human organism.

Oxidizers used in pyrotechnics also include the following explosive (or semiexplosive) substances capable of intramolecular combustion:

a) polynitro compounds - trinitrotoluene (trotyl), hexagene, hexanitroethane, etc.;
b) ammonium salts - ammonium perchlorate NH_4ClO_4 and ammonium nitrate NH_4NO_3.

However, their use frequently involves a marked increase in the explosive properties of the compositions, and in the sensitivity of the compositions to mechanical impulses.

Studies made in the last few years [155] have established that water can act as the oxidizer in compositions containing metals such as magnesium or aluminum as the fuels; these studies are of practical importance because they alter the concept of the role played in combustion processes by the water of crystallization, which may be present in some oxidizers, i.e., salts (concerning the use of crystal hydrate salts, see [41]).

In industrial pyrometallurgical processes, pyrotechnical compositions used for the preparation of pure metals include oxides of a series of metals such as chromium, vanadium, etc., as the oxidizers.

Ellern [9] notes that occasionally, such oxides as Cu_2O, Bi_2O_3, WO_3, peroxides such as sodium peroxide Na_2O_2, and even silver peroxide Ag_2O_2, an unstable substance that decomposes when the temperature is raised only slightly, are used in pyrotechnics as oxidizers. He points out that salt oxidizers such as bromates, iodates, periodates and ferrates have been of only theoretical interest thus far. However, it is well known that lead iodate $Pb(IO_3)_2$ can be used in antihail compositions (see Chapter 22) to prepare PbI_2.

2. PHYSICOCHEMICAL PROPERTIES OF OXIDIZERS

The most important properties of oxidizers that can be quantitatively evaluated are:

1) density (at room temperature);
2) melting point;
3) temperature of intensive decomposition (if sufficiently reliable data are available);
4) equation of the reaction of thermal decomposition;
5) heat of the decomposition reaction;
6) percentage content, calculated on the basis of par. 4, of active oxygen, (i.e., used for the oxidation of the fuel) in the oxidizer;
7) hygroscopicity;
8) toxicity;
9) melting point and boiling point of the decompositions products.

The melting point of the oxidizer is of interest for the pyrotechnist, since an intensive decomposition of the oxidizer takes place in most cases only at a temperature equal to or slightly above its melting point.

The concept of the "temperature of intensive decomposition," referred to in many papers and monographs, is very indefinite; it should be replaced by the value of the temperature at which the partial pressure of oxygen above the oxidizer would be equal to some fixed value (for example, 5 or 50mm Hg), but unfortunately, for most oxidizers such data are as yet unavailable.

Many chemical handbooks give equations of thermal decompositions reactions of oxidizers. These data should be taken into account, but in addition, it should be considered that under the conditions of combustion of pyrotechnic compositions (high temperature + presence of reducing agent = fuel), the equation of decompositions becomes more complete, since the entire (or almost entire) oxygen contained in the oxidizer is consumed by the oxidation of the fuel. If the equation of the decomposition is known, the heat of the decomposition reaction is calculated by using Hess' law (see Ch. 6).

The possibility of a quantitative evaluation of hygroscopicity will be discussed a little later in this chapter.

Table 2.1 lists the properties of oxidizers most frequently used in pyrotechnics (the properties of NH_4ClO_4 and NH_4NO_3, used mainly in rocket compositions, are discussed in Ch. 20). Column 5 of Table 2.1 gives the equations of the most probable decompositions reactions of oxidizers under the combustion conditions of pyrotechnic compositions.

In a slow low-temperature decomposition, potassium chlorate decomposes with the formation of perchlorate

$$4KClO_3 = 3KClO_4 + KCl. \qquad (2.4)$$

However, under the conditions of the combustions process, the products of its decomposition will be only potassium chloride and oxygen

$$2KClO_3 = 2KCl + 3O_2. \qquad (2.5)$$

Potassium chlorate decomposes vigorously only above its melting point (370°C); addition of catalysts markedly lowers its decomposition temperature. The strongest catalytic effect is that of manganese dioxide and cobalt oxide Co_2O_3.

Alekseyenko [16] discussed the comparative thermal stability of chlorates of various metals and concluded that the thermal decomposition of chlorates should be easier the greater the charge and smaller the radius of the metal cation entering into the salt.

A marked decrease in the thermal stability of chlorates is observed in the series Na, Mg, Al, and also in the series Ca, Mg, Be. (these considerations are also cited by Alekseyenko in a discussion of thermal stability of nitrates.)

On strong heating (500-600°C), potassium and sodium perchlorates decompose into chlorides of the corresponding metals and oxygen (see Table 2.1) on next page.

According to Gmelin's data [42, No. 22], the decomposition of $KClO_4$ accelerates in the presence of compounds of Fe, Co, Ni, and W. The addition of copper salts has a particularly strong effect on the decomposition of $KClO_4$. The process of decomposition of alkali metal or alkaline earth nitrates takes place in stages [47], for example:

$$2NaNO_3 = 2NaNO_2 + O_2 - 47 kcal \qquad (2.6)$$

$$2NaNO_2 = Na_2O_2 + N_3 + O_2 - 56 kcal \qquad (2.7)$$

$$Na_2O_2 = Na_2O + 0.5O_3 - 18 kcal \qquad (2.8)$$

Summing up (2.6), (2.7), and (2.8), we obtain

$$2NaNO_3 = Na_2O + N_2 + 2.5O_2 - 121 kcal \qquad (2.9)$$

When nonmetals such as carbon, sulfur, phosphorus or an organic fuel are used as the fuels in the compositions, the decomposition of nitrates ends in the formation of metal oxides (in this case, Na_2O); in cases where the combustion temperature of the composition is not high, the combustion products contain a significant amount of nitrites, as for example in the combustion of a mixture of sodium nitrate and lactose. It should be noted, however, that nitrites of alkali and alkaline earth metals are thermally less stable than the corresponding nitrates; see Voskresenskaya and Berul' [23]. When very energetic reductants such as the metals magnesium or aluminum are used as the fuels, a more extensive decomposition of the nitrate oxidizer may also take place. Thus, the combustion reaction of a mixture of barium nitrate and magnesium may be expressed by the equation

$$Ba(NO_3)_2 + 6Mg = Ba + N_2 + 6MgO + 646 kcal \qquad (2.10)$$

Table 2.1 PROPERTIES OF OXIDIZERS

FORMULA	MOLECULAR WEIGHT	DENSITY G/CM^3	MELTING POINT °C	EQUATION OF DECOMPOSITION REACTION UNDER CONDITIONS PREVAILING DURING COMBUSTION OF COMPOSITION	HEAT OF FORMATION - KCAL/MOLE OF OXIDIZER	HEAT OF FORMATION - KCAL/MOLE OF DECOMPOSITION PRODUCT	HEAT OF DECOMPOSITION ACCORDING TO THE REACTION EQUATION: K.CAL/MOLE	PERCENTAGE OF ACTIVE OXYGEN	AMOUNT OF OXIDIZER EVOLVED BY DECOMPOSITION OF 1G OF OXYGEN	COMPOSITIONS IN WHICH USED
1	2	3	4	5	6	7	8	9	10	11
$KClO_3$	123	2.3	360	$2KClO_3 = 2KCl + 3O_2$	96	106	+10	39	2.55	Smoke & less often in signals
$KClO_4$	139	2.5	610	$2KClO_4 = 2KCl + 4O_2$	103.6	104.2	+1.2	46	2.17	Incendiary & rocket comps, less in signals
$NaClO_4$	122	2.5	482 with decomposition	$2NaClO_4 = 2NaCl + 4O_2$	92	98	+11	52	1.9	Illuminations - hygroscopic
$NaNO_3$	85	2.2	308	$2NaNO_3 = Na_2O + N_2 + 2.5O_2$	111	101	-121	47	2.13	Illuminations - yellow, hygroscopic
KNO_3	101	2.1	336	$2KNO_3 = K_2O + N_2 + 2.5O_2$	119	87	-151	40	2.53	Ignition compositions
$Sr(NO_3)_2$	212	2.9	645	$Sr(NO_3)_2 = SrO + N_2 + 2.5O_2$	231	142	-89	38	2.65	Tracers & red light
$Ba(NO_3)_2$	261	3.2	592	$Ba(NO_3)_2 = BaO + N_2 + 2.5O_2$	237	133	-104	30	3.27	Illumination, tracer, incendiary, green light
Fe_3O_4	232	5.2	1527	$Fe_3O_4 = 3Fe + 2O_2$	266	0	-266	28	3.34	Thermites & thermite incendiary comp.
MnO_2	87	5.0	(>530)	$MnO_2 = Mn + O_2$	125	0	-125	37	2.72	Same + catalyst in match comp
	169	5.0		$MnO_2 = MnO + .5O_2$	125	93	-32	18	5.44	
BaO_2			(~700)	$BaO_2 = BaO + .5O_2$	150	133	-17	9	10.6	Ignition comp.

Note: figures in parentheses - decomposition temperature.

It should be noted that the reaction of magnesium with barium oxide is associated with the evolution of a small amount of heat:

$$BaO + Mg = Ba + MgO + 11 \text{ kcal} \qquad (2.11)$$

When magnesium reacts with sodium oxide, much more heat is evolved:

$$Na_2O + Mg = MgO + 2Na + 44 \text{ kcal} \qquad (2.12)$$

The formation of sodium metal in the combustion of $NaNO_3 + Mg$ mixtures containing over 60% Mg was established experimentally by burning these mixtures in a bomb calorimeter.

Shargorodskiy and Shor [38] obtained decomposition thermograms of Be, Ca, Sr, and Ba nitrates. Analysis of the thermograms established the decomposition temperature of these compounds: 175-200°C for $Be(NO_3)_2$, 480-500°C for $Ca(NO_3)_2$, 580-600°C for $Sr(NO_3)_2$, and 555-600°C for $Ba(NO_3)_2$.

Hogan, Gordon, and #Campell [45] report that a mixture of equivalent parts of $Ba(NO_3)_2$ and $KClO_4$ melts at 520°C.

It follows from Ellern's work [9] that salt oxidizers decompose in the following order of rising decomposition temperatures (for the same cation): permanganates, chlorates, nitrates, perchlorates, bichromates, chromates. The decomposition temperatures of sodium salts are lower than those of potassium salts with the same anion.

The combustion of iron-aluminum and manganese-aluminum thermites forms the free metal iron or manganese

$$MnO_2 + 4Al = 3Mn + 2Al_2O_3. \qquad (2.13)$$

However, in the combustion of compositions containing no active metals (Mg, Al), manganese oxide gives up only one-half on the oxygen it contains (see Table 2.1).

Barium peroxide BaO_2 readily gives up only one-half of the oxygen, converting into BaO.

Columns 6 and 7 list the heat of formation of oxidizers and products of their decomposition.

Column 8 gives a very important characteristic of an oxidizer: the amount of heat absorbed or evolved during the decomposition of the oxidizer.

To obtain the maximum amount of heat from the combustion of a composition, it is of course advantageous to use oxidizers whose decomposition requires the minimum amount of heat. On the other hand, it should be noted that such compositions are usually the most sensitive to mechanical actions. Chlorate compositions are particularly sensitive to mechanical actions and possess pronounced explosive properties, since the decomposition of chlorates evolves a very substantial amount of heat. For example, the heat evolved by the decomposition of $KClO_3$ is sufficient to heat up this salt from room temperature to the temperature of its melting, i.e., 370°C.

Even more exothermic are the decomposition reactions of sodium chlorate and particularly barium chlorate:

$$2NaClO_3 = 2NaCl + 3O_2 + 25 \text{ kcal} \qquad (2.14)$$

$$Ba(ClO_3)_2 = BaCl_2 + 3O_2 + 28 \text{ kcal} \qquad (2.15)$$

Because of their sensitivity to friction and pronounced explosive properties, compositions including barium chlorate find practically no uses at the present time (see also Ch. 16).

Oxidizers whose decomposition proceeds with the evolution of heat may manifest explosive properties under certain conditions even without being mixed with fuels, i.e., by themselves, as pure substances. For example, potassium, sodium or barium chlorate heated above the melting point explode under hard impact.

Potassium perchlorate $KClO_4$ taken by itself has no explosive properties, as its decomposition takes place with an extremely slight evolution of heat, namely, 0.6 kcal/mole, which corresponds to four small calories per 1 g. [Note 1: According to the data of S. M. Skuratov et al. [22], [33], the decomposition reaction of $KClO_4$ is associated with the evolution of 2.5 kcal/mole of heat; however, this figure does not agree with the data of other investigators.]

Sodium perchlorate must be recognized as being more dangerous than potassium perchlorate, since its decomposition is associated with a greater evolution of heat, namely:

$$2NaClO_4 = 2NaCl + 4O_2 + 10,8 kcal \qquad (2.16)$$

It is evident from Table 2.1 that alkali metal and alkaline earth nitrates decompose with a considerable absorption of heat: this substantially decreases the total heat balance of pyrotechnic compositions prepared with their participation.

The consumption of heat by the decomposition of the oxidizer is even more pronounced when oxides of iron and manganese are used.

The consumption of heat due to the decomposition of peroxides, in this case barium peroxide, is comparatively slight. However, the oxygen content of BaO_2 is low (see Table 2.1), and this markedly decreases its quality as an oxidizer.

The facility with which the oxidizer gives up oxygen and the amount of heat necessary to decompose the oxidizer are closely interrelated. Thus, chlorates give up oxygen in the course of combustion of a pyrotechnic composition much more readily (at a lower temperature) than nitrates; the latter in turn give it up more readily than many oxides. It should be noted that to date, oxides have been used in compositions mixed only with magnesium, aluminum and certain other compounds evolving large amounts of heat during combustion.

Column 9 of Table 2.1 shows the percentage of active oxygen, or the number of grams of oxygen liberated by the decomposition of 100 g of oxidizer. The numerical data of this column were calculated on the basis of equations of oxidant decomposition reactions listed in column 5. It should be noted that what is of greatest interest is not the total amount of oxygen present in the oxidant, but the amount of oxygen which actively participates in the reaction and is directly consumed by the oxidation of the fuel; these data are cited in column 9.

As is evident from Table 2.1, the quantity of oxygen given up by the oxidizers amounts to no more than 52% of the weight of the compound.

It is of interest at this point to draw a comparison with the liquid oxidizers nitric acid and tetranitromethane (TNM):

When they decompose according to the reactions

$$2HNO_3 = H_2O + N_2 + 2.5O_2. \qquad (2.17)$$

$$C(NO_2)_4 = C + 2N_2 + 4O_2 \qquad (2.18)$$

the active oxygen content of HNO_3 is calculated to be 63%, and for TNM, 65%.

In hydrogen peroxide, the total oxygen content is high, 94%, but the content of active oxygen is 47%.

The data of column 10 are used in calculations of the percentage content of the components in binary mixtures (see Ch. 5).

Table 2.2 lists data on the melting and boiling points of certain decomposition products of oxidizers; these data can be used to get an idea of the presence or absence of a gaseous phase and liquid slags during the combustion of the compositions, an idea of the intensity of smoke formation during the combustion, etc.

Table 2.2

Compound	KCl	NaCl	SrO	BaO	Fe	Mn	MnO
Melting point, °C	768	800	2450	2190	1527	1242	1785
Boiling point, °C	1415	1465	--	--	2740	1900	--

The alkali metal oxides Na_2O and K_2O melt at the red heat temperature, i.e., approximately 800°C.

3. HYGROSCOPICITY OF OXIDIZERS

Many salts which satisfy most of the requirements on oxidizers are highly hygroscopic. Particularly hygroscopic are soluble magnesium, calcium, sodium, and ammonium salts.

The amount of hygroscopic water absorbed by salts depends on the humidity and temperature of air, on the individual properties of the salt, and on the surface area of contact between the salt and the moist air.

If the relative humidity of air in a closed space is higher than that above the saturated salt solution indicated in handbooks, the salt will absorb the humidity from air, forming a saturated solution, and the salt crystals deliquesce in air. If however the relative humidity of air in the closed space is lower than that above the saturated slat solution, the salt will dry, and in some cases the process of "efflorescence" of crystal hydrates, i.e., loss of water of crystallization, will be observed. Thus, sodium nitrate deliquesces at a relative humidity of air above 77%, and dries at a lower humidity.

The relative humidity of air over a saturated salt solution can be calculated by dividing the pressure of water vapor above the saturated salt solution by the pressure of water vapor above pure water at the same temperature. Water vapor pressure is usually expressed in mm Hg.

The data available in the literature on "hygroscopic points" of salt oxidizers (i.e., data on the relative humidity of air above saturated solutions of oxidizers) at 20°C are listed in Table 2.3.

In cases where no data at all are available on the vapor pressure above a saturated salt solution, the hygroscopicity of the salt can be estimated from its solubility in water. Although this estimate will be inaccurate, as a rule, other conditions being equal, the hygroscopicity of the salt may be assumed to be greater the higher its solubility in water.

As is evident from Table 2.3 (next page), the least hygroscopic salts are $Ba(NO_3)_2$, $KClO_4$, $KClO_3$, followed by NH_4ClO_4, $Ba(ClO_3)_2$, $Pb(NO_3)_2$, and KNO_3.

Salts above the saturated solutions of which the vapor pressure is lower than above a saturated potassium nitrate solution are considered to be hygroscopic.

In practice, it is difficult to work with salts above the solutions of which the relative humidity is lower than 75-80%.

Because of their high hygroscopicity, such oxidizers as $Mg(NO_3)_2$, $Ca(NO_3)_2$, $Mn(NO_3)_2$, $Al(NO_3)_3$, $Sr(ClO_3)_2$, and $Ba(ClO_4)_2$ are not used in pyrotechnics at all. For the same reason, sodium nitrate is used in pyrotechnic compositions only under conditions of reliable insulation of the compositions from the moisture of air.

Information on the hygroscopicity of Strontium salts can be found in [29].

Table 2.3 HYGROSCOPICITY & SOLUBILITY OF OXIDIZERS

Oxidizer	$100P/P_o$ at $T=20°C$	Solubility of anhydrous salt in water in 1 g of solution at temperature deg. C			
		0	20	50	100
Chlorates:					
$KClO_3$	97	3	7	17	36
$Ba(ClO_3)_2$	94	17	26	37	51
$NaClO_3$	74	45	50	59	67
Perchlorates:					
$KClO_4$	94	0.7	1.7	5	18
NH_4ClO_4	96	11	18	30	--
$NaClO_4$	69-73	--	66	71	75
Nitrates:					
$Ba(NO_3)_2$	99	4.8	8	15	25
KNO_3	92.5	12	24	46	71
$Pb(NO_3)_2$	94	27	34	44	56
$Sr(NO_3)_2$	86	28	42	48	50
$NaNO_3$	77	42	47	--	63
NH_4NO_3	67	54	64	78	91

Heavy metal salts find practically no uses as oxidizers in pyrotechnics; this is due to their scarcity, the relatively small amount of oxygen they contain, and the fact that in mixtures of these salts with such metals as magnesium, zinc, etc., a displacement reaction can readily take place in the presence of a small amount of moisture, for example:

$$Pb(NO_3)_2 + Mg = Pb + Mg(NO_3)_2.$$

In general, in the case of pyrotechnic compositions containing powders of active metals (magnesium, zinc, etc.), to obtain a sufficient chemical stability of the composition during storage, the oxidizers used should be chiefly the salts of metals located in the electromotive series (see Table 2.4) above the metals used in the composition.

Table 2.4 STANDARD ELECTRODE POTENTIALS E_0 AT 25°C (ELECTROMOTIVE SERIES)

Electode	E_0, V	Electode	E_0, V	Electode	E_0, V	Electode	E_0, V
Li/Li^{1+}	+3.01	Mg/Mg^{2+}	+2.38	$Ti\backslash Ti^{1+}$	+0.34	Cu/Cu^{1+}	-0.51
K/K^{1+}	+2.92	Al/Al^{3+}	+1.66	Sn/Sn^{2+}	+0.14	Hg/Hg^{1+}	-0.80
Ba/Ba^{2+}	+2.92	Mn/Mn^{2+}	+1.05	Pb/Pb^{2+}	+0.13	Hg/Hg^{2+}	-0.80
Sr/Sr^{2+}	+2.89	Zn/Zn^{2+}	+0.76	$Pt(H_2)/H1$	+0.00	Ag/Ag^{1+}	-0.80
Ca/Ca^{2+}	+2.84	Fe/Fe^{2+}	+0.44	Cu/Cu^{2+}	-0.34	Au/Au^{1+}	-1.70
Na/Na^{1+}	+2.71					$Pt(F^{1-})/F_2$	-2.8

In compositions including magnesium, among soluble salts (as is evident from the electrochemical series), only compounds of potassium, barium, and sodium can be practically used as oxidizers (lithium salts are expensive, and calcium salts are too hygroscopic).

4. TECHNICAL REQUIREMENTS IMPOSED ON OXIDIZERS

The following requirements must be met by oxidizers used in pyrotechnics in regard to the quantity of impurities they contain:

1. Maximum content of the main substance (usually, not less than 98-99%).
2. Minimum moisture content (not more than a few tenths of 1%).
3. Minimum content of impurities of hygroscopic salts & heavy metal salts.
4. The reaction of aqueous solutions of salt oxidizers should be pH neutral.
5. Absence of fuel mixtures and mixtures of solid substances (sand, glass, etc.), which increase the sensitivity of the composition to mechanical actions.
6. Absence of impurities which decrease the chemical stability of the composition during storage (for example, impurities of bromates in chlorates) or impurities decreasing the special effect of the composition.

In addition, the oxidizer powder should be sufficiently pulverized.

As an example, here are the technical specifications for commercial barium nitrate (GOST 1713-53):

1. Pure $Ba(NO_3)_2$ — minimum 99%.
2. Moisture — maximum 0.5%
3. Water-insoluble residue — maximum 0.25%
4. Chlorides in terms of $BaCl_2$ — maximum 0.1%
5. Reaction of aqueous solution to litmus — neutral
6. External appearance and color — fine crystals ranging from white to light yellow in color.

Chlorides are undesirable impurities in oxidizers since they are fairly hygroscopic, as evidenced by the following data on their "hygroscopic points" (in percent at 20°C): KCl 86, NaCl 77, NH_4Cl 80, $MgCl_2 \cdot 6 H_2O$ 33.

The effect of adding chlorides on the hygroscopicity of a salt can be realized from the following data:

Salt composition	$KClO_3$ chemically pure	$KClO_3$ +0.05% KCl	$KClO_3$ +0.1% KCl
Moisture content of salt in % after storing for one day at t=15-20°C	0.1	0.3	0.9

5. PRODUCTION OF OXIDIZERS

CHLORATES. Presently, chlorates are produced both electrochemically and chemically.

Of greatest industrial importance are potassium and sodium chlorates. In the chemical method of production of potassium chlorate, the starting materials are slaked lime, chlorine, and potassium chloride. The reactions of these compounds in water are as follow:

$$6Ca(OH)_2 + 6Cl_2 = Ca(ClO_3)_2 + 5CaCl_2 + 6 H_2O \quad \text{and} \quad Ca(ClO_3)_2 + 2KCl = 2KClO_3 + CaCl_2.$$

After the solution has cooled, potassium chlorate crystals separate and are filtered off and dried. An important drawback of this process is the large amount of waste: about 3 tons of $CaCl_2$ per ton of $KClO_3$.

The advantage of the electrochemical method of producing clorates lies in the absence of waste and simplicity of production. The starting material, i.e., the alkali metal chloride, is found in nature. The material for producing $NaClO_3$ is NaCl solution. The process is carried out in cells without a diaphragm. The chlorine liberated at the anode reacts with the cathodic alkali, sodium chlorate $NaClO_3$ being formed in the solution.

In the case of electrolysis of KCl solution, the process is carried out until the $KClO_3$ content of the solution reaches 200 g/l; on cooling, $KClO_3$ crystals separate out. After the raw material, i.e., potassium chloride, is added to the mother liquor, the later is processed again.

Perchlorates are obtained by electrochemical oxidation of chlorates of the corresponding metals in accordance with the overall equation

$$ClO_3 + H_2O = ClO_4 + H_2.$$

Sodium perchlorate is usually obtained in this manner.

Potassium perchlorate, which is much more sparingly soluble in water than $NaClO_4$, is obtained by means of the exchange reaction

$$KCl + NaClO_4 = KClO_4\downarrow + NaCl.$$

NITRATES. Sodium nitrate used to be extracted from saltpeter deposits in Chile, whence it was exported to all parts of the world to be used as fertilizer. At the present time, $NaNO_3$ is obtained almost exclusively by synthetic means, i.e., reaction of nitrogen oxides with soda solution.

Potassium nitrate KNO_3 is found in very small amounts in nature. It is obtained from potassium chloride by exchange reaction with $NaNO_3$.

Strontium and barium nitrates can be obtained by dissolving the natural salt $SrCO_3$ (strontianite) or $BaCO_3$ (witherite) in nitric acid.

More complex is the process of preparation of Ba and Sr nitrates from their natural salts $BaSO_4$ (barite) and $SrSO_4$ (celestine). The chemistry of this process is discussed below with the aid of the example of barium salts.

When barite is calcined with carbon in an internally fired furnace, the following reaction takes place:

$$BaSO_4 + 4C = BaS + 4CO,$$

then the sulfide is treated with hydrochloric acid:

$$BaS + 2HCl = BaCl_2 + H_2S.$$

The barium chloride formed is converted into $Ba(NO_3)_2$ by reaction with ammonium nitrate solution:

$$BaCl_2 + 2NH_4NO_3 = Ba(NO_3)_2 + 2NH_4Cl.$$

Because of its limited solubility in cold water, barium nitrate is easy to separate from the solution.

A detailed description of the methods of production of the salts listed here can be found in textbooks and monographs by Vol'fkovich [21], Pozin [30], Billiter [18], Shraybman [40], Miniovich [26], and Smirnov [34].

Barium peroxide BaO_2 is obtained by heating porous barium oxide BaO in a stream of hot air (around 500°C). Barium oxide is obtained by calcining barium carbonate or nitrate. Barium peroxide combines with water to form the octahydrate $BaO_2 \cdot 8H_2O$.

Iron scale Fe_3O_4 is found in huge quantities in nature in the form of the mineral magnetite. To remove moisture and organic impurities from iron scale, the latter is subjected to roasting at 400-500°C and is then ground up.

Manganese dioxide MnO_2 is found in nature in the form of the mineral pyrolusite. It can also be obtained artificially from manganese salts or from potassium permanganate.

Chapter 3

FUEL SUBSTANCES

This chapter will discuss the properties of fuels that do not act as binders in pyrotechnic compositions. A description of the properties of binders is given in the following chapter (Ch. 4).

1. SELECTION AND CLASSIFICATION OF FUELS

In selecting a fuel for a pyrotechnic composition, it is necessary to consider all the requirements placed on the composition, one of the decisive factors being the achievement of the best special effect.

The best special effect in incendiary, illumination, tracer and photoillumination compositions is achieved at a high temperature of their combustion; it is therefore necessary to use a fuel of high calorific value. Fuels of high calorific value should also be used in solid pyrotechnic fuel.

For smoke compositions, a high combustion temperature is undesirable in many cases. Frequently, a fuel with a medium calorific value is selected, or an incomplete combustion of the fuel is carried out (for example, combustion of carbon in CO).

A major role in the selection of a fuel is also played by the physico-chemical properties of the products of its oxidation. First of all, it is necessary to know the state of aggregation of the oxidation products at room temperature and at the combustion temperature of the composition. For illumination compositions, it is important to observe a certain optimum relation between the amount of gaseous products that strongly affect the flame dimensions (and hence, the size of the luminous area) and the concentration (density) of solid and liquid particles of the emitters in the flame. Both an excess and a shortage of gaseous products cause a decrease in light intensity.

The combustion products of rocket compositions (solid pyrotechnic fuel) should contain the maximum amount of gases and the smallest possible amount of solids (for more detail see Ch. 20, Sec. 1). In smoke compositions, the amount of gases formed by the combustion should be very substantial, since this causes a timely expulsion of particles of smoke-generating substances into the atmosphere. The combustion products of thermite should remain liquid for a certain period of time, since otherwise the required incendiary effect will not be achieved. In selecting the fuel for flame compositions, it is necessary to consider the luminous radiation intensity of the products of its oxidation, and also the energy distribution of this radiation over the spectrum.

A major role in the selection of a fuel is also played by the ease with which its oxidation can be carried out. For example, silicon or graphite are very hard to oxidize, even when the most active oxidizers are used (combustion in pure oxygen or mixing with one of the most active oxidizers, chlorate), so that their use as fuels is extremely limited.

Aluminum, in finely divided powder or dust, burns fairly vigorously by combining both with the oxygen of the oxidizer and under certain conditions with atmospheric oxygen (in the case of combustion of aluminum dust suspended in air).

Magnesium, which is an easily oxidizable substance, can burn completely in atmospheric oxygen even without being finely divided. The capacity to burn by combining with atmospheric oxygen is very desirable for a fuel in pyrotechnic compositions.

Some fuels are oxidized with excessive ease, so that their mixtures with oxidizers are extremely sensitive to impact and friction, or have an excessively low ignition temperature. An example is white phosphorus, which not only can not be mixed with any oxidizer, but must not be stored in air, to avoid self-ignition. The use of mixtures of red phosphorus with oxidizers is also very limited: mixtures of red phosphorus with chlorates undergo self-ignition almost inevitably, and mixtures with other oxidizers (for example, nitrates) have an excessively high sensitivity to impact and friction.

It is desirable that the smallest possible amount of oxygen be required for the combustion of a unit of weight of fuel, since a high oxidizer content of a composition is disadvantageous; it leads to a decrease in the amount of fuel, and hence, to a decrease in the amount of liberated heat. In addition, in selecting a fuel, a sufficient chemical stability of the composition during storage should be ensured. It should be considered that certain combinations of fuels with oxidizers, for example, a mixture of chlorates with sulfur (for more detail, see Ch. 12) are insufficiently stable chemically.

Thus, the fuels employed in pyrotechnic compositions should meet the following principal requirements:

1) have a heat of combustion ensuring the best special effect of the composition;
2) be readily oxidized by the oxygen of the oxidizer or by atmospheric oxygen;
3) form combustion products that provide for the best special effect of the composition;
4) require the minimum amount of oxygen for their combustion;
5) be chemically and physically stable in the temperature range from -60 to +60°C, and as stable as possible toward weak solutions of acids and alkalis;
6) be nonhygroscopic;
7) be easy to pulverize;
8) have no toxic effect on the human organism;
9) be readily available.

Fuels used for the preparation of pyrotechnic compositions may be divided into the following categories:

INORGANIC FUELS

1. Metals of high calorific value: magnesium, aluminum; zirconium and titanium are used much less frequently.
2. Magnesium-aluminum alloys: aluminum-magnesium (AM), magnesium-silicon, aluminum-silicon, etc.
3. Metals of medium calorific value: zinc, iron, antimony, etc.
4. Nonmetals: phosphorus, carbon (in the form of carbon black or wood charcoal), sulfur.
5. Inorganic compounds.
   a) Sulfides: phosphorus sesquisulfide (P_4S_3), antimony sulfide (Sb_2S_3), pyrite (FeS_2), etc.
   b) Other inorganic compounds: phosphides, silicides, carbides, etc.

ORGANIC FUELS

1. Pure hydrocarbons: benzene, toluene, naphthalene, etc.
2. Mixtures of aliphatic and carbocyclic series: gasoline, kerosene, petroleum, fuel oil, paraffin, etc.
3. Carbohydrates: starch, sugar (lactose, beet sugar), sawdust.
4. Organic substances of other classes: urotropin, stearic acid, (stearin), dicyandiamide, metaldehyde, etc.

2. HIGH-ENERGY FUELS

The largest amount of heat evolved by combustion resulting from combination with atmospheric oxygen as well as oxygen of the oxidizer is evolved by the following twelve elements:

Metals - lithium, beryllium, magnesium, calcium, aluminum, titanium, and zirconium, [Note: High-energy metals also include Nb, V, Sc, and Y, but in our view their discussion as pyrotechnic fuels is premature. Ellern [9] gives recipes for "exotic compositions" with niobium and tantalum].

Nonmetals - hydrogen, boron, carbon, silicon, and phosphorus.

Table 3.1 lists data on the amount of heat evolved by the combustion of these elements.

The quantities of heat Q_1, Q_2, and Q_3 serve to characterize the calorific value.

The amount of heat Q_1 evolved in the combustion of 1 g of element to the oxide can serve as a measure of the calorific value of the fuel provided that it burns by combining with atmospheric oxygen.

Table 3.1 HEAT OF FORMATION OF OXIDES ACCORDING TO THE DATA OF REFERENCES 98, 94

ELEMENT		OXIDE			HEAT OF FORMATION KCAL		
SYMBOL	ATOMIC WEIGHT A	FORMULA	MOLECULAR WEIGHT M	PER MOLE OF OXIDE Q	$Q_1 = \dfrac{Q}{m \times A}$	$Q_2 = \dfrac{Q}{m}$	$Q_3 = \dfrac{Q}{n}$
METALS							
Li	6.9	Li$_2$O	30	143	10.4	4.8	48
Be	9.0	BeO	25	142	15.8	5.7	71
Mg	24.3	MgO	40	144	5.9	3.6	72
Al	27.0	Al$_2$O$_3$	102	400	7.4	3.9	80
Ca	40.1	CaO	56	152	3.8	2.7	76
Ti	47.9	TiO$_2$	80	224	4.7	2.8	75
Zr	91.2	ZrO$_2$	123	260	2.9	2.1	87
NONMETALS							
H	1.0	H$_2$O	18	68.4	34.2	3.8	23
C	12.0	CO$_2$	44	94	7.8	2.1	31
B	10.8	B$_2$O$_3$	70	302	14.0	4.3	60
Si	28.1	SiO$_2$	60	208	7.4	3.5	69
P	31.0	P$_2$O$_5$	142	367	5.9	2.6	52

NOTE: 1 kilocalorie (kcal) = 4.1868 kilojoules (kJ).

Q_1 is calculated by dividing the gram-molecular heat of formation of the oxide Q by the quantity mxA, where A is the atomic weight, and m is the number of atoms of the element comprising the oxide molecule.

The amount of heat Q_2 evolved by the formation of 1 g of oxide may serve to some extent as a measure of the calorific value of the composition in which a given fuel burns by combining with the oxygen of the oxidizer.

The amount of heat Q_3 obtained by dividing the gram-molecular heat of formation of the oxide Q by the number of atoms n in the oxide molecule permits an approximate evaluation the combustion temperature of the element, since as a first approximation, the combustion temperature is proportional to the amount of heat per 1 g-atoms, i.e., is proportional to Q_3.

The highest combustion temperature is developed by metals: zirconium, aluminum, magnesium, and calcium. Hydrogen, carbon, phosphorus, and lithium have a much lower combustion temperature. At the present time, of these elements, only aluminum and magnesium are used on a wide scale, while phosphorus and charcoal (C) are employed on a somewhat smaller scale.

The chief high-energy pyrotechnic fuel is aluminum, a metal present in large amounts in the earth's crust (8.8%) and extracted in very large quantities: in 1960, the world production of aluminum was more than 4 million tons.

Magnesium is in second place; its world production in 1944 (during the war) was about 210 thousand tons, and in 1958, about 110 thousand tons. In recent years, the USSR has become the number 1 magnesium-producing country in the world. The growth of the magnesium industry is considerably affected by the steadily increasing demand for magnesium alloys, which are used in aircraft construction and rocket engineering; magnesium alloys with Al, Zn, Mn, and Zr are of technical importance, and magnesium is also used in the production of titanium metal [68].

Beryllium is a scarce element in the earth's crust; the extraction of its ore (beryl) did not exceed 10,000 tons in 1959; as a fuel in pyrotechnics, beryllium has been insufficiently studied. A major obstacle to its practical application is its very significant toxicity [69], particularly in the finely dispersed state. The high boiling point of beryllium (about 2400°C) makes its vaporization during burning difficult. Beryllium is difficult to burn. For the properties of beryllium, see [69].

Zirconium [62] is an expensive and scarce material whose content in the earth's crust is low (0.02%); compositions containing zirconium have a high combustion temperature and rate. At the present time, zirconium is used only in gasless and incendiary compositions in small-size articles (for more detail, see Ch. 21). Among its advantages are the small amount of oxygen consumed by its combustion and its high corrosion resistance. [108].

Titanium. At 400-600°C, titanium reacts easily not only with atmospheric oxygen, but also with atmospheric nitrogen. The titanium content of the earth's crust is appreciable (0.6%). In the USA, titanium is referred to as a "child of the war" (1941-1945). The world production of titanium in 1956 was about 20,000 tons. Although titanium is still expensive, the prospects for the development of the titanium industry are excellent [53, 63]. There are patents for the use of titanium in pyrotechnic compositions. For example, the following compositions of a smoke mixture is given: Ti 14%, C_2Cl_6 43%, ZnO 43% (DBP patent 1063507).

In the American patent No. 2929697 (1960), a rocket powder composition containing $KClO_4$, raphite, Ti powder (8%) and an organic binder is recommended.

Calcium is not expensive and not scarce. Its content in the earth's crust is high (3.6%), but its production is still lower than that of magnesium, since it has no extensive applications.

The use of powdered calcium as a fuel involves serious difficulties, since calcium at room temperature reacts vigorously with atmospheric moisture and oxygen. Calcium may be used in the form of alloys with other metals.

The lithium content of the earth's crust is low (0.006%). The use of lithium in pyrotechnics, or even alloys including a significant percentage of lithium, is difficult, since it is a metal which reacts very actively with atmospheric moisture and oxygen [64].

Hydrogen is used only in the combined state in the form of organic compounds.

The possibility of using metal hydrides in pyrotechnics is problematic because of the low chemical stability of these compounds [70]; among the drawbacks of hydrides one should also include their low density.

PRINCIPLES OF PYROTECHNICS

The boron content of the earth's crust is low (0.001%); however, the world production of borates and boric acid amounts to hundreds of thousands of tons per year. At one time, the use in rocket technology of boron hydride fuels, which have higher calorific values than ordinary hydrocarbons, was widely advertised in the USA. However, in 1959, a report appeared to the effect that their production has been discontinued because of their high toxicity, transportation difficulty, and also because of their high cost (see, for example, Maxwell and Young [227]).

The use of elemental boron in pyrotechnic compositions involves certain difficulties: the high melting point (2050°C) and boiling point (2550°C) of boron, as well as the low softening temperature (about 450°C) of boron oxide B_2O_3. The latter fact interferes to some extent with the production of high temperatures during the combustion of boron in pyrotechnic compositions, since a substantial part of the heat evolved during its combustion is expended in melting and then in vaporizing boron oxide. Indications given in the literature to the effect that crystalline boron is capable of rapid oxidation do not as yet signify that boron will soon find practical applications in pyrotechnic compositions. As was indicated in [82], [228], the oxidation rate of boron considerably depends on the possibility of a rapid removal of the B_2O_3 oxide film coating its particles. For more detail on the properties of boron, see Samsonov's monograph [65].

The use of elemental silicon in pyrotechnics is hindered by its difficult oxidizability; in the form of alloys with magnesium, aluminum, and other metals, its application in illuminating compositions, for example, may give satisfactory results. Concerning the use of silicon as an additive to explosives, see the paper of Sartorius [81].

In addition to the calorific value of a fuel (see Table 3.1), of great importance in pyrotechnics in many cases is the amount of heat evolved by the combustion of a unit volume of fuel (Q_4 kcal/cm^3). Such data are listed in [9]. We will present them here in somewhat modified form (see Table 3.2).

Table 3.2 AMOUNT OF HEAT (in kcal) EVOLVED BY THE COMBUSTION OF 1g (Q_1) and 1 cm^3 (Q_4) OF CERTAIN FUELS

SUBSTANCE	Q_1	Q_4	SUBSTANCE	Q_1	Q_4
Be	15.8	29	Nb	2.5	21
Al	7.4	20	Mo	1.9	19
Mg	5.9	10	Fe	1.8	14
Ca	3.8	6	Mn	1.7	12
Ti	4.7	200	Ce	1.7	11
Zr	2.9	18	Ta	1.4	23
B	14.0	33	Zn	1.3	9
Si	7.4	18	W	1.1	21
P	5.9	11	Ni	1.0	9

It is evident from Table 3.2 that in calorific value per unit volume, the leading position is that of boron (the calculation was carried out by using a density of 2.3 g/cm^3), while the metals Zr, Nb, Mo, Ta, and W have a lower calorific value comparable to that of aluminum and much higher than that of magnesium.

For the substances considered in Table 3.1, Table 3.3 lists the physicochemical properties mentioned in Sec. 1 of this chapter.

Table 3.3 PHYSICOCHEMICAL PROPERTIES OF FUELS AND THEIR OXIDES

SYMBOL	FUEL					OXIDE		
	DENSITY G/CM³	IGNITION TEMPERATURE OF POWDER IN AIR, °C	°C OF MELTING POINT	°C OF BOILING POINT	AMOUNT OF FUEL G, BURNED IN 1g OF OXYGEN	FORMULA	°C OF MELTING	°C OF BOILING
METALS								
Li	0.5	180	181	1370	0.87	Li_2O	(2600)	
Be	1.8	>800	1284	(2400)	0.56	BeO	2530	(4120)
Mg	1.7	550	650	1105	1.52	MgO	~2800	--
Al	2.7	>800	660	2500	1.12	Al_2O_3	2030	(~3000) see note
Ca	1.5	~600	850	1420	2.50	CaO	2600	(3500)
Ti	4.5	300-600	1660	(3260)	1.50	TiO_2	1920	(~3000) see note
Zr	6.5	180-200	1868	(4750)	2.85	ZrO_2	2700	(4300)
NONMETALS								
H	0.07 (liq.)	--	-259	-253	0.12	H_2O	0	100
C Graphite	2.2	--	>3000	--	0.38	CO_2	-78	Sublimes
B Crystalline	2.3	>900	2050	(2550)	0.45	B_2O_3	~450	(1500)
Si	2.3	>900	1430	2600	0.88	SiO_2	1713	(~2950)

Note: With decomposition.
The boiling temperature of oxides within parentheses, 3000°C and above, are not reliable.

The ignition temperature of metal powder considerably depends on the size and shape of the powder particles as well as on the quality of the oxide film coating them. The higher the dispersity of the metal powder, the lower the ignition temperature.

Schichter [161], who made a sampling of the foreign literature, points out that the ignition temperature of titanium powder may vary from 300 to 600°C, and some specimens of finely dispersed zirconium powder will ignite at room temperature. He reports that moistened zirconium powder burns more vigorously than dry powder (the $2H_2O + Zr$ system is capable of burning and exploding), and quenching of burning zirconium is permissible only by dousing with powdered CaF_2 or CaO, since H_2O, CCl_4, and even $CaCO_3$ react vigorously with zirconium. Ellern [9] points out that a number of serious incidents have occurred in work with finely dispersed zirconium powder (2-5 microns). The ignition temperature of finely dispersed zirconium powder has been observed to be about 85°C. It is transported and, insofar as possible, treated under water. Fractions of Zr powders have an ignition temperature of the order of 180-200°C. Zirconium always contains traces of hydrogen, and its ignition temperature is lower the less hydrogen it contains. Zr powder with a particle size of 10 microns or more is estimated [9] to be fairly safe to handle.

The largest amount of oxygen for oxidations is required by hydrogen, followed by carbon for combustion in CO_2 (see Table 3.3). It follows that compositions in which the fuels are organic substances should contain a large amount of oxidizer and correspondingly a very small amount of fuel.

One of the main factors determining the chemical stability of metals is their standard electrode potential. These data are listed in Table 2.4.

The oxidizability of metals by gaseous oxygen is determined by the quality of the oxide film coating the metal. In the case of an oxide whose vaporization is difficult, the ease of oxidation of the metal depends on the properties of the oxide film formed.

According to the rule of Pilling and Bedworth [80]; [170, p. 43], if the volume of the oxide formed is smaller than that of the metal it replaces, the outside film has a loose, cellular structure and can not reliably protect the metal from further oxidation. If, however, the ratio of the volume of the oxide to that of the metal is greater than unity, the oxide film formed has a compact, continuous structure, effectively isolates the metal from the action of gaseous oxygen, and hence, prevents further oxidation of the metal.

The coefficient α of Pilling and Bedworth is calculated from the formula:

$$\alpha = \frac{M_{ox} \times D_{me}}{D_{ox} \times A_{me}} n,$$

D_{ox} is the density of the oxide,
D_{me} is the density of the metal,
A_{me} is the atomic weight of the metal,
n is the number of metal atoms in the oxide formula.

Table 3.4 lists the values of α for different metals and silicon [242, 80].

Table 3.4 RATIO OF VOLUME OF OXIDE TO VOLUME OF METAL

Na	0.55	Ba	0.78	Pb	1.31	Zn	1.59	Fe	2.06
K	0.45	Ca	0.64	Cd	1.32	Ni	1.68	Mn	2.07
Li	0.58	Mg	0.81	Sn	1.33	Cu	1.70	Co	2.10
Sr	0.69			Zr	1.45	Ti	1.73	Cr	3.92
				Al	1.45	Si	2.04		

As is evident from Table 3.4, for light metals, i.e., alkali, alkaline earth metals, and magnesium, $\alpha < 1$, and for heavy metals and aluminum, $\alpha > 1$. As was pointed out by Pilling and Bedworth, the value of α determines the behavior of metals in high-temperature corrosion: if $\alpha < 1$, the metal corrodes easily and rapidly.

It appears that the small value of for magnesium ($\alpha = 0.81$) is one of the reasons for the rapid combustion rate of magnesium compositions.

Samsonov and Portnoy [66] noted that "for very high values of α, the oxide layer undergoes considerable stresses, cracks, and loses its protective properties; for this reason, the best protective properties are displayed by layers for which does not appreciably exceed 1."

PRINCIPLES OF PYROTECHNICS

Most important for pyrotechnists are the physicochemical properties of two metals: magnesium and aluminum.

MAGNESIUM (according to the data of V. M. Gus'kov [56], Kh. Strelets, A. Tayt et al. [68], see also the handbook [79]). The heat of fusion and boiling of magnesium is 2.1 and 30.5 kcal/g-atom, respectively (for the melting and boiling points, see Table 3.3). The heat of sublimation at the melting point is 34.0±1.5 kcal/g-atom. The atomic heat capacity is 5.86 cal/g-atom deg at 0°C, 6.70 at 300°C, 7.58 at 650°C, for liquid magnesium 8.10, and for gaseous 4.97. The thermal conductivity is 0.37 cal/cm sec deg (at 20°C). The saturated vapor pressure in Hg is 1 at 622°C, 5 at 702°C, 20 at 750°C, and 100 at 909°C [73].

Magnesium is very active chemically, but up to about 350°C an oxide film protects it from oxidation to a certain extent (see the papers of Makolkin and Vernidub [61]). On heating to a higher temperature, the oxidation of magnesium speeds up rapidly. At 500°C, according to the data of Pilling and Bedworth [80], the oxidation rate of magnesium is 0.52 mg/cm^2 hr. Magnesium in the form of lumps and plates ignites in air at 600-650°C, and powdered magnesium, at about 550°C. On burning in air, magnesium forms magnesium oxide MgO and partly the nitride Mg_3N_2.

Concerning the mechanism of combustion of magnesium ribbon in oxygen, air, and other gaseous oxygen mixtures, see Coffin [143] and Kirschfeld [146]. The paper of Delavo indicates that the addition to air of 1% by volume of SiF_4 or BF_3 is sufficient to extinguish the flame of burning magnesium.

Magnesium oxide is a light white powder (density, 3.6 g/cm^3); thoroughly calcined magnesium oxide loses its ability to combine with water and dissolves in acids. The volatility of MgO is appreciable at about 2000°C; its melting point is about 2800°C. Most of the boiling temperature values reported in the literature range from 3000 to 3600°C. Since the melting and boiling temperatures increase in the homologous series Ba, Sr, Ca, Mg [7, p. 36], and the boiling temperature of MgO will not be lower than 3100°C. The latent heat of sublimation of MgO is estimated at 150 kcal/mole [68]. Magnesium oxide is widely employed in the production of refractories.

Magnesium nitride Mg_3N_2 is a grayish-green solid substance easily decomposed by water: the combination of magnesium with nitrogen is associated with a much lower evolution of heat than that of magnesium with oxygen. The heat of formation of Mg_3N_2 from the elements is 115 kcal/mole (for 3MgO, 144 x 3 = 432 kcal).

ALUMINUM [201, 55]. The heat of fusion and boiling of aluminum are 2.5 and 69.6 kcal/g-atom, respectively. The atomic heat capacity is 6.0 cal/g-atom deg at 0°C; 7.36 at 660°C, and that of liquid aluminum 7.19 at 1000°C. The vapor pressure in Hg is 2 x 10^{-5} at 660°C, 1 at 1284°C, 5 at 1421°C, 20 at 1555°C, and 100 at 1749°C [73]. The thermal conductivity is 0.52 cal/cm sec deg at 20°C.

Aluminum is chemically active, but under ordinary conditions (including in the powdered state), its oxidation is prevented by a thin but strong oxide film coating its surface. When heated, powdered aluminum burns vigorously in air. At the temperature of red heat, it combines actively with sulfur, forming the compound Al_2S_3. At 800°C, aluminum combines with nitrogen to form the nitride AlN, consisting of white crystals with a fusion temperature of 2200°C (at a nitrogen pressure of 4 atmospheres).

Aluminum is trivalent in its compounds, but at high temperatures, compounds of univalent aluminum exist (see Ch. 17).

Aluminum oxide is a white powder with the following density: corundum 3.96, alumina 3.42 g/cm^3. The temperature dependence of the molar heat capacity in the range 100-1400°C for Al_2O_3

is expressed by the formula $C_p = 23.86 + 0.00673t$. The fusion temperature of Al_2O_3 is 2050°C. At high temperatures (of the order of 3000°C), Al_2O_3 dissociates significantly with the detachment of oxygen, forming lower oxides AlO, and in a reducing atmosphere Al_2O, so that the boiling temperature of Al_2O_3 given in handbooks is very tentative.

Of the metal alloys used in pyrotechnics particular attention should be given to magnesium-aluminum alloys. The intermetallic compound Mg_4Al_3 (54% magnesium by weight) has a heat of formation of +49 kcal/mole; its density is 2.15, and melting point, 463°C. This alloy is advantageously different from the corresponding mixtures of magnesium and aluminum in that it has a lesser tendency to corrode, and its brittleness is greater, so that it can be ground up easily. It is well known that intermetallic compounds are generally characterized by a significant brittleness and a hardness greater than that of the constituent components. Aluminum-magnesium alloys containing 85-90% magnesium have been named "electron" alloys.

3. TECHNICAL REQUIREMENTS PLACED ON METAL POWDERS

The following requirements are usually placed on metal powders in pyrotechnics:

1. Maximum content of active (unoxidized) metal (from 90 to 98% for different metals and different grades of powders).
2. A content of iron and silicon impurities not in excess of a few tenths of 1% (with the exception of special alloys containing these elements as integral components).
3. Minimum amounts (usually, traces) of heavy metal impurities (copper, lead).
4. Content of fats - not more than a few tenths of one percent; in some cases, none are allowed to be present.
5. Moisture content, not more than tenths of one percent.
6. The powder is standardized according to the degree of pulverization.

A silicon impurity is dangerous if silicon is present in the form of compounds (sand, glass) that increase the sensitivity of the compositions. The presence of copper, lead and iron impurities (particularly, the first two) in the metal powder may favor the formation of a galvanic couple in a moistened composition, and this will unquestionably accelerate the process of dissociation of the composition in storage.

A high fat content slows down the combustion process and may promote an increase in sparking.

The requirements placed on aluminum powder with respect to its content of the active metal and impurities are as follows:

Active aluminum	not less than 96%
Aluminum oxide	not more than 3%
Iron impurities	not more than 0.8%
Silicon impurities	not more than 0.6%
Copper and lead impurities	traces
Fats	not more than 0.5%
Moisture	not more than 0.2%

The requirements placed on magnesium powder with respect to its content of active metal and impurities are as follows:

Active metal	not less than 98%
Magnesium oxide	not more than 1%
Iron	not more than 0.35%
Silicon	not more than 0.15%
Moisture	not more than 0.2%
Copper	not allowed
Chloride salts	traces
Fats	not allowed

Pulverization (screen number indicates the number of openings per linear centimeter):

Passes through screen No. 30	100%
Passes through screen No. 60	not more than 10%

4. PRODUCTION OF METAL POWDERS

Several methods are used to prepare metal powders:

1) mechanical grinding;
2) spraying of liquid metals;
3) reduction of oxides;
4) electrolysis.

Mechanical grinding is carried out in ball or hammer mills. This method is most convenient when brittle metals or alloys are ground up.

Spraying of a metal melt (a very economical method) is used to prepare powders of comparatively low-melting metals, including aluminum and zinc. Several versions of this method exist: 1) spraying with compressed air or gas, 2) centrifugal spraying, 3) granulation by pouring the metal into water.

The method of thermic reduction of oxides is used, for example, to prepare titanium and niobium powders. Electrolysis of molten media produces powders of high-melting metals, zirconium, etc. The shape of the powder particles depends on the method of their preparation. When metal melts are sprayed, the powder particles are in the shape of droplets or spheres.

When metal oxides are reduced, the powders obtained are porous and loose. In powders obtained electrolytically, the shape is irregular, branched (dendrites). The particle size of metal powders depends on the manner of their preparation, and varies over very wide limits, from 500 microns (coarse powders), to 50-20 microns (fine powders) and down to 5-2 microns (superfine powders). Obviously, a metal powder is finer the higher is its reactivity and lower its density (g/cm^3) when poured freely (or shaken).

Magnesium is very reactive chemically, and therefore magnesium powder is obtained most frequently by mechanical grinding. To prepare magnesium powder, magnesium bars are converted on special cutting machines into fine shavings. Further comminution is carried out in an inert gas atmosphere in special ball mills. The production of magnesium powder is dangerous because of the high chemical activity of the finely divided metal. Another method of obtaining magnesium powder by cooling magnesium vapor in a condenser has also been described.

Aluminum powder is obtained by spraying the molten metal with compressed air or by centrifugal spraying [34]. Aluminum dust is obtained by comminution of aluminum powder (or foil sheets) in ball or hammer mills. This process (of significant increase in the surface area of aluminum) involves a risk of its oxidation. To avoid this possibility, before the comminution, to the aluminum powder is added about 1% of lubricating additives (stearin or paraffin). After the comminution it is removed by extraction or heating in a vacuum.

PRINCIPLES OF PYROTECHNICS

In Germany, during the period 1939-45, about 25,000 tons of aluminum powder was produced per year, and the annual production of powders of magnesium and magnesium alloys was about 10,000 tons [54].

Information on the methods of preparation and properties of metal powders may be found in [84], [54], and [76]. Dispersity is found in [57],[74] and[75].

5. INORGANIC FUELS OF MEDIUM CALORIFIC VALUE

In cases where a large evolution of heat and a high combustion temperature are not required, for example, in smoke or fuse compositions, the use as fuels of zinc, antimony, iron, manganese, sulfur, and possibly some other elements, as well as certain inorganic compounds, is possible. Not only elements, but alloys of metals and inorganic compounds can be used as fuels in pyrotechnics. Obviously, other things being equal, the heat of combustion of a compound will be greater the lower its heat of formation from the elements. Thus, the following selection principle takes form:

<u>The fuels used in pyrotechnics may be compounds of the elements listed in Tables 3.1 and 3.5, having a low heat of formation (Q, kcal/g).</u>

The heat of formation of oxides of certain elements of comparatively low calorific value (from the standpoint of the characteristics Q_1 and Q_2) is given in Table 3.5.

Table 3.5 HEAT OF FORMATION OF OXIDES

ELEMENT		OXIDE		HEAT OF FORMATION KCAL			
SYMBOL	Atomic Weight	FORMULA	Molecular Weight	Per Mole of Oxide	$Q_1 = \dfrac{Q}{m \times A}$	$Q_2 = \dfrac{Q}{m}$	$Q_3 = \dfrac{Q}{n}$
Na	23.1	Na_2O	62	99	2.1	1.6	33
K	39.1	K_2O	94	85	1.1	0.9	29
Cr	52.0	Cr_2O_3	152	273	2.6	1.8	55
Mn	54.9	MnO	71	93	1.7	1.3	46
Fe	55.8	Fe_2O_3	160	195	1.7	1.2	28
Co	58.9	CoO	75	57	1.0	0.8	28
Ni	58.7	NiO	75	58	1.0	0.8	29
Zn	65.4	ZnO	81	83	1.3	1.0	41
Ga	69.7	Ga_2O_3	187	256	1.8	1.4	51
Ge	72.6	GeO_2	105	128	1.8	1.2	43
As	74.9	As_2O_5	230	219	1.5	1.0	31
Mo	95.5	MoO_3	144	180	1.9	1.3	45
Sb	121.8	Sb_2O_5	324	230	0.9	0.7	33
S	32.1	SO_2	64	71	2.2	1.1	24
C	12.0	CO	29	26	2.2	0.9	13

See Table 3.1 for explanation of symbols

Compounds with covalent bonds usually have a a low heat of formation. In discussing such compounds from the standpoint of pyrotechnics, attention should be turned to compounds of metals with nonmetals of groups III, IV, V, and VI of the periodic system, i.e., borides, carbides, silicides, phosphides, and sulfides. Many of these compounds have not yet been tested by pyrotechnists, while others have unsuitable properties, being either too easy (many phosphides) or too difficult (silicides of certain metals) to oxidize.

Among binary inorganic compounds, the following sulfides have practical applications: pyrite FeS_2, antimony sulfide Sb_2S_3, and some others. Their properties are listed in Table 3.6.

Table 3.6 PHYSICOCHEMICAL PROPERTIES OF SULFUR COMPOUNDS

FORMULA	MOLECULAR WT.	DENSITY °C	MELTING POINT °C	BOILING POINT °C	COLOR	HEAT OF FORMATION KCAL/MOLE	COMBUSTION PRODUCTS ADDITIONAL OF SO_2	AMOUNT OF HEAT, KCAL, DUE TO COMBUSTION OF 1g OF SUBSTANCE	AMOUNT OF SUBSTANCE, g, BURNING IN 1g OF OXYGEN
P_4S_3	220	1.3	172	407	yellow	77	P_2O_5	3.9	0.86
FeS_2	120	5.0	--	--	gray	36	Fe_2O_3	1.7	1.36
As_2S_2	214	3.5	320	565	orange	19	As_2O_3	1.3	1.92
As_2S_3	246	3.5	310	707	yellow	20	As_2O_3	1.4	1.71
Sb_2S_3	340	4.6	548	--	black	36	Sb_2O_3	1.1	2.36

6. ORGANIC FUELS

Liquid hydrocarbons (gasoline, kerosene, petroleum, etc.) are used in incendiary mixtures burning in atmospheric oxygen.

Table 3.7 and the text below give the characteristics of various petroleum products.

Table 3.7 PROPERTIES OF PETROLEUM PRODUCTS

PETROLEUM PRODUCT	DENSITY	TEMP °C BOILING POINT	TEMP °C FLASH POINT
Petroleum ether	0.64-0.67	50-60	-58
Gasoline:			
a) light	0.67-0.72	60-100	--
b) heavy	0.72-0.76	100-120	--
Ligroin	--	120-160	from +5 to +15
Kerosene:			
a) light	0.78-0.84	150-300	from +28 to +45
b) heavy	0.80-0.87	--	--
High ign. temp. kerosene	--	270	100
Spindle oil	0.88-0.89	--	160
Machine & Cylinder oil	0.90-0.92	--	180-220
Fuel oil	0.91	300	100-140
Petroleum	0.78-0.92	--	30-90
Benzene	0.88	80	from -12 to +10

The amount of heat Q_1 evolved by the combustion of 1 g of the product is 11.2 kcal for gasoline, 10.8 kcal for petroleum, and 10.0 kcal for benzene. Gasoline and kerosene freeze at a temperature below -80°C, benzene freezes at +6°C. The hydrogen content of gasoline is 14-15%, and that of benzene, 7.7%.

Turpentine is obtained from the resinous secretions of conifers; its chief component is the unsaturated hydrocarbon pinene ($C_{10}H_{16}$). Turpentine differs from saturated hydrocarbons in its marked ease of oxidation; it ignites readily on contact with concentrated nitric acid.

If the evolution of a large amount of heat markedly decreases the special effect, various carbohydrates are usually employed as fuels. The amount of oxidizer added to them should ensure the combustion of the carbon of the organic substance to carbon monoxide CO only.

The properties of certain carbohydrates are indicated below.

Starch $(C_6H_{10}O_5)_n = [C_6H_7O_2(OH)_3]_n$. The heat of formation is 1.55 kcal per g, density 1.6; practically insoluble in cold water, it dissolves much better in hot water; in an acid medium, starch hydrolyzes to grape sugar: $(C_6H_{10}O_5)_n + nH_2O = nC_6H_{12}O_6$ (d-glucose)

Lactose $C_{12}H_{22}O_{11} \cdot H_2O$. Density 1.5; water is driven off only at 125°C; it melts at about 200°C (with decomposition). At room temperature, 17 g of lactose dissolves in 100 g of water; on heating, the solubility in water increases sharply; it is very sparingly soluble in ethanol. The heat of formation is 651 kcal per 1 g-mole.

Beet (cane) sugar $C_{12}H_{22}O_{11}$. Density 1.6; melts around 160°C with slight decomposition. At room temperature, 190 g of sugar dissolves in 100 g of water. It is sparingly soluble in ethanol.

Wood (sawdust) consists mostly of cellulose $(C_6H_{10}O_5)_n$, whose content may reach 2/3 of the weight of dry wood. The heat of formation of cellulose is 1.55 kcal per g; air-dried wood yields about 3-3.5 kcal per g, and cellulose, 4.2 kcal per g on burning.

Among other organic fuels used in pyrotechnics should be mentioned stearin, naphthalene, paraffin, urotropin, and dicyandiamide.

Stearin or stearic acid $C_{17}H_{35}COOH$. A handbook of pyrotechnics published toward the end of the 19th century [Note: P. F. Simonenko, Pyrotechnics, St. Petersburg, 1894.] states: "stearin (acidum stearinicum) is one of the components of tallow. It is used in the form of shavings, since it is very difficult to grind. Sticks of stearin are used for this purpose. The stearin must not have the odor of tallow or soil the hands. It is used in the compositions of certain fires."

The melting point of stearic acid $C_{18}H_{36}O_2$ is 71°C, and the boiling range, 359-383°C (at atmospheric pressure); its density is 0.94. Stearic acid is a surface-active agent, and is therefore strongly adsorbed on the surface of the component powders of a composition. The heat of formation of stearic acid from the elements is 223.8 kcal/mole [111].

At the present time, stearin is used in the manufacture of solid fuels and as a fuel (and simultaneously, a plasticizer) in many flame compositions (illuminating and rocket compositions, etc). Commercial stearin is a mixture of stearic and palmitic acids. The melting of palmitic acid $C_{16}H_{32}O_2$. is 62°C, and its boiling point, 268°C.

Naphthalene $C_{10}H_8$. the melting point is 80°C, boiling point 218°C, and density (at 20°C), 1.16. The heat capacity of solid naphthalene at 20°C is 0.30 cal/g deg, and that of liquid naphthalene at 87°C, 0.30 cal/g deg; the latent heat of fusion is 34.6 cal/g, and the latent heat of vaporization, 75.5 cal/g. The heat of formation of naphthalene from the elements is 16 kcal/mole. In flame compositions, naphthalene is used as the fuel; in compositions of white smokes, which have a low combustion temperature (less than 500°C), it burns partially and partially sublimes, playing the part of an additional smoke-forming substance.

Paraffin is a mixture of saturated hydrocarbons. In stoichiometric calculations, the conventional formula $C_{26}H_{54}$ can be used. It is a white or slightly yellowish mass without taste or odor. It is obtained by processing certain grades of petroleum, brown coal, or peat. The melting range is 44-58°C, and the density, 0.88-0.91 g/cm³. It is chemically less reactive than naphthalene or stearin. The heat of combustion of paraffin (at the melting point, 52.5-53°C) is 11.19 kcal/g [111].

Urotropin or hexamethylenetetramine, $C_6H_{12}N_4$. It is obtained by condensation of formaldehyde with ammonia. It contains considerable nitrogen, and therefore its combustion gives a colorless, nearly soot-free flame; this property is particularly valued when it is used in signal light compositions.

Urotropin is the chief component of commonly used "solid alcohol" pellets (used for heating water, food, etc.). It is soluble in water: at room temperature, 167 g of urotropin dissolves in 100 g of water; it is sparingly soluble in absolute ethanol, and even less in benzene. It is decomposed by acids. It is thermally stable - distils under vacuum at 230-270°C almost without decomposition.

Urotropin is an endothermic compound, i.e., has a heat of formation from the elements of -30 kcal/mole, which is even more negative than that of naphthalene (see above). It is widely employed in medicine; it serves as the raw material for the production of the powerful explosive hexogen.

Dicyandiamide - DCDA $C_2N_4H_4$, $NH_2C(=NH)NH-C=N$. It is used as a fuel and simultaneously as a flash eliminator in smoke compositions. It is obtained by dimerization of cyanamide in the presence of NH_3. The melting point is 209°C, and the density, 1.40 g/cm^3. It is poorly soluble in cold water (2.2 g per 100 g of H_2O) and nonhygroscopic. It is soluble in acetone, hot water, and hot alcohol, and insoluble in benzene. When heated above 180°C, it dissociates by releasing ammonia, to form melamine $C_3N_3(NH_2)_3$. The heat of formation of DCDA from the elements is slightly negative, namely, −4.5 kcal/mole [111].

During World War II, DCDA was added as a flash eliminator to artillery powder charges (cp. [51], Vol. V, p. 827). In industry, DCDA is used for the production of amino plastics.

Chapter 4

METHODS OF COMPACTION OF COMPOSITIONS AND TESTING OF THE STRENGTH OF PELLETS. ORGANIC POLYMER BINDERS.

1. METHODS OF COMPACTION OF COMPOSITIONS. USES OF BINDERS

The compaction and shaping of pyrotechnic compositions may be carried out in different ways: pressing, screw conveying, pouring, and in some cases, packing by hand.

In the manufacture of pyrotechnics by the screw extrusion method, the working tool is an Archimedes screw, called a screw conveyer. The conveyer is used to feed the composition to the shell of the articles, and at the same time, to compact the compositions inside this shell.

Compaction of the compositions is achieved by applying from the outside to the objects being loaded (on the side opposite to that where the composition is supplied) a force of constant magnitude white up to a certain limit prevents the object from moving away from the screw conveyer during the latter's operation [156].

The use of a screw conveyer is an industrial method of loading. However, its use in pyrotechnic production is difficult in view of the sensitivity of pyrotechnic compositions to mechanical actions. The use of a screw conveyer is also difficult because many pyrotechnic compositions do not have the degree of plasticity required in this method of loading.

Less sensitive to mechanical impulses and sufficiently plastic are some compositions of masking smokes (see Ch. 18); for such compositions, loading by the screw conveyer method is permissible.

Loading by pouring is used in the case of yellow phosphorus, its alloys and solutions. Incendiary ammunition with liquid, condensed or solidified fuels (TG, napalm, etc.) is also loaded only by pouring. In the past few years, loading by pouring has been used for composite powders containing not less than 20% of organic substances (see Ch. 20).

The main obstacle to the application of this method to loading ammunition with other types of pyrotechnic compositions is the high melting temperature of the main components of pyrotechnic compositions, i.e., inorganic oxidizers and metallic fuels.

The content of low melting components (with melting points below 120-150°C) in pyrotechnic compositions containing an oxidizer usually does not exceed 10-15%; for this reason, it is impossible in most cases to use this method of suspending high melting components in a liquid melt used, for example, in loading high-percentage amatols (60/40) by pouring.

The most frequently used method of loading of pyrotechnic compositions is that of cold pressing on hydraulic or mechanical presses.

Extrusion at higher temperature (60-100°C) might improve the quality of loading, but this mode of operation would be too dangerous, since raising the temperature markedly increases the sensitivity of the compositions to impact and friction.

As already pointed out, compaction of a composition is necessary in order to make its combustion slower and more uniform. A compacted (pressed composition) must have a high mechanical stability. This is particularly necessary in cases where the products (torches, pellets, segments, etc.) are subjected to significant destructive mechanical forces when they are used under combat conditions. A product made of a compacted pyrotechnic composition should have a sufficient mechanical strength to prevent cracking, chipping, or falling off of fragments of the composition either in the course of shooting or during combustion of the composition. It is not always possible or expedient to achieve a high strength of pressed compositions only by using high pressures during pressing.

In order to increase the strength of pyrotechnic products, binders are often introduced into the compositions. [Note: Organic binders simultaneously act as fuels: their introduction also slows down the process of combustion of many compositions and increases their chemical stability.] The binders most frequently used are artificial and natural resins, rubber, and some other organic substances; however, in some cases, the use of inorganic substances such as sulfur, etc. is also possible for this purpose. In some cases, the introduction of binders is aimed at imparting the necessary strength to granular pyrotechnic compositions; the binder should ensure the necessary strength of the grains (granules) during transportation, storage, and action of the product.

2. STRENGTH TESTING OF PELLETS

The majority of pyrotechnic compositions are loaded into the ammunition in the form of compacted products of the required shapes.

The method most frequently used at the present time for testing the strength of products made of pyrotechnic compositions is the determination of the force (in kg/cm^2) required to crush a test cylinder (pellet) of compacted pyrotechnic composition. This is done on special testing machines (Fig. 4.1).

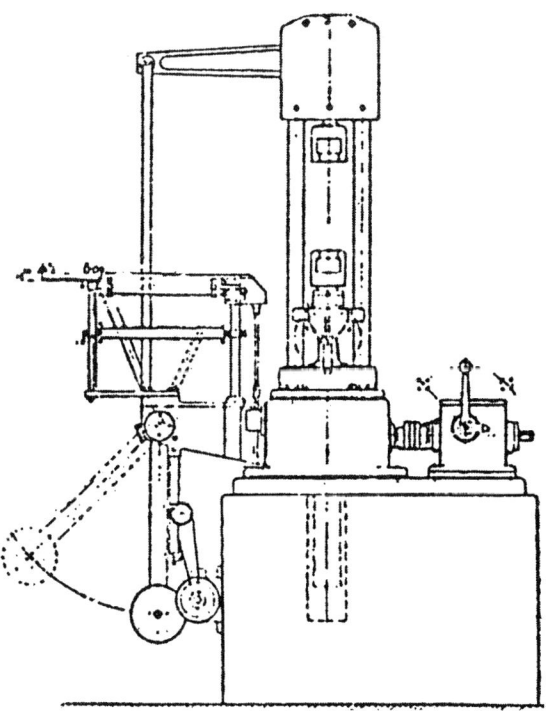

Fig. 4.1. IM-4R testing machine.

In its design, the IM-4R machine is one of a number of testing machines with a mechanical drive and a lever-pendulum force-measuring mechanism. The machine has automatic plotting instruments which automatically trace out load-strain curves during the tests. [Note: For more detail on the design and operation of the testing machines, see (83, p. 49-56). a special attachment, the reverser, is used with the IM-4R machine. When the IM-4A machine is employed, it is not necessary to use a reverser.]

The tested specimen of compacted composition (usually, a pellet 20 mm in diameter and 30 mm high) is placed between two small steel plates, and, as these plates are slowly brought together at a constant rate, the force required to crush the pellet is determined.

The testing machines make it possible accurately to determine the crushing force on the pellet, ensure constant testing conditions, chiefly a constant rate of load increase, and make it possible to record a curve reflecting the character of deformation of the pellet with increasing load.

The compessive strength of the tested specimen α is calculated from the formula:

$$\alpha = \frac{P_{max}}{S} \, kg/cm^2$$

where P_{max} is the force in kg required for complete crushing of the pellet; S is the cross-sectional area of the specimen (pellet) in cm^2. [Ed. note: P_{max} is the gravitational force produced by the mass measured in kg. Kg is not a unit of force.]

3. FACTORS AFFECTING THE STRENGTH OF PRESSED COMPOSITIONS

The strength of a pressed article depends on:
1) the properties of the main oxidizer-fuel mixture;
2) the properties of the binder and its amount in the composition;
3) the degree of comminution of the components and process of manufacture of the composition (order of mixing of the components, duration of mixing, type of mixing apparatus, etc.);
4) the method of introducing the binder into the composition; in dry form, in the form of a solution (lacquer);
5) the concentration of the binder solution, etc.;
6) the specific pressure of the pressing and holding time during pressing;
7) the height of simultaneously pressed column, the pressing area, and their ratio.

Unquestionably, the crystalline system and hardness of the crystals of pressed components greatly affect the strength of the pressed product. Compositions consisting of components of great hardness do not undergo pressing too well and, other things being equal, yield articles of lower strength.

The strength of the article increases with rising pressing pressure (see Table 4.1) [Note: This increase reaches a determined range. For several compositions (containing aluminum powder), the phenomenon of "repressing", that is a lessening of the compressive strength during high pressure pressing, is observed], but the compressive strength α of the article does not usually exceed 20-25% of the specific pressing pressure.

In one-sided pressing, the height of a simultaneously pressed column of the composition must not exceed the diameter of the pressed product by a factor of more than 1.5-2. Otherwise, the transfer of pressure from layer to layer does not take place to the required extent, and the lower portion of the pressed column remains insufficiently compacted. In two-sided pressing, this ratio may be increased substantially.

Table 4.1 STRENGTH OF ARTICLE VERSUS PRESSING PRESSURE
(Pellets 20mm dia. and 20mm height were tested)

Composition Formula %	Specific Pressing Pressure kg/cm²	Compressive Strength α	
		Yield Point kg/cm²	% of Pressing Pressure
Ba(NO$_3$)$_2$ 80	1000	80	8.0
Al (powder) 20	2000	176	8.8
	3000	288	9.6
Ba(NO$_3$)$_2$ 55	1000	216	21.6
Al (powder) 45	2000	384	19.2
	3000	633	21.1

The pressure p_n inside the composition at a distance h from the die may be calculated from the formula [84]:

$$p_n = p \times e^{-A \times h}$$

where p is the pressure at the die;
  e is the base of natural logarithms;
  A is the proportionality constant, which changes in approximately inverse proportion to the diameter of the product being pressed (it is determined experimentally).

In some cases, a sufficient strength of the pressed article may also be obtained without adding any special binders to the composition. Usually, however, in order to obtain strong articles, it is necessary to resort to the introduction of binders.

Binders also increase the plasticity of the compositions; as the specific pressing pressure decreases, the density and strength of compositions containing a binder decrease much less than those of compositions containing no binder.

4. CLASSIFICATION OF BINDERS; THEIR PHYSICOCHEMICAL PROPERTIES

Note: The properties of binders used in rocket compositions are given in Ch. 20.

The following binders are most frequently used in pyrotechnic compositions:

1) synthetic resins - iditol, bakelite, etc.;
2) resins of natural origin and product of their processing: colophony, resinates, shellac;
3) drying oils;
4) glue - dextrin.

In addition, the use of various asphalts and bitumens as binders is possible. In some cases, solutions of nitrocellulose and rubber in suitable solvents are employed.

The solvent for nitrocellulose may be an alcohol-ether mixture, acetone, and for rubber, benzene, benzine, etc.

The characteristic properties of resins are:

1) insolubility in water;
2) solubility in organic solvents;
3) capacity to form films on drying of the resin solution;
4) stickiness of the solution;
5) absolute resistance to decay (in contrast to adhesives of animal origin).

Iditol is an artificial resin (of "lacquer-resin" type), obtained by condensing excess phenol with formaldehyde in the presence of an acid catalyst (for example, HCl).

The condensation reaction in its initial stage takes place in the following direction:

$$CH_2O + 2C_6H_5OH = CH_2(C_6H_4OH)_2 + H_2O.$$

On polymerizing, the initial condensation product, dihydroxydiphenylmethane, converts into a resin on heating.

In pyrotechnic calculations, the conventional formula $C_{13}H_{12}O_2$ is used for iditol.

The heat of formation of a commercial sample of iditol, calculated on the basis of its heat of combustion, is 0.74 kcal/g, and its density 1.25-1.30. Iditol is soluble in ethyl alcohol. Commercial samples of iditol always contain certain quantities of free phenol, which explains their reddish color. The technical requirements placed on iditol amount chiefly to the standardization of its softening temperature (not below 90-97°C for different grades), establishment of its phenol content limits (from 0.1 to 3.0% for different grades), and testing for the absence of colophony impurities therein.

These resins are soluble in alcohols and practically insoluble in hydrocarbons and mineral oils. They are stable to water, acids, ammonia, and weak solutions of alkali. They are decomposed by a strong alkali solutions.

Bakelite is a synthetic resin produced by condensing phenol with excess formaldehyde in the presence of an alkaline catalyst (for example, ammonia). In the case of Bakelite formation, the reaction takes place according to the equation:

$$CH_2O + C_6H_5OH = C_6H_4(OH)CH_2OH.$$

The condensation reaction then takes place on heating to form Bakelite A:

$$2C_6H_4(OH)CH_2OH = (C_6H_4CH_2OH)_2O + H_2O.$$

Bakelite A $C_{14}H_{14}O_3$ has a softening temperature of 75-100°C, and is soluble in ethanol.

Bakelite A is usually formed at a temperature no higher than 100°C. The conversion of Bakelite from form A via form B to form C, however, takes place at a higher temperature: 120-150°C or even higher (up to 180°C).

Bakelite A is usually introduced into pyrotechnic compositions in the form of a powder or alcoholic lacquer; the conversion of Bakelite into form C (Bakelization) is accomplished by heating the pressed composition [Note: Thermite incendiary compositions.] to about 150°C; the density of Bakelites is 1.20-1.29.

On heating to a higher temperature, Bakelite A leads to the formation of bakelite B, which is sometimes assigned the conventional formula $C_{43}H_{38}O_7$. On further heating, Bakelite B polymerizes to Bakelite C.

Bakelite C is a nonmelting substance (but softens on heating), insoluble in most ordinary organic solvents. It is chemically stable and has a high mechanical strength.

In calculations, it is most convenient to use the conventional formula $C_{12}H_{11}O_2$ (which almost coincides in ultimate composition with the formula $C_{43}H_{38}O_7$). Bakelite is used as a binder in compositions and products required to have a special mechanical strength.

Shellac is a natural resin secreted by trees growing in Southern Asia. Its empirical formula is $C_{16}H_{24}O_5$; and its density, 1.05-1.20. Shellac is soluble in ethyl alcohol, and almost insoluble in petroleum ether; it is imported, and may be replaced by domestic binders.

Colophony is obtained from resins of conifer trees. The main components of colophony are the unsaturated cyclic acids, abietic and pimaric acids. Abietic acid has the formula $C_{20}H_{30}O_2$. The density of colophony is 1.0-1.1; the softening temperature should not be below 65°C. Colophony is soluble in ethanol, ether, benzene and its homologs, and partly in petroleum ether; it dissolves in drying oil on heating.

Resinates is the term applied to products of the reaction of colophony with hydroxides or salts of the corresponding metals. Calcium resinate is obtained by melting colophony with slaked lime at 230-240°C. However, this reaction does not go to completion; this is evident from the fact that calcium resinate is characterized by a definite acid number (not above 80), although much lower than for colophony (160-180). The softening temperature of calcium resinate is 120-150°C. The composition of calcium resinate may be expressed with a certain approximation by the formula $(C_{19}H_{29}COO)_2Ca$. Petroleum ether or a 1:1 alcohol-petroleum ether mixture is used as the solvent for calcium resinate.

Among resinates of other metals, use may be made of strontium resinate, which has certain advantages in composition where it is necessary to obtain a red flame color.

Ellern [9] indicates that the following monomers, which after polymerization become binders, can be used in pyrotechnic compositions: methyl methacrylate, vinyledene chloride, styrene, acrylonitrile, vinyl acetate. His paper also gives values of the heats of polymerization, which in terms of a mole of monomer are, respectively, 13.0, 14.4, 16.1, 17.3, and 21.3 kcal, corresponding to 130-325 small calories per gram of substance.

Data on the properties of synthetic resins and organic polymers can also be found in the books of Losev and Petrov [87], Petrov and Levin [89], Losev and Trostyanskaya [88], and Lazarev and Sorokin [86], in the compendium New Materials in Technology [90], and in the literature listed in the second edition of our book [7, p. 53].

Among organic binders not included in the class of resins we should also mention drying oil and dextrin.

Linseed oil (GOST 7931-56) [GOST = Soviet State Union Standard] is the product of polymerization and partial oxidation of flaxseed or hempseed oil. It is a viscous liquid of light yellow or light brown color. The density is in the range for different grades of 0.920-0.938, and the viscosity according to the VU viscometer at 20°C, in the range 12-16 cp. Drying oil is obtained by blowing air through flaxseed (hempseed) oil heated to 160°C. To speed up its drying, siccatives or desiccants, which speed up the drying process, are added. Commonly used siccatives are resinates or linoleates of manganese, lead, or cobalt. In pyrotechnics, the best grades of drying oil are used without siccatives; occasionally, a solution of colophony in drying oil is used as the binder; for the properties of drying oil, see [85].

Dextrin is a gum obtained by heating starch with dilute acids. The formula of dextrin is $(C_6H_{10}O_5)$, and its density, 1.04. Dextrin dissolves readily in cold and hot water.

Table 4.2 compares certain physicochemical properties of organic fuels as binders and as substances having no binding properties.

To simplify the calculations, Table 4.2 also gives the conventional empirical formulas for high molecular compounds; these formulas were obtained on the basis of ultimate analysis of the corresponding substances, and their significance is of practical value only.

The data of column 4 are used to calculate the content of components in binary mixtures (see Ch. 5). The ultimate composition of organic fuels and their oxygen content in particular play a major part in the selection of the fuel in flame compositions.

Table 4.2 PHYSICOCHEMICAL PROPERTIES OF ORGANIC FUELS

Name & Formula	Density g/cm^3	Conventional molecular wt.	Amount of substance in g burning in 1 g of oxygen		Ultimate Composition		
			To CO & H_2O	To CO_2 & H_2	C	H	O
1	2	3	4		5		
Iditol $C_{13}H_{12}O_2$	1.3	200	0.74	0.42	79	6	15
Bakelite $C_{12}H_{11}O_2$	1.3	187	0.75	0.42	78	5	17
Shellac $C_{16}H_{24}O_5$	1.1	269	0.80	0.47	65	8	27
Colophony $C_{20}H_{30}O_2$	1.1	302	0.57	0.36	79	10	11
Calcium resinate $C_{20}H_{29}O_2Ca$	.1.2	643	0.61	0.38	75	9	10
Drying Oil $C_{16}H_{26}O_2$	0.93	250	0.58	0.36	77	10	13
Starch $C_6H_{10}O_5$	1.6	162	1.69	0.85	45	6	49
Lactose $C_{12}H_{24}O_{12}$	1.5	360	1.88	0.94	40	7	53
Paraffin $C_{26}H_{54}$	0.91	366	0.43	0.29	86	14	0
Stearin $C_{18}H_{36}O_2$	0.94	284	0.52	0.34	76	13	11
Naphthalene $C_{10}H_8$	1.14	128	0.57	0.33	94	6	0
Dicyandiamide $C_2N_4H_4$	1.40	84	1.31	0.88	29	4	0
Urotropin $C_6H_{12}N_4$	--	140	0.73	0.48	51	9	0
Wood charcoal C_6H_2O	1.5-1.7	90	0.94	0.48	81	2.17	
Carbon (graphite)	2.2	12	0.75	0.38	100	--	--
Flame-coloring salts							
Sodium oxalate $Na_2C_2O_4$	2.3	1.34	--	8.37	18	--	48
Strontium oxalate SrC_2O_4	--	176	--	10.98	14	--	36

Among inorganic binders, elemental sulfur is sometimes used in pyrotechnics. In cold pressing, its introduction into the composition barely affects the latter's strength. However, in the case of hot pressing at a temperature close to the melting point of sulfur, its presence increases the strength of the composition.

5. AMOUNT OF BINDER INTRODUCED INTO THE COMPOSITION

It is desirable to introduce the binder into the pyrotechnic composition in the amount of not more than 10-12% (this does not apply to rocket compositions).

It should be recalled that all binders of organic origin simultaneously play the part of a fuel in the composition, and hence, require the corresponding consumption of oxidizers for their combustion.

Adding a binder, particularly colophony and calcium resinate, markedly slows down the combustion of the compositions. Therefore, in order to obtain a composition with the necessary combustion rate, in practice, over 10-12% of binder is sometimes introduced into the basic oxidizer-metal mixture.

When the binder content of the composition is raised above 10-12%, the strength of the product increases insignificantly.

Organic binders affect the increase in the strength of the composition the most when they are added into the compositions in the form of a solution in a suitable solvent (such as lacquer).

Table 4.3 shows the dependence of the strength of the pressed composition on the amount of binder introduced at a pressing pressure of 3000 kg/cm².

Table 4.3 EFFECT OF THE AMOUNT OF BINDER ON THE STRENGTH OF THE COMPOSITION (Pellets 20mm in diameter and 20mm high were tested)

Composition formulas, %			Compressive Strength, α kg/cm²
$Ba(NO_3)_2$	Aluminum powder	Iditol	
80	20	0	288
78	20	2	408
76	20	4	460
74	20	6	530
72	20	8	562

Chapter 5

PRINCIPLES OF FORMULATION AND CALCULATION OF PYROTECHNIC COMPOSITIONS

The basic assumptions for the calculation of binary and multicomponent mixtures were formulated in the second half of the 19th Century by the Russian pyrotechnist P. S. Tsytovich [92]. In the calculation, he proceeded from the assumption that the entire fuel present in binary mixtures is burned up by the oxidizer oxygen; the possibility that atmospheric oxygen participates in the combustion of the composition was not considered. Ternary and multicomponent mixtures were assumed by Tsytovich to consist of two or more binary mixtures. The relative proportions of the binary mixtures forming the multicomponent composition was established by testing many variants, from which the composition producing the best pyrotechnic effect was selected.

1. CALCULATION OF BINARY MIXTURES

To obtain the formula of a binary mixture, it is necessary to write the equation of the combustion reaction, and on its basis, to calculate the weight relations between the oxidant and fuel.

Example 1. The combustion reaction of a binary mixture containing potassium perchlorate and magnesium may be expressed by the equation

$$KClO_4 + 4Mg = KCl + 4MgO. \quad (5.1)$$

To 139 g of potassium perchlorate there should correspond 24.3 x 4 = 97.2 g (rounded off to 97 g) of magnesium. Adding up the values obtained, 139 g + 97 g = 236 g, we obtain the total amount of the composition in grams. Hence it is easy to find that 100 g of the composition will contain

$$KClO_4 \; \frac{139 \times 100}{236} = 58.9 \text{ g}; \qquad Mg \; \frac{97 \times 100}{236} = 41.1 \text{ g}.$$

Rounding off the figures obtained, we get the composition formula 59% $KClO_4$, 41% mg.

However, seldom, even in the case of binary mixtures, can one write the composition of the end products of a reaction with full confidence that the formula obtained will correspond to the reality. This can be done only when data on the chemical analysis of the combustion products are available; if they are not available, one can speak only of a probable equation of the combustion reaction. In this connection, it is necessary to discuss the oxidizer decomposition equations which are most probable in a given specific case, and the most probable oxidation products of the fuel.

Table 2.1 gives the most probable oxidizer decomposition equations, and Table 3.3, the oxidation products of inorganic fuels. By using them, in many cases one can write the probable equation of the combustion reaction of a binary mixture and find the formula of the composition.

Example 2. Set up the equation of the combustion reaction and find the formula of a binary mixture consisting of barium nitrate and magnesium.

$$Ba(NO_3)_2 = BaO + N_2 + 2.5O_2. \quad (5.2)$$

and in Table 3.3, the indication that the oxidation product of magnesium is MgO.

Using the oxygen evolved by the decomposition of the oxidizer to oxidize magnesium, we obtain the equation of the reaction

$$Ba(NO_3)_2 + 5Mg = BaO + N_2 + 5MgO \qquad (5.3)$$

and determine the formula of the composition:

> barium nitrate 68%
> magnesium 32%

This composition may be used as a photomixture.

It is well known that in the presence of insufficient oxygen, magnesium may react with nitrogen, and therefore another equation of the combustion reaction (without the participation of atmospheric oxygen) may be written:

$$Ba(NO_3)_2 + 8Mg = BaO + 5MgO + Mg_3N_2. \qquad (5.4)$$

Hence we obtain the following composition formula:

> barium nitrate 57%
> magnesium 43%

In carrying out the calculations and setting up the proposed equations of the combustion reaction, it is necessary to be guided by thermochemical consideration. For example, in the case at hand, it is necessary to consider that the combination of magnesium with nitrogen evolves approximately 1/3 the amount of heat liberated by the combination of magnesium with oxygen, and therefore, it is not always advantageous to direct the reaction toward the formation of nitrides. [Note: A significant gain in the amount of heat will be obtained only when the Mg_3N_2 formed can be oxidized by atmospheric oxygen to MgO.]

Tsytovich showed long ago that in setting up the combustion equations of composition containing coal or organic fuels, the calculation can be made 1) for the total oxidation of the fuel with the formation of carbon dioxide and water, or 2) when there is a decreased oxidizer content, for the formation of carbon monoxide and water.

Example 3. Set up the equation of the combustion reaction of mixtures of Xpotassium nitrate with iditol.

Taking the simplified formula $C_{13}H_{12}O_2$ for iditol, we obtain two equations:

$$12KNO_3 + C_{13}H_{12}O_2 = 6 K_2O + 6N_2 + 13 CO_2 + 6H_2O. \qquad (5.5)$$

$$34KNO_3 + 5 C_{13}H_{12}O_2 = 17 K_2O + 17N_2 + 65CO + 30H_2O \qquad (5.6)$$

Formula according to (5.5)
Potassium nitrate 86%
Iditol 14%

Formula according to (5.6)
Potassium nitrate 77%
Iditol 23%

Equations (5.5) and (5.6) correspond to the reality only to a certain extent; as is well known, the combustion of mixtures of potassium nitrate with organic fuels also forms certain quantities of potassium nitrite and peroxide, and the combination of potassium oxide with carbon dioxide forms potassium carbonate.

The selection of the coefficients in the calculation of mixtures (particularly with organic fuels) requires a considerable amount of time. For this reason, Demidov [91] suggested the use of tables which show how many grams of oxidizer is required to produce 1 g of oxygen and what amount of fuel can be oxidized by 1 g of oxygen.

These tables were prepared as follows.

It is well known that under combustion conditions of the compositions, potassium perchlorate decomposes according to the equation

$$KClO_4 = KCl + 2O_2. \quad (5.7)$$

It follows that in order to obtain 1 g of oxygen, it is necessary to use up

$$139 \div 64 = 2.17 \text{ g of } KClO_4$$

The oxidation of aluminum takes place according to the equation

$$2Al + 1.5O_2 = Al_2O_3.$$

Hence, 1 g of oxygen can oxidize $54:48 = 1.12$ g of aluminum. Such data for oxidizers and fuels are listed in Tables 2.1, 3.3 and 4.2.

Example 4. Find the formula of a composition containing potassium perchlorate in iditol if combustion of iditol to carbon dioxide and water is to be achieved. For potassium perchlorate in Table 2.1, we find the figure 2.17, and for iditol in column 4 of Table 4.2, the figure 0.42. These figures express the amounts in weight of the components of the mixture (in g):

$KClO_4$	2.17
Iditol	0.42
Total mixture	2.59

or in percent

$KClO_4$	83.8%
Iditol	16.2%

2. FORMULATION AND CALCULATION OF TERNARY AND MULTICOMPONENT MIXTURES

In many cases, ternary mixtures may be considered to consist of two binary mixtures containing one and the same oxidizer.

However, the presence of two different fuels in a composition markedly alters the direction of the reaction, and such an approach then becomes unacceptable. For example, in the case the barium nitrate-aluminum-sulfur, a reaction takes place between the aluminum and the sulfur, forming aluminum sulfide Al_2S_3, which can then burn to Al_2O_3 and SO_2.

Included among the combustion products of such a composition may be BaO, Al_2O_3, $Ba(AlO_2)_2$, BaS, $BaSO_4$, Al_2S_3, SO_2, N_2, etc.

The composition of the combustion products depends not only on the relative amounts of the components in the composition, but also on the conditions of combustion of the composition: on the pressure of the ambient medium, initial temperature, density of the composition, diameter of the flame, etc.

In tentative calculations for ternary compositions including an oxidizer, a metal fuel and an organic fuel (binder), the following method may be employed.

Example 5. Find the formula of the ternary mixture barium nitrate-magnesium-iditol when iditol burns completely to carbon dioxide and water. Setting up the reaction equations or using Demidov's tables, we find the content of the components in two binary mixtures:

  Barium nitrate .... 68%   Barium nitrate 88%
  Magnesium .... 32%   Iditol 12%

Considering that a 4% iditol content of the composition will provide for an adequate mechanical strength of the illumination pellet, we choose the ratio of the binary mixtures as 2:1. On the basis of this condition, we make a calculation that reduces to some simple arithmetic and requires no explanation:

$$\text{Barium nitrate} \quad \frac{68 \times 2}{3} + \frac{88 \times 1}{3} \approx 75\%$$

$$\text{Magnesium} \quad \frac{32 \times 2}{3} \approx 21\%$$

$$\text{Iditol} \quad \frac{12 \times 1}{3} = 4\%$$

It is obvious that a haphazardly chosen ratio of the two binary mixtures is not optimal, and the formula obtained is tentative and should be refined experimentally by taking into consideration the special requirements imposed on illuminating compositions.

In certain compositions of signal lights, the salt coloring the flame (in a formal approach without considering the intermediate combustion products) does not participate in the combustion of the composition.

Example 6. Find the formula of a red light composition including potassium chlorate, strontium carbonate, and iditol. The combustion reaction of this composition may be approximately expressed by the equation

$$10KClO_3 + nSrCO_3 + C_{13}H_{12}O_2 = 10KCl + nSrO + (13+n)CO_2 + 6H_2O \qquad (5.8)$$

Practice shows that in order to obtain a flame of satisfactory red color, it is sufficient to take 20-25% of strontium carbonate, and to divide the remaining 80-75% between potassium chlorate and iditol on the basis of the above reaction equation.

In the presence of 25% of strontium carbonate in the composition, we obtain the formula:

  Potassium chlorate 64.4%
  Strontium carbonate 25%
  Iditol 10.6%

3. CALCULATION OF COMPOSITIONS WITH A NEGATIVE OXYGEN BALANCE

In many cases, the special effect of pyrotechnic compositions is not decreased, but on the contrary, increased if not only the oxidizer oxygen, but also atmospheric oxygen participates in the combustion process. This is due to the fact that for many types of compositions (illuminating, incendiary, etc.,), the special effect increases with increasing heat of combustion of the composition. Other things being equal, the heat of combustion of a composition will be greater, the higher is its content of completely burning fuel (whose combustion is also due at least partly to combination with atmospheric oxygen).

An adequate special effect due to pyrotechnic compositions including an excess of fuel is usually obtained in cases where the fuel is an easily oxidizable substance capable of burning partly by combining with atmospheric oxygen.

The most typical example of a fuel whose combustion in a composition can be partly due to combination with atmospheric oxygen is magnesium. In many cases, it is useful to formulate magnesium-base compositions where only one-half of the magnesium is oxidized by combining with the oxidizer oxygen, the combustion of the other half being due to combination with atmospheric oxygen.

On the contrary, compositions including difficult-to-oxidize fuels (coarse particles of aluminum or silicon) should contain an amount of oxidizer sufficient for its complete combustion, since in this case the fuel can no longer burn up by combining with atmospheric oxygen. The amount of fuel that can burn by combining with atmospheric oxygen is determined experimentally.

Pyrotechnic compositions containing excess oxidizer above the amount necessary for complete oxidation of the fuel are referred to as compositions with a positive oxygen balance; the excess oxygen not participating actively in the combustion process of the composition is definitely detrimental in all cases. For all practical purposes, compositions with a positive oxygen balance are not used in pyrotechnics.

[Note: an exception are compositions for "oxygen candles" - see (294) and Ch. 22.]

Compositions containing only the amount of oxidizer necessary for complete combustion of the entire fuel (or fuels) contained in the composition are referred to as compositions with a zero oxygen balance.

Compositions containing the oxidizer in an insufficient amount to ensure the complete combustion of the fuel are referred to as compositions with a negative oxygen balance.

Most pyrotechnic compositions currently employed have a negative oxygen balance. The effectiveness of the action of such compositions greatly depends on the extent to which atmospheric oxygen can practically participate in the combustion process.

The term "oxygen balance (n) of a composition" is applied to the amount of oxygen in grams whose addition is necessary for complete oxidation of all the fuel elements in 100 g of composition. The oxygen balance in the presence of which the best special effect is obtained in the composition is known as the optimum oxygen balance.

The ratio of the amount of oxidizer contained in the composition to the amount of oxidizer necessary for complete combustion of the total fuel contained in the composition is known as the oxidizer availability coefficient of the composition (k).

In calculating binary mixtures of magnesium or aluminum with alkali or alkaline earth nitrates, the concepts of "active" and "total" oxygen balance are employed.

The term "active" oxygen balance refers to the evolution by the oxidizer, during the combustion of the composition, of only weakly bound, so-called active oxygen, for example:

$$Sr(NO_3)_2 + 5Mg = SrO + N_2 + 5MgO. \qquad (5.9)$$

In the case of the "total" oxygen balance, the entire oxygen contained in the oxidizer is taken into account, and the equation is set up on the basis of the assumption that the metal contained in the oxidizer is obtained in free form at least in the flame zone. The reaction equation given below was set up on the basis of the "total" oxygen balance:

$$Sr(NO_3)_2 + 6Mg = Sr + N_2 + 6MgO. \qquad (5.10)$$

Compositions with a "total" oxygen balance obviously have a negative oxygen balance.

Given below is an example of a calculation of compositions with a negative oxygen balance. In the calculation, the required oxygen balance (in grams) is specified.

Example 7. Calculate the binary mixture potassium chlorate-magnesium on the condition that its oxygen balance n=-20g O_2.

In Tables 2.1 and 3.3, we find the numbers 2.55 and 1.55 respectively, for potassium chlorate and magnesium. We calculate that 20 g of oxygen oxidize 20 x 1.52 = 30.4 g of magnesium.

The remaining 69.6 g of composition should be calculated in the usual manner for a zero oxygen balance.

The potassium chlorate content of the composition is found to be

$$\frac{2.55 \times 69 \times 6}{2.55 + 1.52} = 43.6\%$$

and the composition will contain 100-43.6-30.4 = 26.0% of magnesium. The oxidizer oxygen will burn 56.4-30.4=26.0% of magnesium.

The oxidizer availability coefficient of the composition will in this case be k=26.0÷56.4 = 0.46. A similar calculation can be carried out for multicomponent mixtures.

In many cases, it is necessary to determine by calculation the validity of an already existing composition formula. In particular, calculation of the oxygen balance n and coefficient k makes it possible to estimate the degree to which it is necessary to bring the composition in contact with atmospheric oxygen during its combustion; calculation of these values permits one to explain the causes of sparking of the composition, etc.

The order of calculation of n and k is shown in the following example:

Example 8. Calculate n and k for a yellow light composition having the following formula (in percent):

          Potassium chlorate 60
          Sodium oxalate 25
          Shellac 15

We find from Tables 4.2 and 2.1 that 2.55 g of potassium chlorate is necessary to burn 0.47 g of shellac or 8.37 g of sodium oxalate to carbon dioxide and water. Hence, the combustion of 15 g of shellac requires

$$\frac{2.55 \times 15}{0.47} = 81.5 \text{ g of potassium chlorate,}$$

and the combustion of 25 g of sodium oxalate requires

$$\frac{2.55 \times 25}{8.37} = 7.6 \text{ g of potassium chlorate.}$$

Hence, the oxidizer availability coefficient of the composition for the given composition will be

$$k = \frac{60}{81.5 + 7.6} = 0.67,$$

and the value of the oxygen balance n for the given composition will be

$$n = \frac{60 - (81.5 + 7.6)}{2.55} = -11.4 \text{ g of } O_2$$

4. CALCULATION OF METAL CHLORIDE COMPOSITIONS

In metal chloride compositions, the role of the oxidizer is played by a chlorinated organic compound, and that of the fuel, by an active metal powder.

The oxidizer in compositions of this type should be taken in an amount such that the chlorine it contains is sufficient for the complete oxidation of the metal. For example, for each g-atom of zinc, it is necessary to take enough oxidizer so that it contains 2 g-atoms of chlorine. On this basis, the equation of the reaction between hexachloroethane and zinc may be written as follows:

$$C_2Cl_6 + 3Zn = 2C + 3ZnCl_2. \quad (5.11)$$

To simplify the calculations, use may be made of table 5.1, which indicates what amount of oxidizer yields 1 g of chlorine on decomposing, and the amount of metal combining with 1 g of chlorine.

Table 5.1

Chlorinated Organic Compound (Oxidizer)	Molecular Weight	Quantity Producing 1 g of Chlorine	Metal Fuel	Atomic Weight	Quantity of Metal Combining w/1 g Chlorine
Carbon tetrachloride CCl_4	154	1.08	Zinc	65.4	0.92
Hexacloroethane C_2Cl_6	237	1.11	Aluminum	27.0	0.27
Hexachlorobenzene C_6Cl_6	285	1.34	Magnesium	24.3	0.34
Hexachlorocyclohexane (hexachlorane) $C_6H_6Cl_6$	291	1.37	Zirconium	91.2	0.64
Polyvinyl chloride $(C_2H_3Cl)_n$	60.5	1.76	Iron	55.8	0.53 ($FeCl_3$ is formed)
			Iron	55.8	0.79 ($FeCl_2$ is formed)

PRINCIPLES OF PYROTECHNICS

Example 9. Calculate the percent content of the components in the binary mixture hexachloroethane-aluminum, using Table 5.1.

$$\begin{array}{ll} \text{Hexachloroethane} & 1.11 \text{ g} \\ \text{Aluminum} & 0.27 \text{ g} \\ \hline \text{Total mixture} & 1.38 \text{ g} \end{array}$$

or in percent

$$\text{Hexachloroethane} \quad \frac{1.11 \times 100}{1.38} = 80.5\%$$

$$\text{Aluminum} \ldots\ldots\ldots\ldots 100 - 80.5 = 19.5\%$$

A more complex calculation is that of ternary mixtures containing chlorinated organic compounds.

Example 10. Calculate the composition of a green light containing barium nitrate-hexachloroethane-magnesium.

Practice shows that in order to obtain flames of good color, the compositions of green and red lights with nitrate oxidizers should contain at least 15% of chlorinated organic compounds. Hence the additional condition: the composition should contain 15% of hexachloroethane, whose carbon should be oxidized at least to CO.

Solution. The combination with 15% C_2Cl_6 will require (see Table 5.1):

$$\frac{15 \times 0.34}{1.11} = 4.6\% \text{ of magnesium.}$$

When decomposing, hexachloroethane forms

$$\frac{15 \times 24}{237} = 1.5\% \text{ of carbon,}$$

which as stated must be oxidized to CO. Using Tables 2.1 and 4.2, we find the figures 3.27 and 0.75, and calculate the amount of barium nitrate necessary for oxidizing carbon:

$$0.75 \div 3.27 = 1.5 \div x;$$
$$x = 6.5\% \text{ Ba(NO}_3)_2$$

We know that 100 g of composition should contain

$$15 \text{ g } C_2Cl_6 + 4.6 \text{ g Mg} + 6.5 \text{ g Ba(NO}_3)_2 \qquad (5.12)$$

We find how many grams of composition correspond to the binary mixture $Ba(NO_3)_2 + Mg$:

$$z = 100 - 15 - 4.6 - 6.5 = 73.9 \text{ g.}$$

Using Tables 2.1 and 3.3, we find that 73.9 g of mixture contains:

$$\frac{3.27 \times 73.9}{3.27 + 1.52} = 48.7 \text{ g of Ba(NO}_3)_2 \text{ and } 25.2 \text{ g of Mg.}$$

Summarizing the calculations, we obtain the formula:

Barium nitrate .... 48.7 + 6.5 = 55.2%
Hexachloroethane 15%
Magnesium .... 25.2 + 4.6 = 29.8%

The solid combustion products of this composition will be MgO, BaO, and $MgCl_2$; however, the reverse exchange reaction taking place in the flame $BaO+MgCl_2 = BaCl_2+MgO$, forms barium chloride, which gives a green color to the flame.

When the organic binder iditol is introduced into this mixture, we obtain a formula of the composition $Ba(NO_3)_2$ 59.5%, Mg 20.5%, C_2Cl_6 15% and iditol 5% (cf. also [7, p.62]).

5. CALCULATION OF COMPOSITIONS WITH A FLUORINE BALANCE

In principle, this type of calculation is similar to that of metal chloride compositions.

The role of oxidizer is played in this case by fluorine compounds (fluorides of relatively inactive metals or fluorinated organic compounds), and the role of fuel, by active metal powders. Enough oxidizer should be taken so that it suffices for the complete oxidation of the metal.

To simplify the calculations, use may be made of Table 5.2, which indicates what amount of oxidizer yields 1 g of fluorine on decomposition, and the amount of metal combining with 1 g of fluorine.

Table 5.2

Oxidizer - Fluorine Compound	Molecular Weight	Quantity Producing 1 g of Fluorine	Metal Fuel	Atomic Weight	Quantity of Metal Combining w/1 g Fluorine
CuF_2	104	2.74	Be	9.0	0.24
AgF	127	6.68	Mg	24.3	0.64
PbF_2	245	6.45	Al	27.0	0.47
$(C_2F_4)_n$	100	1.32	Zr	91.2	1.20
Teflon	(monomer)				
XeF_4*	207	2.72			

*Xenon tetrafluoride is a recently synthesized solid compound stable at room temperature.

Example 11. Calculate the binary mixture of teflon and zirconium.

Solution. Combination with 1.32 g of teflon will require 1.20 g of zirconium. The teflon content of the mixture will be 1.32÷2.52 = 52.4%, and the zirconium content, 47.6%.

The combustion of compositions containing an excess of oxidizer may involve the evolution of free fluorine in certain cases. The formation of free fluorine will be most likely when the oxidizers used are fluorides of variable valence metals CoF_3, MnF_3, PbF_4, etc., for example:

$$6CoF_3+2Mg = 2MgF_2+6CoF_2+F_2. \quad (5.13)$$

The properties of fluorine compounds are described in the monographs of Simons [205] and Ryss [203]; see also the brochure of Knunyants and Fokin [200] and Ch. 17 of this book.

Chapter 6

HEAT OF COMBUSTION OF PYROTECHNIC COMPOSITIONS

The amount of heat evolved by the combustion of pyrotechnic compositions substantially determines their special effect. The heat of combustion of pyrotechnic compositions may be determined in two ways:

1) by calculation using Hess' law,
2) experimentally, by burning the compositions in a bomb calorimeter.

1. CALCULATION OF THE HEAT OF COMBUSTION

The calculations are based on Hess' law, which is stated as follows: the amount of heat evolved by a chemical reaction depends only on the initial and final state of the system and is independent of the path taken by the reaction.

A different formulation of this law may be used for the calculations, namely: "if a system passes directly from state 1 to state 3 once, and another time via several intermediate states, the heat of the direct conversion is equal to the sum of the heats of the intermediate reactions."

In other words:

$$Q_{1,3} = Q_{1,2} + Q_{2,3}$$

where $Q_{1,3}$ is the amount of heat emitted or absorbed when the system passes from state 1 to state 3; $Q_{1,2}$ and $Q_{2,3}$ and the amounts of heat emitted or absorbed by the corresponding transitions of the system from state 1 to state 2 and from state 2 to state 3.

It follows from Hess' law that the heat of formation of the combustion products of a pyrotechnic composition from the elements is equal to the sum of the heats of formation of the components of the composition, to which should be added the amount of heat evolved by the combustion of the composition. Consequently, the heat of the combustion reaction is determined as the difference between the heat of formation of the combustion products and the heat of formation of the components of the composition:

$Q_{2,3}$	=	$Q_{1,3}$	–	$Q_{1,2}$
Heat of combustion of pyrotechnic composition		Heat of formation of the combustion products		Heat of formation of the components of the composition

The combustion of pyrotechnic compositions usually takes place at a comparatively slow rate and mostly in an open space; on this basis, it may be assumed that the combustion takes place at constant pressure. Since thermochemical reference tables usually give the heats of formation of compounds at constant pressure, the above formula is similarly used to calculate the heat of reaction $Q_{2,3}$ at constant pressure Q_p.

This differentiates the calculation of the heat of combustion of pyrocompositions from the calculation of the heat of explosion of an explosive, used to calculate the heat of reaction at constant volume Q_v. The transition from Q_p to Q_v is achieved by use of the formula

$$Q_v = Q_p + 0.57n,$$

where n is the number of moles of gasses formed in the reaction (cf. [7]).

In the calculations, the standard heats of formation of the substances from the element (ΔH_{298}) should be found in the works of Rossini [98], Britske, Kapustinskiy, et al. [93], Karapet'yants [94], Glushko (ed.) [233], and the monograph of Kubaschewski and Evans [95]; a list of the heats of formation of many substances, taken from the works of Medard, is given in the monograph of Andreyev and Balyayev [111].

Table 6.1 gives the values of the heat of formation of components of pyrotechnic compositions and certain products of their combustion.

Data on the heat of formation of certain oxidizers and products of their decomposition are given in Table 2.1, and the heat of formation of the oxides, in Tables 3.1 and 3.5 (see Ch. 3).

As an example, here is a calculation of the heat of combustion of the mixture

$$3\ Ba(NO_3)_2 + 10 Al = 3 BaO + 3 N_2 + 5\ Al_2O_3.$$

The heat of formation of the combustion product (taken from [98])

$$5\ Al_2O_3 \ldots 400.5 = 2000$$
$$3 BaO \ \ \ldots 133.3 = \ \ 399$$
$$\overline{}$$
$$2399\ \text{kcal}\ (Q_{1,3})$$

The heat of formation of the components of the composition

$$3\ Ba(NO_3)_2 \ldots 237 \times 3 = 711\ \text{kcal}\ (Q_{1,2}).$$

The heat of the combustion reaction

$$Q_{2,3} = 2399 - 711 = 1688\ \text{kcal}$$

The sum by weight of $Ba(NO_3)_2$ and aluminum

$$M = 261 \times 4 \times 3 + 27.0 \times 10 = 1054.$$

The heat of combustion of the composition in kcal/g

$$q = \frac{1688}{1054} = 1.601\ \text{kcal/g}$$

This method of calculation is fairly accurate, but it does not give a clear representation of the energy contribution of the fuel and oxidizer, taken separately, to the total heat balance of the composition. This analysis can be carried out by using several other methods of calculation. This will be illustrated by the example of the mixture $Ba(NO_3)_2 + Mg$. the stoichiometric calculation gives the ratio (p. 106) of the components: $Ba(NO_3)_2$ 68%, Mg 32%. Using Table 3.1, we find that the combustion of 0.32 g of magnesium evolves $0.32 \times 5.9 = 1.88$ kcal.

We find from Table 2.1 that the decomposition of 261 g of Ba(NO$_3$)$_2$ requires 104 kcal. We calculate that the decomposition of 0.68 g of Ba(NO$_3$)$_2$ requires consumption of 0.27 kcal. Comparing the data, we obtain the heat of combustion of the mixture

$$q = 1.88 - 0.27 = 1.61 \text{ kcal/g}.$$

In the case at hand, the decomposition of the oxidizer consumes $0.27 \times 100 \div 1.88 = 14\%$ of the amount of heat evolved by the combustion of magnesium. Using this method to calculate the heat of combustion of iron-aluminum thermite Fe$_3$O$_4$ 75%, Al 25%, we find that its heat balance $q = 1.82 - 0.86 = 0.9x$ kcal/g, i.e., in this case, the decomposition of the oxidizer consumes as much as 47% of the heat evolved by the combustion of aluminum.

Table 6.1 HEAT OF FORMATION (ΔH_{298}) OF COMPONENTS OF PYROTECHNIC COMPOSITIONS AND CERTAIN PRODUCTS OF THEIR COMBUSTION*X

Compound	Heat of Formation kcal/g-mole	Compound	Heat of Formation kcal/g-mole	Compound	Heat of Formation kcal/g-mole
NaF	136	SrO$_2$	154	Starch	227
Na$_3$AlF$_6$	758	Sr$_3$N$_2$	92	Lactose	651
Na$_2$SiF$_6$	669	SrSO$_4$	92	Eethanol	67
NaHCO$_3$	226	SrCl$_2$	198	Ethyl ether	65
Na$_2$CO$_3$	271	BaCO$_3$	285	Shellac	227
Na$_2$C$_2$O$_4$	315	Ba$_3$N$_2$	90	Iditol	149
NaNO$_2$	88	ZnCl$_2$	99	Trotyl	16
Na$_2$SO$_4$	332	AlN	75	Hexogen	-21
K$_2$CO$_3$	282	AlCl$_3$	167	Urotropin	-30
K$_2$SO$_4$	342	Al$_2$S$_3$	140	Hexachloro-ethane	54
Cu(OH)$_2$	107	Pb(NO$_3$)$_2$	107	Naphthalene	-16
CuCO$_3$	143	FeCl$_3$	94	Anthracene	-32
CuCNS	(-10)	NH$_3$	11	Carbon disulfide	-21
CuCl	32	NH$_4$Cl	75	Pyroxylin	+656
CuCl$_2$	53	HCl	22	Collodion	+639
MgCO$_3$	267	CCl$_4$	25	Liq. methane	18
Mg$_3$N$_2$	115	MgS	84	Benzene	-13
MgCl$_2$	153	MgF	264		
SrCO$_3$	290	SrC$_2$O$_4$	(288)		

*The numerical data are given with the accuracy required for technical calculations; data for shellac and iditol are calculated on the basis of experimental data on the heat of combustion of these substances calculated by the author.

Neglecting the substances whose composition, like that of the "electron" magnesium alloy, involves atmospheric oxygen, we find that the highest heat of combustion is that of compositions of photomixtures, followed by illuminating, tracer and rocket compositions; a much smaller amount of heat is evolved by the combustion of compositions of signal lights, and finally, the smallest amount of heat is produced by the combustion of smoke compositions. The formulas of ignition compositions are so numerous, and the components included in a composition so diverse, that the heat obtained from their combustion varies over very wide limits.

If the process of their combustion involves the participation of atmospheric oxygen, compositions with a negative oxygen balance yield a larger amount of heat than compositions made up of the same components but taken in stoichiometric proportions.

Let us take as an example a composition consisting of 44% $KClO_3$ and 56% Mg with oxygen balance n=-20 g of O_2; the combustion reaction of this composition may be reprensented by the equation

$$KClO_3 + 6.5Mg + 1.75O_2 = KCl + 6.5MgO.$$

The heat of combustion of this composition

$$q = \frac{144 \times 6.5 + 106 - 96}{123 + 24.3 \times 6.5} = 3.37 \text{ kcal/g}$$

In comparison with the composition shown in Table 6.2, consisting of the same components but taken in stoichiometric proportions, we will obtain a 45% increase in the heat of combustion.

Table 6.2 lists data on the heat of combustion of certain pyrotechnic compositions; the components of these compositions are taken in stoichiometric proportions; the heat of combustion is calculated from Hess' law.

Table 6.2 HEAT OF COMBUSTION OF PYROTECHNIC COMPOSITIONS & SOLID ROCKET FUEL (SRF)

Composition Formula %	Heat of Combustion kcal/g	Use
$KClO_4$ - 83, Be - 17	3.19	No practical use
$KClO_4$ - 60, Mg - 40	2.24	Photoflash
$KClO_3$ - 63, Mg - 37	2.29	Photoflash
$Ba(NO_3)_2$ - 68, Mg - 32	1.65	Photoflash
$KClO_4$ - 66, Al - 34	2.45	Incendiary
$NaNO_3$ - 60, Al - 40	2.00	Illuminating (w/o binder)
Fe_2O_3 - 75, Al - 25	0.96	Thermite
$Ba(NO_3)_2$ - 75, Mg - 21, Iditol - 4	1.23	Illuminating
$Ba(NO_3)_2$ - 63, Al - 27, Sulfur - 10	1.40	Illuminating
$Sr(NO_3)_2$ - 69, Mg - 25, Calcium resinate - 6	1.48	Tracer
Nitroglycerine colloidal powder	1.23	Rocket fuel
Mixture [228]: NH_4ClO_4 - 90, Resin - 10	1.26	Rocket fuel
NH_4ClO_4 - 80, Resin - 20	1.01	Rocket fuel
$Ba(ClO_3)_2 \cdot H_2O$ - 88, Iditol - 12	0.99	Green light
$KClO_3$ - 57, $SrCO_3$ - 25, Shellac - 18	0.61	Red light
C_2Cl_6 - 81, Al - 19	0.96	Masking smoke
C_2Cl_6 - 17, $KClO_4$ - 22, Zn - 61	0.52	Masking smoke
$KClO_3$ - 35, Lactose - 25, Rhodamine dye 40	0.38	Red smoke*
KNO_3 - 75, Coal (carbon) - 15, Sulfur - 10	0.66	Black smoke
Mg - 90, Al - 10	6.10	"Electron" alloy (burns in atmospheric oxygen)

*In practice, decomposition of the dye takes place to some extent during the combustion of the composition.

2. EXPERIMENTAL DETERMINATION OF THE HEAT OF COMBUSTION

To determine the heat of combustion of pyrotechnic compositions experimentally, a certain weighed amount of composition is burned in a bomb calorimeter. The amount of heat evolved is determined as the product of the heat capacity of the system (water + instrument) and the difference between the final and initial temperature of the water into which the calorimeter is immersed. The heat capacity of the system, also known as the "water number" of the calorimeter, is determined by special experiments; substances of known heat of combustion, for example, benzoic acid, are burned in the bomb calorimeter.

The volume of the bomb calorimeter is usually 300-400 cm^3, and the weight of the water in the calorimetric vessel is about 3 kg (the water is weighed to within 1 g).

In most cases, the pyrotechnic composition is burned in an atmosphere of air, and more seldom in an atmosphere of nitrogen or argon. There is no point in burning a pyrotechnic composition in an oxygen atmosphere, since the heat of combustion of the composition for the complete oxidation of all of its fuel elements can almost always be calculated with sufficient accuracy on the basis of Hess' law.

If it is necessary to determine the heat of combustion of a composition having a negative oxygen balance with the stipulation that it burns in air, care is taken to provide the bomb calorimeter with the maximum amount of air per unit weight of the composition, and to this end, the weighed sample of the composition is reduced to the limit allowed by the accuracy of the determination: the amount taken is usually equal to 0.5-1 g, in such a way that the water temperature rise in the calorimetric vessel is at least 0.3°C. It should be considered that a 300 cm^3 bomb filled with air ($p=1$ atm) contains 0.1 g of oxygen, which is sufficient for the oxidation of only 0.5 g of a composition having an oxygen balance $n=-20$ g of O_2.

The fact that in many cases atmospheric oxygen actively participates in the process of combustion of pyrotechnic compositions having a negative oxygen balance has been ascertained on the basis of many experimental data.

Only certain smoke compositions burn in a nitrogen atmosphere (atmospheric oxygen had an oxidizing effect, burning a smoke-forming substance partly or even completely, and thus distorting the results of the determination). With respect to compositions containing magnesium or aluminum, nitrogen is no longer an inert gas, since these metals react with it, forming nitrides Mg_3N_2 or AlN.

If it is necessary completely to eliminate the influence of the ambient medium on the combustion of compositions containing magnesium or aluminum, the combustion should be carried out in an atmosphere of an inert gas, for example, argon.

To ignite many pyotechnic compositions, it is completely sufficient to bring them in contact with a fine nichrome wire heated to redness with an electric current. If this method is not sufficient to ignite the composition, a small amount (hundredths of a gram) of an ignition composition, whose heat of combustion must be determined beforehand, is poured on the main composition to be burned. The introduction of the ignition composition lowers the accuracy of the calorimetric determination because of the possibility of a chemical reaction between the combustion products of the composition being tested and the ignition composition. For this reason, an ignition composition should be used only in extreme cases.

Concerning the procedure employed in calorimetric experiments, see Popov's monograph [97].

3. RELATIONSHIP BETWEEN THE USE OF COMPOSITIONS AND THEIR HEAT OF COMBUSTION

On the basis of the calculated data given in Table 6.2 as well as experimental data, a relationship can be established between the function of compositions and the amount of heat evolved by their combustion (Table 6.3).

Table 6.3 RELATIONSHIP BETWEEN THE FUNCTION OF COMPOSITIONS AND THEIR HEAT OF COMBUSTION

Function of Compositions	Heat of Combustion, kcal/g
Photoflash mixtures	1.7 - 3.0
Illumination & tracer compositions	1.5 - 2.0
Rocket compositions (solid rocket fuels w/o complete burning in air: a) colloidal nitroglycerin powders b) mixed powders (tentative data)	 0.8 - 1.3 1.0 - 1.5
Incendiary compositions (w/oxidizer)	0.8 - 3.0
Night signal compositions	0.6 - 1.2
Masking smoke compositions	0.4 - 1.0
Colored signal smoke compositions	0.3 - 0.6

Chapter 7

GASEOUS COMBUSTION PRODUCTS OF PYROTECHNIC COMPOSITIONS

The formation of a certain amount of gaseous substances by the combustion reaction is typical of nearly all types of pyrotechnic compositions. Among compositions used for practical purposes, iron-aluminum thermite appears to be the only one giving no gaseous products on combustion. [Note: However, at the reaction temperature (~2400°C) and in the case of thermite, a part (comparatively small) in intermediate products and reaction products are in the vapor state.]

In illuminating compositions, tracer compositions and signal lights, and in photomixtures, the formation of gaseous substances is required in order to obtain a flame during the combustion and thus to increase the amount of emitted luminous energy. The formation of gaseous products during the combustion of solid pyrotechnic fuels is a necessary condition for producing the thrust of a rocket engine.

In smoke compositions, the presence of the gaseous phase in the reaction products of combustion is required in order to achieve the expulsion of particles of various smoke-forming substances from the reaction sphere into the atmosphere.

In the combustion of incendiary compositions, the formation of gaseous products is also desirable, since this substantially expands the fire center produced.

In addition to gaseous products, the combustion of pyrotechnic compositions also yields a certain amount of solid reaction products. An exception is solid pyrotechnic fuel, as well as certain fuel compounds or their mixtures, which burn by reacting with atmospheric oxygen. Thus, liquid hydrocarbons such as gasoline, kerosene, naphtha, etc., used in pyrotechnic incendiary devices, yield very few solid products on burning.

The proportion between the amount of gaseous and solid products of the combustion reaction is determined by the ultimate use of the composition and the special effect requirements imposed on it. For example, for illuminating and tracer compositions, the gaseous reaction products comprise 15 to 25% of the weight of the composition. Smoke compositions evolve a large amount of gaseous products on burning.

The amount of gaseous products obtained by burning 1 g of pyrotechnic composition is usually expressed, not in terms of weight, but in terms of the volume which they occupy at standard conditions (0°C and 760mm Hg). This volume of gaseous products is known as the specific volume and is denoted by v_0.

Usually, when carrying out the calculations, to the volume occupied by the gases (CO_2, CO, etc.) formed in the course of the combustion reaction is added the volume occupied at standard conditions by the water vapor formed.

If the volume v_0 is known, the volume v_t of the gaseous products at the temperature of the combustion reaction is calculated from the commonly known Gay-Lussac formula:

$$v_t = v_0 (1+0.00366\ T).$$

In calculating v_t, it is necessary to consider not only the true gases and water vapor, but also all the substances which are in the vapor state at the temperature of the combustion reaction.

An example is a mixture of potassium chlorate and aluminum; the combustion reaction takes place according to the equation

$$KClO_3 + 2Al = KCl + Al_2O_3.$$

Since the combustion temperature of this mixture is about 3000°C, and potassium chloride boils at 1415°C, all of it will be in the vapor state at the reaction temperature. This explains the phenomenon that in some cases, the combustion of this mixture may occur at a rate approaching the rate of an explosion.

Gaseous products of the combustion reaction of pyrotechnic compositions are formed mainly as a result of oxidation or decomposition of components containing hydrogen, nitrogen, sulfur, and chlorine. The first two elements enter into the composition of organic compounds used as components in the manufacture of pyrotechnic compositions.

Nitrogen is contained in organic nitro or amino compounds, and also in nitrates, which are the most common oxidizers in pyrotechnics. Sulfur is used in the form of sulfur compounds (Sb_2S_3, pyrite, etc.), and is sometimes introduced into pyrotechnic compositions in the elemental state as well.

Chlorine is contained in chlorinated organic compounds; free chlorine and hydrogen chloride are formed by the combustion of compositions containing ammonium perchlorate NH_4ClO_4. The decomposition and oxidation of compounds containing these elements form H_2O, CO_2, CO, N_2, SO_2, H_2, Cl_2, HCl, and in some cases, the nitrogen oxides NO and NO_2.

Table 7.1 shows the volume occupied at standard conditions by 1 g of the above gases.

Table 7.1 VOLUME OF 1g OF GAS AT STANDARD CONDITIONS IN ml

Gas	ml/g	Gas	ml/g
H_2	11200	HCl	614
H_2O	1247	CO_2	509
CO	800	SO_2	350
N_2	800	Cl_2	315

As is evident from Table 7.1, for the same weight, the largest volume in the gaseous state is occupied by hydrogen, followed by water vapor, and carbon monoxide CO. On the basis of these data, one can reach the practically important conclusion that in order to obtain a large amount (by volume) of gaseous combustion products, it is necessary to use as fuel components, organic substances containing a large amount of hydrogen; the amount of oxidizer introduced into the composition should be calculated so that the fuel burns only to H_2O and CO, or, if required by thermochemical considerations, to H_2O and CO_2. Among solid fuels, a large amount of hydrogen is contained in paraffin, stearin, etc.

The specific volume of the gaseous reaction products may be determined in two ways.

1. Experimentally - during the combustion of the pyrotechnic composition in the bomb calorimeter. The gaseous products for measuring their volume are transferred from the bomb calorimeter to a gasometer [154]. One can also operate otherwise, i.e., use a mercury manometer to measure the pressure of the gases directly in the bomb calorimeter and use the value obtained

(allowing for the temperature of the gases in the bomb) to calculate the volume of the gases. In determining the specific volume, it is necessary to consider that the combustion products after cooling will have a somewhat different composition than during the reaction. In some cases, this distorts the results of the determination. For example, cooling of the combustion products involves condensation of the metal chlorides formed during the combustion of perchlorate (or chlorate) compositions; in addition, reactions of formation of carbonates from CO_2 and metal oxides may take place; all this causes a decrease in v_0.

It should be noted that the volume occupied by water vapor does not enter into the experimentally determined value of the specific volume of the gases.

2. By calculation - on the basis of the equation of the combustion reaction of the pyrotechnic composition. The equation of the reaction can be set up in two ways:

a) tentatively - on the basis of available experimental data on the direction of the combustion reaction of other compositions similar in formulation to those being studied;

b) more accurately - on the basis of results of chemical analysis of gaseous and solid combustion products and determination of the amount of water formed by the reaction.

The specific volume is calculated from the formula

$$v_0 = \frac{22.4 \times n \times 1000}{m},$$

where n is the number of g-moles of gaseous substances evolved by the combustion (sum of the coefficients of the gaseous substances on the right side of the reaction equation); m-weight of the reacting composition in g.

Example 1. Calculate v_0 of the following illumination composition: 75% barium nitrate, 21% magnesium, 4% iditol.

The combustion reaction may be approximately expressed by the equation

$14.6\ Ba(NO_3)_2 + 43.2Mg + C_{13}H_{12}O_2 = 14.6BaO + 14.6N_2 + 43.2MgO + 13CO_2 + 6H_2O$, whence

$$v_0 = \frac{22.4(14.6 + 13 + 6) \times 1000}{14.6 \times 261 + 43.2 \times 24.3 + 200} = 149\ cm^3/g\ of\ composition.$$

Table 7.2 lists the values of v_0 calculated for various compositions; the composition formulas were taken from Table 6.2. A certain relationship can be established between the function of the composition and the specific volume of the gaseous products of its combustion (see Table 7.2).

It should be noted that the specific volume of gaseous products for the pyrotechnic compositions currently in use (with the exception of solid rocket fuel) is much smaller than for the main explosives; thus, for hexogen, v_0 is 908 cm^3/g, for tetryl, 750 cm^3/g, and for trotyl, 688 cm^3/g.

Table 7.2 SPECIFIC VOLUME OF GASEOUS COMBUSTION REACTION PRODUCTS OF CERTAIN COMPOSITIONS

Use of Composition	Formulation in %	Gaseous Reaction Products	v_0 cm³/g	Rate of gases in % of weight of composition
Photoflash	$Ba(NO_3)_2$ - 68 Mg - 32	N_2	58	7
Photoflash	$Ba(NO_3)_2$ - 74 Al - 26	N_2	61	8
Iron-aluminum thermite	Fe_3O_4 - 75 Al - 25	--	0	0
Illuminating	$Ba(NO_3)_2$ - 75 Al - 25 Iditol - 4	N_2 CO_2 H_2O	144	21
Green signal	$Ba(ClO_3)_2 \cdot H_2O$ - 88 Iditol - 12	CO_2 H_2O	330	43
Red signal	$KClO_3$ - 57 $SrCO_3$ - 25 Shellac - 18	CO H_2O	375	40
Red smoke	$KClO_3$ - 35 Lactose - 25 Rhodamine - 40	CO H_2O	365	39
Smoke powder	KNO_3 - 75 Wood charcoal - 15 Sulfur - 10	CO CO_2 H_2O N_2	280	--

It should be pointed out in conclusion that for solid rocket fuels, other more exact methods of calculation of the amount of gaseous reaction products are employed (see Ch. 20). The numerical data given in Table 7.3 should be treated only as tentative ones in view of the insufficiency of both literature and experimental material.

Table 7.3 RELATIONSHIP BETWEEN THE FUNCTION OF COMPOSITIONS AND THE VALUE OF v_0

FUNCTION OF COMPOSITION	v_0 cm³/g
Photoflash	50 - 100
Illumination and tracer	100 - 300
Night signal	300 - 450
Smoke (except metal chloride mixtures)	300 - 500
Incendiary (w/oxidizer)	0 - 300
Solid rocket fuel	600 - 850

Chapter 8

COMBUSTION TEMPERATURE OF PYROTECHNIC COMPOSITIONS

The determination of the combustion temperature of pyrotechnic compositions is of major importance, since it provides a criterion for evaluating existing compositions and facilitates the creation of new, improved formulations.

The combustion temperature of pyrotechnic compositions may be determined in two ways:

1) calculation by means of formula (8.7). This involves the commonly known assumption that the temperature of a reaction is equal to its heat divided by the total heat capacity of the reaction products; however, this does not take into account the losses to the surroundings, and in particular, radiation losses, which are fairly substantial for pyrotechnic flames.
2) direct measurement (with the aid of optical pyrometers or thermocouples).

1. CALCULATION OF THE COMBUSTION TEMPERATURE OF COMPOSITIONS

It should be pointed out at the very outset that a flame has different temperatures in different zones. Calculation gives only the upper limit of the possible temperature, in other words, the maximum flame temperature.

The determination of the combustion temperature of compositions by calculation is complicated by the lack of accurate numerical data on the heat capacity of many compounds at high temperatures (above 2000°C); for many compounds, the latent heats of fusion and vaporization either have not been determined at all, or have been determined with insufficient accuracy.

The only calculated values of the combustion temperature are the upper (practically inaccessible) limit, since in actuality, the combustion temperature should be considerably lower because of the heat loss due to thermal dissociation of the combustion products and also the heat loss to the surroundings.

Combustion-temperature calculation results of satisfactory accuracy, obtained by the method described below involving the use of formula (8.7), can be obtained when the temperature being sought does not exceed 2000-2500°C. Otherwise, only tentative data are obtained (for example, in the calculation of the combustion temperature of compositions containing over 10-15% of magnesium, aluminum, or other metals of high calorific value).

Considering that in the combustion of pyrotechnic compositions, the gaseous reaction products are usually able to expand, the calculations are made by using heat capacity values at constant pressure C_p. The molar heat capacities of gases at constant volume C_v and constant pressure are related as follows:

$$C_p = C_v + R. \qquad (8.1)$$

where R is the gas constant, equal to 1.98 cal/g mole deg.

The molar heat capacity of a gas C_v may be calculated from the formula

$$C_v = \frac{nR}{2} \qquad (8.2)$$

For monatomic gases, the number 1 is equal to 3, for diatomic ones, 5, and for polyatomic ones, 6.

Using formulas (8.1) and (8.2), we obtain the values of C_p in cal/g-mole deg:

    For monatomic gases 4.95
    For diatomic gases 6.93
    For polyatomic gases 7.92

Experimental data do not always confirm the theoretical calculation of heat capacity for diatomic gases and particularly polyatomic ones. Table 8.1 gives values of molar heat capacity C_p obtained experimentally. For elements at high temperatures (1000°C and above) in the solid state, it may be assumed in accordance with Dulong & Petit's Law that their gram-atomic heat capacity has a constant value equal to approximately 6.4 cal/deg.

Table 8.1 VALUE OF AVERAGE MOLAR HEAT CAPACITIES FOR GASES, C_p

Temperature Range, °C	N_2, O_2, CO	H_2	H_2O	CO_2
0-100	7.0	6.9	8.0	9.1
0-500	7.1	7.0	8.3	10.3
0-1000	7.3	7.1	8.8	11.3
0-1500	7.5	7.4	9.5	11.9
0-2000	7.7	7.6	10.3	12.3
0-3000	8.0	7.8	12.8	12.7

Kopp's Law, according to which the heat capacity of a compound is equal to the sum of the atomic heat capacities of its component elements, applies to a certain extent to compounds which are in the solid state at high temperatures.

Table 8.2 gives the experimental values of the heat capacity for certain substances which are of interest to a pyrotechnist. The difference between C_p and C_v for solids is very slight and may be neglected for practical purposes.

Table 8.2 AVERAGE MOLECULAR HEAT CAPACITY OF SOLIDS C_p

SUBSTANCE	TEMPERATURE, °C	C_p	SUBSTANCE	TEMPERATURE, °C	C_p
Fe	0-100	6.2	NaCl	600	13.8
Fe	20-1500	9.6	$Al+_2+O_3$	30-300	23.0
Cu	20-100	5.9	$Al+_2+O_3$	30-1100	27.7
Cu	20-1500	9.4	$Al+_2+O_3$	30-1500	28.1
NaCl	20-785	13.6	$Al+_2+O_3$	20-2030	28.5
MgO	20-1735	12.1	$Al+_2+O_3$	570-600	29.4
MgO	20-2370	14.0	Al_2O_3	965-973	33.2
MgO	20-2780	14.3	$BaCl_2$	100	19.6
KCl	400	13.3	$BaCO_3$	1000	31.8

Concerning the heat capacity of liquids at high temperatures (above 1000°C), it is difficult to indicate any definite relationships. It should be noted only that the heat capacity of a substance in the liquid state is usually higher than in the solid state. The molar heat capacity of high-melting liquids is sometimes approximately assumed to be equal to 1.3 of the solids.

Latent heat of fusion and vaporization. Richards and Compton found that the following relation applies to most elements:

$$\frac{Q_s}{T_s} = 0.002\text{—}0.003 \qquad (8.3)$$

where Q_s is the heat of fusion in kcal/g-atom;
T_s is the melting point in °K.

This relation is not sufficiently accurate for many elements.

In our view, the latent heat of fusion for many inorganic compounds can be approximately calculated from the empirical formula

$$\frac{Q_s}{T_s} = 0.002 \times n, \qquad (8.4)$$

where n is the number of atoms in the molecule of a compound.

The value of the latent heat of vaporization of one and the same substance is not constant, but decreases as the temperature at which vaporization takes place rises. Thus, for water, the gram-molecular heat of vaporization is 10.8 kcal at 0°C, and 9.7 kcal at 100°C.

The heat of vaporization of liquids at high temperatures reaches high values. The relationship between the heat of vaporization (Q_k kcal/mole) and the boiling point of a liquid at 760 Hg (T_k in °K) is expressed by Trouton's formula

$$\frac{Q_k}{T_k} = 0.02 \qquad (8.5)$$

For many (particularly, high-boiling) inorganic compounds, the latent heat of vaporization may, in our view, be calculated with a higher accuracy than from Trouton's formula by using the empirical formula

$$\frac{Q_k}{T_k} = 0.011 \times n \qquad (8.6)$$

where n is the number of atoms in the compound.

Obviously, the heat of vaporization of compounds is much higher than their heat of fusion.

The values of Q_k and Q_s for high-melting and high-boiling inorganic compounds are so appreciable that their consideration is absolutely necessary in calculating the combustion temperatures.

The maximum combustion temperature is calculated from the formula

$$t = \frac{Q - \sum(Q_s + Q_k)}{\sum C_p} \qquad (8.7)$$

where Q is the amount of heat evolved during the combustion reaction;

ΣC_p, is the sum of the heat capacities of the reactions products;

$\Sigma(Q_s+Q_k)$ is the sum of latent heats of fusion and vaporization of the combustion products;

t is the desired combustion temperature in °C.

Given below are two examples of calculation of the combustion temperature of compositions by use of this formula.

Example 1. Calculate the combustion temperature of a red light composition consisting of 65% $KClO_3$, 20% $SrCO_3$, and 15% $C_{13}H_{12}O_2$ resin (iditol). The combination reaction of this composition may be approximately expressed by the equation

$$7.1 KClO_3 + 1.8 SrCO_3 + C_{14}H_{12}O_2 = 7.1 kcl + 1.8 SrCO_3 + 6 H_2O_v + 4.3 CO_2 + 8.7 CO.$$

This involves an inaccuracy, since the dissociation reaction $SrCO_3 = SrO + CO_2$, as well as the reactions $2kcl + SrO = SrCl_2 + K_2O$ and $2SrCl_2 = 2SrCl + Cl_2$ take place in the flame (see Ch. 16).

The heat of combustion is calculated from Hess' law, and the heat of formation of iditol is assumed to be 0.74 kcal/g:

For H_2O	6 x 57.4 = 344		
For CO_2	4.3 x 94 = 404	For $KClO_3$	7.1 x 96 = 682
For CO	8.7 x 26 = 226	For $C_{13}H_{12}O_2$	0.74 x 200 = 148
For KCl_2	7.1 x 106 = 752		830 kcal
	1726 kcal		

The amount of heat evolved by the reaction is

$$Q = 1726 - 830 = 896 \text{ kcal}.$$

It is assumed that the combustion temperature will be about 1500°C, and so we take into account the latent heats of fusion and boiling of KCl.

In handbooks [98] we find:

$$Q_{s\,KCl} = 6.3 \text{ kcal at } 768°C$$

$$Q_{k\,KCl} = 40 \text{ kcal at } 1415°C$$

Hence $\Sigma(Q_s+Q_k) = 46.3 \times 7.1 = 329$ kcal

and $Q - \Sigma(Q_s+Q_k) = 896 - 329 = 567$ kcal

The capacity C_p of gases and water vapor in the range 0-1500°C (see Table 8.1) is

For H_2O	6 x 9.5 = 57.0
For CO_2	4.3 x 11.9 = 51.2
For CO	8.7 x 7.5 = 65.2
	173.4

According to the Kopp law, we assume the heat capacity of KCl and $SrCO_3$ to be equal to 12.8 and 32.0 cal/deg mole, respectively.

Hence
12.8 x 7.1 = 90.9
32.0 x 1.8 = 57.6
148.5

PRINCIPLES OF PYROTECHNICS

and finally

$$\Sigma C_p = 173.4 + 148.5 = 321.9 \text{ cal/deg}$$

whence

$$t = \frac{567 \times 1000}{321.9} = 1760°C.$$

If we took into account the thermal dissociation $SrCO_3 = SrO + CO_2 - 54$ kcal, which partly takes place in the flame, the calculation would yield a value of the temperature of the order of 1600°C.

Example 2. Calculate the combustion temperature of a smoke mixture with the following composition: CCl_4 39%, Zn 34%, $NaClO_3$ 15%, NH_4Cl 9%, SiO_2 (kieselguhr) 3%. We set up the equation of the combustion reaction:

$$0.257 CCl_4 + 0.514 Zn + 0.137 NaClO_3 + 0.164 NH_4Cl + 0.055 SiO_2 =$$
$$0.514 ZnCl_2 + 0.137 NaCl + 0.164 NH_4Cl + 0.055 SiO_2 + 0.103 CO + 0.154 CO_2.$$

In this composition, kieselguhr plays the part of absorber of liquid CCl_4 and does not participate in the combustion reaction. Q = 47.2 kcal.

For $ZnCl_2$ $T_s = 365°C$, $T_k = 732°C$;
For NaCl $T_s = 800°C$, $T_k = 1440°C$.

The sublimation temperature of ammonium chloride (NH_4Cl) is 335°C.

For SiO_2 $T_k = 1470°C$.

For smoke compositions, the reaction temperature seldom exceeds 1000°C; therefore, the latent heat of vaporization of NaCl and heat of fusion of SiO_2 are not considered in further calculations.

Let us determine the initial values: $Q_{s\,ZnCl_2} = 3.8$ (calculated from formula 8.4):

$$Q_{s\,NaCl} = 7.2; \quad Q_{k\,ZnCl_2} = 33;$$

Q of sublimation of NH_4Cl = 39 kcal.

Hence

$$\Sigma(Q_s + Q_k) = 0.514(3.8 + 33) + 0.137 \cdot 7.2 + 0.164 \cdot 39 = 26.5 \text{ kcal.}$$

$$Q - \Sigma(Q_s + Q_k) = 20.7 \text{ kcal.}$$

We calculate the heat capacities:

For $ZnCl_2$	0.514×20	$= 10.28$
For NaCl	0.137×13.6	$= 1.87$
For NH_4	0.164×36	$= 5.90$
For SiO_2	0.055×18	$= 0.99$
For CO	0.103×7.3	$= 0.75$
For CO_2	0.154×11.3	$= 1.64$

$$\Sigma C_p = 21.43$$

$$t = \frac{20.7 \times 1000}{21.43} = 965°C.$$

In 1953, in calculating the combustion temperature of aluminothermic reactions, Venturini [109] introduced certain modifications into the calculation method. For ordinary iron-aluminum thermite, he obtained the value of 3500°K (3227°C), which was inconsistent with the results of experimental determinations (see below). At first glance, Venturini's thermochemical calculation is justified, but his appreciable error lies in the fact that, as the author himself indicated, he did not consider the possibility of vaporization of the products of the combustion reaction.

Vaporization also takes place to some extent in aluminothermic processes. In our view, neglect of this phenomenon caused Venturini to obtain excessively high data.

It should be noted once again that if the combustion temperature is in excess of 2000-2500°C, the data obtained by calculation from formula (8.7) reflect only the order of magnitude, and can therefore be useful only in comparing compositions differing markedly in their formulation.

For rocket fuels, a high degree of accuracy is required in the calculation of the combustion temperature and other characteristics of the combustion products. In this case, the combustion temperature is calculated by means of very laborious computations in which the processes of thermal dissociation and vaporization of the combustion reaction products are considered. The initial data for such calculations are given in the handbook [233].

2. EXPERIMENTAL DETERMINATION OF THE COMBUSTION TEMPERATURE

The combustion temperature of most pyrotechnic flame compositions lies in the range 2000-3000°C.

The flame temperature during the combustion of such compositions is measured more frequently by means of optical methods.

The basis for the use of optical methods in measuring the combustion temperature of pyrotechnic compositions is the premise that the radiation of a pyrotechnic flame follows (at least approximately) the laws of radiation of an absolute blackbody (ABB).

In this case, the radiation should obey:

1) the Stefan-Boltzmann law
$$E = c \times T^4,$$

where E is the total radiation energy;
c is a constant equal to 5.73×10^{-12} W/cm² deg;
T is the absolute temperature in °K.

2) Wien's displacement law: $\lambda_{max} \times T = 2884$

if λ_{max} is measured in microns;

3) Planck's radiation law

Actually, the radiation of the flame of pyrotechnic compositions differs somewhat in character from the radiation of ABB. The radiation of ABB has a continuous spectrum.

The presence of incandescent solid or liquid particles in the flame produces a continuous flame spectrum. In addition, some substances exist in the flame in the gaseous state, giving rise to a discontinuous (line or band) spectrum. For this reason, in most cases, the flame of pyrotechnic compositions has a continuous radiation spectrum with a discontinuous radiation spectrum of the gaseous phase superimposed on it. The comparative intensity of the continuous and discontinuous spectra depends on many factors, including primarily the temperature of the flame and the quantitative ratio of the solid to liquid phase therein.

It follows from the above that in most cases, by measuring the flame temperature with an optical pyrometer, one can obtain only an approximate estimate of the true flame temperature (for white flames, the error amount of approximately ±50 to ±100°C).

The most accessible methods are those of measurement of the brightness and color temperature of a flame. The brightness temperature T_b is the temperature of ABB at which the flame brightness for ($\lambda=0.665\mu$) is equal to the brightness of the radiator studied at the same wavelength. The color temperature T_c is the temperature of ABB at which the chromaticity of its radiation is the same as the chromaticity of the radiator studied.

In this case, the radiation intensity of ABB for two different wavelengths λ_1 and λ_2 is equal to the radiation intensity of a given physical body at the same wavelengths. The experimental determination of the color temperature of a body amounts to determining the ratio of the radiation intensities for two different wavelengths,

$$K = \frac{E_{\lambda_1}}{E_{\lambda_2}}$$

then finding from a table the temperature at which for ABB the ratio of the radiation intensities at wavelengths λ_1 and λ_2 is equal to K.

TYPES OF OPTICAL PYROMETERS

A disappearing-filament optical pyrometer (Fig. 8.1) is a visual pyrometer in which the brightness of the light emitted by the body studied (flame) is measured by comparison with the brightness of a standard incandescent body (lamp filament) at the same effective wavelength $\lambda=0.665\ \mu$.

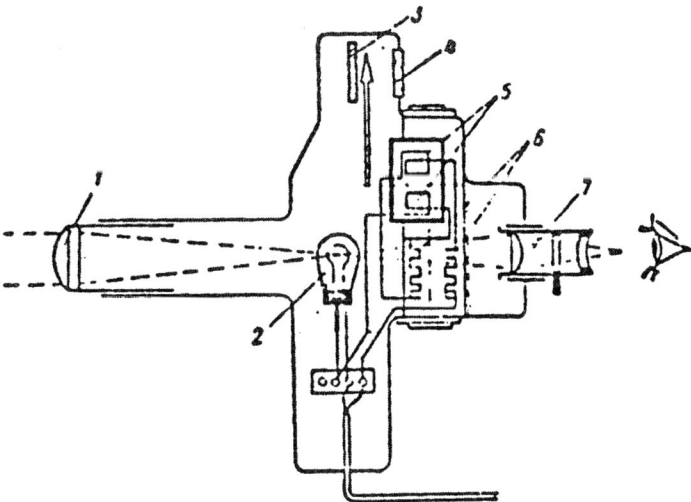

Fig. 8.1. Diagram of a disappearing-filament optical pyrometer. 1 - objective lens, 2 - lamp, 3 - scale, 4 - glass, 5 - movable rollers, 6 - variable resistance, 7 - sliding ocular tube.

In the pyrometer, the lamp filament is placed at the focal point of the objective, which together with the ocular forms an optical system superposing the image of the flame on the image of the lamp filament.

By changing the current intensity, one can change the brightness of the incandescent filament until it is the same as the brightness of the flame. This effect is perceived by the observer as a disappearance of the incandescent filament against the background of the flame; at the same time, a reading is taken on a rheostat scale graduated in degrees of brightness temperature. The mass-produced OPPIR-C17 pyrometer has two measurement scales on which the temperature can be measured in the range 1500-2500°C and 2200-6000°C.

A cinephotopyrometer is an ordinary movie camera provided with a red light filter ($\lambda=0.665\mu$) and a set of ribbon filament lamps mounted in a row next to the object being measured (Fig. 8.2).

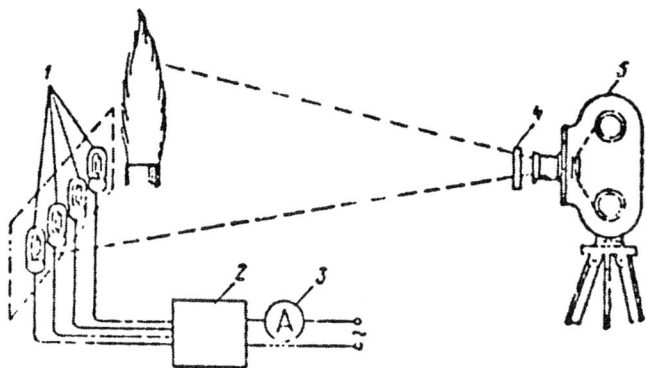

Fig. 8.2. Diagram of cinephotopyrometer. 1 - ribbon filament lamps, 2 - transformer, 3 - ammeter, 4 - light filter, 5 - movie camera.

This method of temperature measurement is based on the principle of photography of a flame in its own light; higher-temperature flames will produce higher blackening densities on the photographic film than lower-temperature flames. By simultaneously photographing ribbon filament lamps of known temperature with the flame, blackening densities s for known temperatures are obtained on the movie film frame. For the range of normal densities, the calibration graph is plotted. Having a certain density, we find the brightness temperature from the graph. This method is of particular interest for pyrotechnists, since it can be used to determine the temperatures in various portions of a flame, and also to record changes of temperature with time (Fig. 8.3).

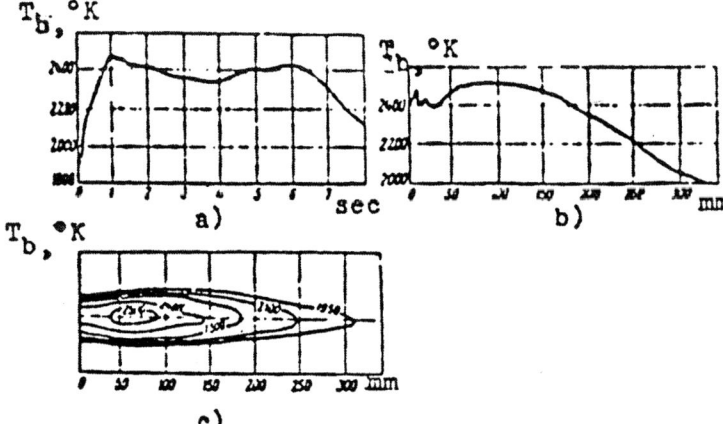

Fig. 8.3. Flame temperature data obtained with a cinephotopyrometer. a - graph of temperature change with time; b - graph of temperature change with flame height, c - graph of temperature fields.

The FEP-0.65 photoelectric pyrometer is a variant of the optical pyrometer especially designed for measuring the temperature of pyrotechnic flames with recording of the measurement data on an oscillograph. The FEP-0.65 is used to measure the mean brightness temperature and emissive power of a flame (the quantity of energy of a definite wavelength radiated by a unit area of the body per unit time at temperature T, i.e., E_λ in cal/m² hr or W/cm² [101]).

To measure the emissive power, behind the flame is placed a ribbon filament lamp with a lens (Fig. 8.4), which projects the ribbon filament on the portion of the flame studied. Between the lamp and the lens is placed a revolving screen (modulator) which masks the light of the lamp periodically at a frequency of 10 cps.

At first, the photocurrent of a single lamp (I_1) is recorded on the oscillograph. During the combustion of a pyrotechnic composition, when the screen masks the lamp, the photocurrent produced by one flame (I_f) is recorded. When the light of the lamp is not blocked by the screen, however, the photocurrent produced by the flame and lamp together (I_{f+1}) is recorded.

From the measured values, the transmissivity of the flame is calculated:

$$\tau_\lambda = \frac{I_{f+1} - I_1}{I_f}$$

then the absorptivity of the flame $\alpha_\lambda = 1 - \tau_\lambda$ which according to Kirchoff's law is equal to the emissive power of the flame E_λ.

In measuring the brightness temperature, it is necessary only to have data on the value of the photocurrent due to the flame (I_f); these data are used to determine the temperature from the graph, the change in temperature with time, the temperature fluctuation, and the maximum and minimum temperature in any region of combustion.

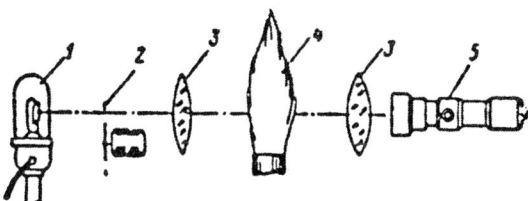

Fig. 8.4. Diagram of the determination of the emissive power of a flame.
1 - ribbon filament lamp, 2 - modulator, 3 - lens, 4 - flame of pyrotechnic composition, 5 - FEP-065.

The color pyrometer (red/blue ratio) is a recording luxmeter measuring the radiation intensity in the red and blue portions of the spectrum simultaneously. The most common version of the design (Fig. 8.5) consists of a unit of two photocells before which is mounted a rotating disc with two rows of holes; light filters (red $\lambda=0.685\mu$ and blue $\lambda=0.456\mu$) are mounted on one row of holes. The photocells are so arranged that the beam strikes both photocells at the same time. A photocell without a filter is used to introduce a correction for the change in luminous flux during the displacement of the light filters.

The color temperature can also be determined by the spectral method on the basis of the continuous spectrum by using those portions of the latter where no bands or lines are superposed on it. A description of the methods of measuring flame temperatures is given in the literature: see the books of Kadyshevich [101], Kirillin and Sheyndlin [102], Dean [105], and the papers of Pokhil [103] and Sobolev [104].

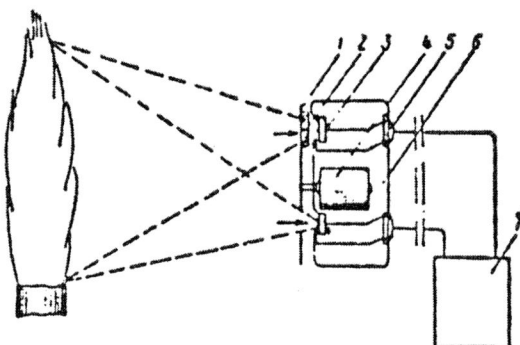

Fig. 8.5. Diagram of a color pyrometer. 1 - rotating disc, 2 - light filter, 3 - photocell, 4 - electric motor, 5 - contact terminals, 6 - housing, 7 - oscillograph.

The temperature of the incandescent solid or liquid slag formed during the combustion of a pyrotechnic composition is determined much more accurately than the flame temperature by an optical pyrometer.

Wartenberg [7, p. 87] used the optical method to determine the temperature of a flowing jet of iron-aluminum thermite and found to be 2400 ±50°C.

Eggert, Eder, and Dziobek [7, p. 87] measured the radiation intensity of magnesium flames in different portions of the spectrum and on this basis calculated the "color temperature" of the flame: the color temperature of the flame during the combustion of magnesium in oxygen was ~3700°C, during the combustion magnesium in air ~3400°C, and the color temperature of the flame of the photoflash $Th(NO_3)_4$ +Mg ~3100°C.

The temperature of combustion of aluminum in oxygen, measured by Quelleron and Scartazini with the aid of a brightness pyrometer was found to be ~3000-3200°C. It follows that a temperature of the order of 3000-3500°C can be obtained during the combustion of Mg and Al powders.

A higher temperature is prevented by the large heat loss due to vaporization and partial decomposition of oxides of these metals.

The highest temperature in the combustion of metal powders is produced by the combustion of zirconium powder in oxygen. Its estimate based on thermodynamic calculations [106] gives 4930°K.

The temperature limit is determined in this case by the boiling temperature of the oxide ZrO_2 at p=1 at (see Table 3.3).

Hence the possibility of reaching very high temperatures is determined not only by the high calorific value of the fuel, but also by the maximum boiling temperature, and also by the high chemical stability of the combustion products (metal oxides).

An even higher combustion temperature (5400°C) was produced by the gaseous mixture H_2+F_2 [223, p. 139].

According to Harrison's approximate estimate [108] (observation of evaporation of gold and platinum) the combustion temperature of titanium in oxygen lies in the range 2950-3500°C.

The relationship between the combustion temperature of elements in oxygen and the atomic number of an element in the periodic system is shown in Fig. 8.6.

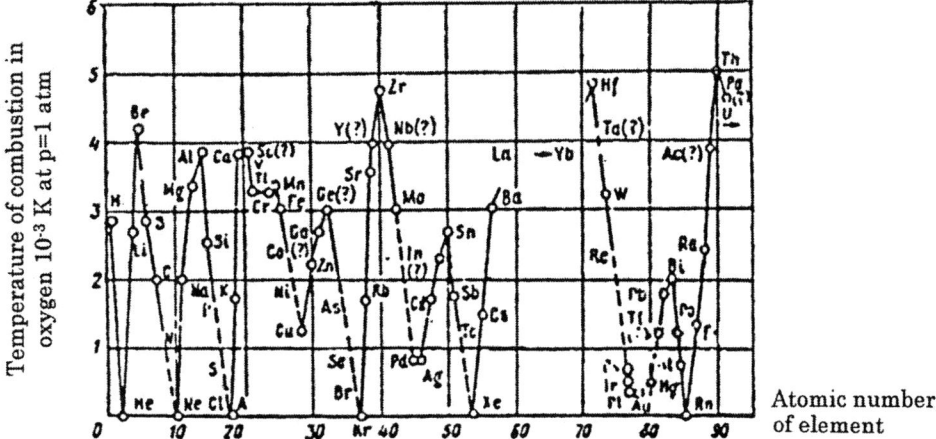

Fig. 8.6. Combustion temperature of elements in oxygen in °K at p=1 at [243].

By using an optical brightness pyrometer to measure the temperature of flames formed during the combustion of match compounds containing about 40% of potassium chlorate, Arditti [7, p. 88] obtained temperature values of 1400-1500°C.

In conclusion, let us also note that the temperature in different zones of a pyrotechnic flame is different, and the term "flame temperature" is usually taken to mean the temperature in the hottest zone of the flame.

The combustion temperature of certain slowly burning smoke compositions, ranging from 360 to 600°C, may be measured with quartz mercury thermometers in which the mercury is under pressure.

A necessary condition for obtaining temperature values close to the true value in these experiments is a slow rate of combustion of the compositions so that the bulb of the mercury thermometer is able to warm up sufficiently.

A chromel-alumel thermocouple can be used to measure temperatures up to 1300°C. To measure higher temperatures, up to 1600°C, use may be made of a thermocouple whose one lead is made of platinum, and the other, of an alloy of platinum with 10-15% rhodium (melting point of platinum, 1771°C).

W-Ir and W-Re thermocouples can be used up to temperatures of 2100 and 2700°C, respectively, but they are of little use in pyrotechnics, since the presence of a reducing or inert medium would then be required.

The Ir-Rh/Ir (40% Ir) thermocouple is graduated and used to 2100°C with an accuracy of 10°. For limited periods of time, it can also be used in air.

In measurements of the combustion temperature of pyrotechnic compositions with the aid of a thermocouple, particular attention should be given to the thermal lag of the pyrometer and care should be taken to reduce it as much as possible. Accurate results can be obtained only when the thermocouple leads are made of unsheathed wire not more than 0.05-0.1 mm in diameter, or of a ribbon of the same thickness; the lag of the recording instrument, a millivolt meter, should also be as small as possible.

Depending upon the diameter of the thermocouple leads (unsheathed) the following results were obtained for the composition of red signal smoke (rhodamine dye):

Diameter of leads in mm	0.5	0.2	0.1
Value for the maximum reaction temperature in °C	338	697	837

Obviously, the reaction temperature is slightly higher than 837°C; as the diameter of the thermocouple leads decreases further, results closer to the actual values can be obtained.

In the measurement of the reaction temperature of two other smoke mixtures for signal smokes by means of an iron-constantan thermocouple with a lead diameter of 0.1 mm, the following data were obtained:

for mixture No. 1   1080°C
for mixture No. 2   1000°C

The following values were obtained by calculation for the combustion temperatures of these mixtures:

for mixture No. 1   1262°C
for mixture No. 2   1070°C

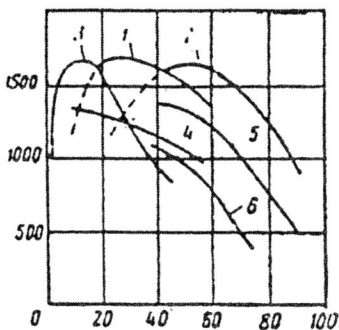

% of fuel in binary mixture

Fig. 8.7. Maximum temperature during the combustion of mixtures (measured with a Pt/Pt-Rh thermocouple).

Using a thermocouple, Hill and Sutton studied the change in combustion temperatures of binary mixtures with changing ratios of the components. Results of this study are in Fig. 8.7.

3. RELATIONSHIP BETWEEN THE USE OF A COMPOSITION AND THE MAXIMUM COMBUSTION TEMPERATURE

On the basis of the experimental material published thus far, it may be concluded that there exists a definite relationship between the function of a composition and the maximum temperature in the combustion zone (Table 8.3).

Table 8.3 FUNCTION OF A COMPOSITION AND MAXIMUM TEMPERATURE IN THE FLAME

Composition	Maximum Temperature in the flame °C
Photoilluminating	2500-3500
Illuminating and tracer	2000-2500
Incendiary w/oxidizer	2000-3500
Rocket (solid fuel)	2000-2900
Signal night	1200-2000
Smoke compositions	400-1200

Chapter 9

SENSITIVITY OF PYROTECHNIC COMPOSITIONS

The initial impulse is taken to mean the minimum amount of energy necessary to stimulate a reaction of combustion (or explosion) in a pyrotechnic composition. The lower this amount of energy, the most sensitive to external actions will be a given pyrotechnic composition.

A rapid reaction in a pyrotechnic composition can be stimulated by using various forms of energy: thermal, mechanical, electrical, radiant, etc., but the nature of the process thus produced in the pyrotechnic composition is different.

To obtain the normal effect from the action of pyrotechnic compositions, use is made in most cases of an initial thermal impulse (flame, quick match, ignition compositions); only in work with photomixtures and certain incendiary compositions is the action of an explosive occasionally applied as the initial impulse (i.e., the combined action of a mechanical and a thermal impulse is employed), so as to produce a deliberate explosion in the pyrotechnic composition.

Sensitivity tests of a pyrotechnic composition are aimed at establishing:

1. The appropriate methods of its preparation, and storage conditions that will guarantee the absence of ignition or explosion.
2. Correct ignition methods that will cause the composition to yield the required special effect.

Sensitivity tests of pyrotechnic compositions consist in the determination of:

1. The self-ignition temperature.
2. Sensitivity to a flame.
3. Sensitivity to shock.
4. Sensitivity to friction.

Less frequent are tests for:

1. Ignitability by special ignition compositions.
2. Sensitivity to impact of bullets (ordinary or incendiary ones.

In designing a new pyrotechnic composition and selecting its components, every pyrotechnist must fully realize what sensitivity the new composition will have. In most cases, this can be anticipated in general terms in advance of the tests.

1. DETERMINATION OF THE SENSITIVITY OF COMPOSITIONS TO THERMAL EFFECTS

The determination of the sensitivity of pyrotechnic compositions to thermal effects consists in determining the temperature of their self-ignition and the sensitivity to a flame.

DETERMINATIONS OF SELF-IGNITION TEMPERATURE

The decomposition reaction of pyrotechnic compositions may take place at a slow rate and at lower temperatures than the temperature of their self-ignition.

"Self-ignition temperature" is the term applied to the lowest temperature to which a composition must be heated so as to cause its spontaneous inflammation, associated with a clearly visible luminous, sound or smoke effect.

According to modern views, the phenomenon of self-ignition consists essentially in the following. At a low temperature, the rate of the exothermic reaction in a pyrotechnic composition is slow, and all of the heat evolved is dissipated into the ambient medium. As the temperature rises, the reaction rate and inflow of heat to the composition increase rapidly, and at a certain instant, the heat inflow begins to exceed the heat outflow; the temperature begins to rise as a result of self-heating, and the rate of the chemical reactions taking place in the composition rises sharply, resulting in self-ignition of the composition.

The self-ignition temperature for one and the same composition is not always a constant, but depends significantly on the conditions of heat inflow and outflow prevailing during its determination; therefore, in practice, the weighed sample of composition and the design of the instrument (method and time of heating) are strictly regulated.

One of the possible methods of determining the self-ignition temperature of pyrotechnic compositions consists in the following.

A special iron bath filled with Wood's alloy is heated until the alloy passes completely into the liquid state. Three glass test tubes, one of which contains a thermometer, are then immersed in the bath to one-third of their height.

The bath is then heated to a temperature close to the expected self-ignition temperature of the composition. When this temperature is reached, 0.1 g of pyrotechnic composition is then poured into each of the empty tubes, and, the bath temperature being kept constant, the time necessary for self-ignition of the composition to take place (induction time) is recorded. If self-ignition fails to occur within 5 min, the test is discontinued, and the word "rejected" is written in the notebook.

A series of such tests conducted at various temperatures, always with fresh samples of the composition, make it possible to establish to within 5°C the minimum temperature below which no self-ignition takes place after 5 min of testing. This temperature is taken as the self-ignition temperature.

If desired, the testing results may be supplemented by plotting a graph showing the dependence of the induction time (ignition delay time) on the testing temperature.

Tests of compositions whose self-ignition temperature is above 400-450°C are carried out by using an electric furnace instead of a bath; in this case, it is entirely sufficient to determine the self-ignition temperature to within 10°C. As an illustration, data on the self-ignition temperature of certain binary mixtures, obtained by using the above-described procedure, are presented below.

Table 9.1 SELF-IGNITION TEMPERATURE OF BINARY MIXTUES IN °C
(Oxidizer & fuel taken in stoichiometric amounts)

Oxidizer	Self-ignition Temperature				
	Sulfur	Lactose	Charcoal	Mg powder	Al dust
Potassium chlorate	220	195	335	540	785
Potassium perchlorate	560	315	460	460	765
Potassium nitrate	440	390	415	565	890

Table 9.2 SELF-IGNITION TEMPERATURE OF BINARY MIX WITH POTASSIUM CHLORATE

Fuel	Self-ignition temperature °C	Fuel	Self-ignition temperature °C
Shellac	250	Antimony	295
Iditiol	345	Graphite	890
Colophony	335		

The self-ignition temperature of binary mixtures whose composition includes chlorates as the oxidizer, for example, potassium chlorate (Tables 9.1 and 9.2), are in most cases significantly lower than the self-ignition temperature of perchlorate or nitrate mixtures (with the same fuels). An exception to this rule are binary mixtures containing metal powders as fuels.

Ellern [9] divides pyrotechnic fuels (without any oxidizer being added) into three groups, according to the value of their flash point, namely:

1. Flash point   between 200 and 350°C
2. Flash point   between 350 and 600°C
3. Flash point   above 700°C

In the first group, he includes Zn, Ti, P_{red}, S, Th, and, as he writes, possibly Nb and Ta. In the second group, Mg (550-540°C), Mn (492°C), antimony and arsenic sulfides, and possibly, zinc, and calcium silicide. Ellern's third group includes aluminum only. It should be assumed that the lower the flash point of the fuel, the lower, as a rule, will be the self-ignition temperature of a pyrotechnic composition that includes this fuel.

In 1943, N. F. Zhirov developed and standardized a new method of determining the self-ignition temperature of pyrotechnic compositions (GOST 2040-43). According to this method, the self-ignition temperature is determined in an electric furnace whose temperature is measured with a thermocouple connected to a millivoltmeter.

In this testing method, in addition to the self-ignition temperature and induction time, one can also find the so-called flash point, whose value is adopted in accordance with GOST as a characteristic of the composition. The accepted flash point is the temperature at which the induction time to self-ignition is equal to zero.

For many compositions, the flash point (GOST 2040-43) is much higher than their self-ignition temperature, determined by the old method in a bath, and there is no total agreement between the data obtained by testing the same compositions by these two methods (Table 9.3).

Table 9.3 IGNITION POINT & FLASH POINT OF BINARY MIXTURES

No.	Formula %	Ignition Point (in bath)	Flash Point, °C
1	Potassium chlorate 86; Iditol 14	445	540
2	Potassium nitrate 86; Iditol 14	460	570
3	Potassium chlorate 63; Magnesium 37	540	670
4	Potassium nitrate 63; Magnesium 37	565	670
5	Sodium nitrate 65; Magnesium 35	--	610
6	Barium nitrate 68; Magnesium 32	510	660
7	Sodium nitrate 66; Aluminum dust 34	--	800
8	Black powder	310	480
9	Ignition composition: Potassium nitrate 75; Mg 10; Iditol 15	445	555

N. F. Zhirov's method, which is accepted at the present time as the basic method of testing, is more complex than the one described earlier, and each determination is more labor-consuming, considering the number of parallel tests required by the standard. However, in working with compositions that are difficult to ignite (incendiary compositions and others), where the purpose of the determination is selection of the appropriate igniter for the composition being tested, the use of N. F. Zhirov's method has undoubted advantages over the old methodology.

DETERMINATION OF SENSITIVITY TO A FLAME

The determination of the sensitivity of a pyrotechnic composition to a fire impulse is aimed at facilitating its matching with an ignition composition that will ensure its dependable ignition under combat conditions. In addition, the results of this test also characterize the degree of inflammability of the composition, this being taken to mean, for example, the possibility of its ignition upon being struck by a random spark.

As a firing pulse in this test, a flame is used, formed in the discharge of a flame from an open straight cut of a burning Bickford fuse. The characteristic of the sensitivity of the composition in this test is the largest distance between the lower section of the Bickford fuse and surface of the composition at which ignition of the composition still takes place.

According to Zhirov, the sensitivity of a composition to a ray of fire is characterized by:

a) an upper sensitivity limit, i.e., that maximum distance between the section of the Bickford fuse and the surface of the composition at which a 100% ignition of the composition takes place;
b) a lower sensitivity limit, i.e., that minimum distance between the section of the Bickford fuse and the surface of the composition at which 100% rejection takes place (the composition does not ignite);
c) the reliability factor, i.e., the slope of the straight line connecting points obtained experimentally on a graph where the abscissa axis represents the distance between the section of the Bickford fuse and the surface of the composition, and the ordinate axis represents the percent ignition.

Given below are the results of tests or certain compositions according to this method (see Table 9.4).

Table 9.4 SENSITIVITY OF COMPOSITIONS TO FLAME IMPULSE
(Data of I.I. Vernidub & V.A. Sukhikh)

Composition	Upper Limit cm	Lower Limit cm	Reliability Factor
Black powder, Grade 1	2	15	0.7
Igniter	3	13	0.8
Red light	0	2	3.0
Illuminating	0	3	1.0

A disadvantage of this method of testing is an incompletely satisfactory reproducibility of the results. The method should be improved: instead of a Bickford fuse, an ignition capsule can be employed.

Of little use for the practical evaluation of the inflammability of compositions are testing methods based on a prolonged action of a gas burner flame on the surface of the composition, where the measure of the inflammability of the composition is the induction period, i.e., the time interval from the start of heating to the instant of inflammation.

For compositions which completely fail to be ignited by a flame from a Bickford fuse, additional, as yet unstandardized tests are conducted to determine their capacity to be kindled by ignition compositions of different igniting capacities. This is done by gradually passing from weaker to stronger ignition compositions.

Unfortunately, no method has been developed thus far for determining the sensitivity of compositions to radiant energy; cf. [119, p. 27].

2. DETERMINATION OF THE SENSITIVITY OF COMPOSITIONS TO MECHANICAL EFFECTS

During the preparation and compaction, no matter how carefully these operations are carried out, pyrotechnic compositions are inevitably subjected to certain mechanical effects. Most frequently, these effects give rise to friction forces, but the possibility of jolts and impacts is not excluded. The possibility of an impact during the transportation of pyrotechnic products also exists.

During artillery shelling, considerable inertial stresses are set up at the instant of discharge. These stresses may give rise to shifts of the pyrotechnic charges and other filler elements present in the shells, and a significant friction may be produced. For this reason, the study of the sensitivity of pyrotechnic compositions to mechanical effects is of great importance.

Testing of compositions for sensitivity to mechanical effects usually involves testing for sensitivity to impact and friction.

It should be noted that occasionally, pyrotechnic compositions more sensitive to friction are also more sensitive to impact, but there is no exact correspondence between these characteristics, nor is there any correspondence between the flash point of compositions and their sensitivity to impact or friction. This is explained by the fact that the capacity for absorbing energy in a given form is different in different pyrotechnic compositions. This capacity depends on the physical properties of the compositions and on the conditions in which the action is sustained. See also [111, p. 266].

DETERMINATION OF SENSITIVITY TO IMPACT

In the determination of the sensitivity of pyrotechnic compositions to impact, the same apparatus is used, i.e., impact testing machines and roller instruments, as in similar testing of explosives (Fig. 9.2). Before they are put into operation, the impact testing machines and roller instruments are inspected in accordance with GOST 4545-48.

According to the specification of GOST 2039-43, a 10 kg load is used in work with pyrotechnic compositions on an impact tester; more sensitive compositions are tested with a 5 kg load. The area of the composition subjected to impact is 0.5 cm², and a 0.05 g sample of the composition is taken.

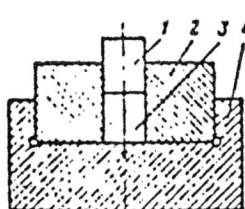

Fig. 9.2 Roller instrument.
1 - Striker; 2 - Directing sleeve; 3 - Anvil; 4 - Pan.

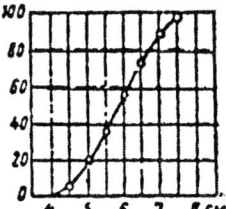

Fig. 9.3 Number of explosions (in %) versus drop height of load.

The following two quantities are indicated as a characteristic of the sensitivity of pyrocompositions to impact according to GOST: 1) the smallest drop height for which 95 ±5% of the explosions are obtained in 25 parallel tests, knows as the <u>upper sensitivity limit</u>, and 2) the largest drop height of the load for which 100% of rejects are obtained, known as the <u>lower sensitivity limit</u>.

Simultaneously, standard compositions are tested under analogous conditions, and the test results of experimental and standard compositions are compared.

In tests of major importance, large numbers of tests are conducted at various drop heights of the load, and a graph is plotted for the number of explosions in percent (from 0 to 100%) versus the drop height of the load (Fig. 9.3).

In cases where the sensitivity curves obtained are more or less symmetric, one can find the drop height of the load for which 50% of the explosions are obtained, then express the sensitivity to impact in terms of the magnitude of the work (in kg·m) referred to a unit surface of the composition over which the impact propagates:

$$A = \frac{p \times h}{S} \frac{kg \times m}{cm^2}$$

where p is the weight of the load in kg;
h is the height of its fall in m;
S is the cross-sectional area of the roller in cm².

Listed below are data thus obtained on the sensitivity to impact of certain binary pyrotechnic compositions in which the oxidizer and fuel are taken in stoichiometric proportions (see Tables 9.5 and 9.6).

Table 9.5 IMPACT WORK IN kg·m/cm² FOR BINARY PYROTECHNIC MIXTURES

Oxidizer	Fuel				
	Sulfur	Lactose	Charcoal	Magnesium powder	Aluminum dust
Potassium chlorate	1.1	1.8	3.2	4.5	4.5
Potassium perchlorate	1.2	2.9	4.2	4.4	5
Potassium nitrate	3.6	5	5	4.6	5

Table 9.6 SENSITIVITY TO IMPACT OF BINARY MIXTURS WITH POTASSIUM CHLORATE

Fuel	Impact Work kg·m/cm²	Fuel	Impact Work kg·m/cm²
Potassium thiocyanate	0.5	Naphthalene	1.3
Realgar	0.6	Potassium ferrocyanide	2.2
Paraffin	1.1	Antimony sulfide	3.5
Sulfur	1.1	Graphite	10

Many papers on possible methods of mathematically expressing the sensitivity of substances to impact have been published. It has been frequently stated that it is necessary to consider the height of rebound of the load (h') and to evaluate the sensitivity of the substance or mixture in terms of the magnitude of the work actually absorbed on impact ($A = p(h-h')/S$).

In addition, it has been pointed out that the percentage of explosions for the same impact work largely depends on the final fall velocity of the load; for example, when a 2 kg load falls from a height of 0.5 m, the percentage of explosions is greater than when a 5 kg load falls from a height of 0.2 m.

New designs of impact testers and instruments have been but most of these complex designs have not found any broad practical application.

Particular mention should be made of the work of N.A. Kholevo, who developed a new design of an instrument named "Instrument No. 2" (Fig 9.2). As is evident from the figure, a distinctive feature of this instrument is the presence of an annular groove cut out inside the matrix. This feature permits the substance being tested to be pushed out from under the roller and into this groove at the instant of impact.

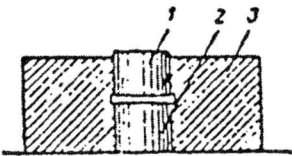

Fig. 9.4. Instrument No. 2. 1 - hammer, 2 - anvil, 3 - directing sleeve.

In practical terms, the use of this new design signifies the introduction of a combined new test involving impact elements as well as friction elements.

In recent years, a series of published papers have dealt with the mechanism of generation of explosions in explosives and with methods of evaluating the sensitivity of explosives to mechanical effects, namely, those of N. A. Kholevo, K. K. Andreyev and co-workers, and F. Bowden [119, pp. 5-130; 111; 116]. To a certain extent, the conclusions reached in these papers can be extended to pyrotechnic compositions.

In particular, Andreyev [119, p. 92] concludes that "testing on an impact tester in its different versions (instrument No. 1 according to GOST 4545-48 and instrument No. 2) has the fundamental disadvantage of mainly characterizing the possibility of formation of reaction centers, not the propagation of the process." In this connection, he points out that "in combination with the determination of combustibility (capacity to burn), the results of testing on V. S. Kozlov's pendulum instrument apparently can characterize the danger posed by explosives subjected to mechanical effects better than other methods can." This statement by Andreyev does not mean that tests on an impact tester are not necessary, but reflect his view that they do not adequately simulate the conditions prevailing in practical handling of explosives.

DETERMINATION OF SENSITIVITY TO FRICTION

Some indications of the degree of sensitivity of compositions to friction may be provided by their crushing in an unglazed porcelain mortar. A mortar 10 cm in diameter and 6 cm high in which a dash of pyrotechnic composition is placed (an amount no greater than 0.02-0.03 g) is usually employed for this purpose. The composition is crushed with an unglazed pestle, and the operator conducting the test should wear protective spectacles (made of plexiglas). This test does not require much time and is easy to carry out, but is obviously subjective in character (depends on the mode of crushing and strength of the experimenter).

Obviously, in principle, an instrument can be built which executes the rotatory and vibratory motion of the pestle in the mortar and which also maintains a constant pressure of the pestle crushing the sample of the composition.

PRINCIPLES OF PYROTECHNICS

In 1938, S. G. Dobrysh designed an instrument for determining the sensitivity of explosives and pyrotechnic compositions to friction. The composition being tested is crushed between two parallel steel surfaces, the lower of which (die) is stationary, and the upper (punch) rotates at 150 rpm. A pressure up to 2000 kg/cm^2 can be applied to the punch. The sample of the composition weighs 0.02 g. To increase the pressure, 0.002 g of fine quartz sand is added to the sample. The friction lasts 10 sec. The sensitivity to friction is estimated quantitatively from the minimum pressure (in kg/cm^2) at which inflammation takes place as the sample is being crushed.

Six tests are carried out for each load. At the same time, experiments with a standard substance are conducted whose sensitivity to friction is already known.

The above-described method requires a considerable improvement, however; many compositions crushed in this instrument do not explode even under a load of 2000 kg/cm^2.

The sensitivity to friction of certain binary mixtures tested on Dobrysh's instrument is shown in Table 9.7. Components of these mixtures were taken in stoichiometric proportions.

Table 9.7 FRICTION SENSITIVITY OF BINARY MIXTURES WITH POTASSIUM CHLORATE

Fuel	Sensitivity to Friction, k/cm^2	Fuel	Sensitivity to Friction, k/cm^2
Potassium ferricyanide	8	Sulfur	65
Potassium thiocyanate	22	Potassium ferrocyanide	85
Lactose	60	Antimony sulfide	90

Bowden and Carton [111, p. 322] designed an instrument for determining the sensitivity of explosives to friction. In this instrument, a sample of explosive is compressed between two parallel horizontal steel surfaces (a roller and a bar). Under a lateral impact, the steel bar moves in a horizontal direction, and the explosive is subjected to a strong friction during this movement.

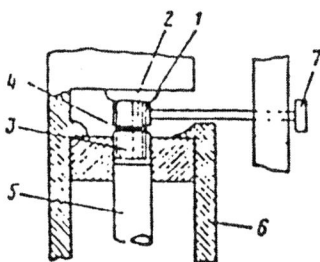

Fig. 9.5. Part of Bowden-Kozlov instrument for determining the sensitivity of explosives to friction. 1 - movable roller, 2 - stop, 3 - immovable roller, 4 - explosive, 5 - piston for transmitting the pressure of the press to roller 3, 6 - sleeve, 7 - rod with head transmitting the impact.

V. S. Kozlov modified this instrument. In the Bowden-Kozlov instrument (Fig. 9.5), a sample of explosive is compressed between two rollers 10 mm in diameter. The impact of the pendulum is transmitted to upper movable roller 1, which then moves in a horizontal direction between stationary stop 2 located above it and roller 3 located below, mounted in sleeve 6. The sensitivity of the explosive in this instrument is determined from the change in the frequency of explosions as the pressing pressure changes. The instrument permits one to increase this pressure to 8000-10,000 kg/cm$_2$. On this instrument, trotyl produces explosions starting at a compression pressure of about 5000 kg/cm$_2$. Since a certain misalignment of the movable roller usually takes place during the impact, we are dealing in this case, not with friction alone, but with a combination of friction and sliding impact.

3. FACTORS AFFECTING THE SENSITIVITY OF COMPOSITIONS TO INITIAL IMPULSES

The amount of energy that must be imparted to a system in order to give rise to a fast chemical reaction is determined, on the one hand, by the internal resistance of the system, and on the other hand, by the possibilities of its own energetics.

The effect of energetics is manifested in the ability of a reaction, begun in a small volume by an external impulse, to develop spontaneously at the expense of the energy evolved during its course. The greater this tendency toward spontaneous development of the reaction, the higher will be the probability of propagation of the reaction even if the initial impulse is comparatively weak.

"In order for a chemical reaction to take place, it is necessary to break the existing chemical bonds and to form new ones," states N. N. Semeenov.

The internal resistance of a system is determined by the amount of energy necessary to break the bonds. This quantity is known as the activation energy of a process. The total amount of energy evolved by a process is equal to the amount of energy produced by the formation of new bonds minus the energy necessary for breaking the initial bonds.

Neglecting for the time being the influence of external physical factors, we can state that the sensitivity of a chemical system to initial impulses is determined primarily by the magnitude of the activation energy and magnitude of the heat of reaction. Both quantities are expressed in the same units, kcal/g-mole. However, this general statement, which is unquestionably valid in many cases, does not permit one to evaluate the sensitivity of individual substances (explosives). [Note: The main difficulties are: 1) the fact that the actual activation energy of the process has very little in common with the activation energy of thermal decomposition, and 2) the influence of various physical factors related to the different rates of absorption of the energy coming from the outside.] The situation becomes even more difficult when a multicomponent system is considered, and, as should be particularly emphasized, a heterogeneous system such as a pyrotechnic composition.

In considering the possibility of stimulating a reaction in a pyrotechnic composition, we should direct our attention first of all to the ease of carrying out individual "elementary" processes: the process of decomposition of the oxidizer, then the process of oxidation of the fuel.

Comprehensive characteristics of these simpler, but essentially still very complex processes, should again be the activation energy and heat of reaction of the process. For practical purposes, however, we must limit ourselves in most cases to the examination of the heat of reaction of the process, and instead of the activation energy, consider the values of the decomposition temperature of the oxidizer, or the flash point or decomposition temperature of the fuel. The corresponding numerical data for various components of a composition may be found in the tables of Ch.2 and 3.

The less energy is required to bring about the decomposition of the oxidizer and the lower its decomposition temperature, the more sensitive the composition will be to initial impulses, other things being equal (mainly, the same fuel). This can be seen in Tables 9.1 and 9.5.

Comparing the properties of compositions with three different oxidizers - potassium chlorate, potassium perchlorate, and potassium nitrate - we see that chlorate compositions have the highest sensitivity, since the decomposition of potassium chlorate takes place at a comparatively low temperature, ~350°C, and it is very important to note that in contrast to the decompositions of the majority of other oxidizers, it is associated with the evolution of an appreciable amount of heat.

A slightly lower sensitivity is exhibited by mixtures with potassium perchlorate oxidizer (see the heat of its decomposition reaction in Table 2.1), and a much lower sensitivity, by mixtures with potassium nitrate as the oxidizer (see the heat of its decomposition).

Such metal oxides as ferric oxide, ferriferrous oxide, etc., decompose with the greatest difficulty and at the highest temperatures. It is evident that compositions prepared with their participation will have the lowest sensitivity to all types of initial impulses. Let us note in particular that iron-aluminum thermite shows very little sensitivity to impact and friction, has an ignition point of about 1300°C, and fails to be ignited not only by a flame from a Bickford fuse, but also when heated by a gas burner flame.

The case of sensitivity of compositions with the same oxidizer but different fuels is much more complex. A low sensitivity is exhibited by compositions that include only fuels of very low calorific value, and an oxidizer requiring a large amount of heat for its decomposition. Such compositions include, for example, binary mixtures of nitrates and sulfur. A low sensitivity also characterizes compositions containing only fuels that are very difficult to oxidize and have a very high inflammation temperature, for example, graphite or silicon.

In discussing the sensitivity of compositions with different fuels, of great importance are the individual physical properties of the fuel, such as the melting and boiling point, thermal conductivity, plasticity, hardness, etc.

It is in this case that the equivalence with respect to the sensitivity of the compositions to various types of initial impulses is disturbed the most, and it becomes necessary to find out on an individual basis how the flash point, sensitivity to friction, etc., change as the fuel changes.

Among individual binary mixtures should be mentioned the extremely high sensitivity of mixtures of potassium chlorate and phosphorus; they explode when lightly crushed with a rubber stopper or upon the slightest impact.

Very sensitive to impact and friction are mixtures of chlorates with sulfur, selenium, metallic arsenic, and sulfides of phosphorus, arsenic, and antimony. Mixtures of chlorates with thiocyanates, ferricyanides, and ferrocyanides, as may be seen from Table 9.6 and 9.7, are particularly sensitive to shock and friction. Mixtures of chlorates with various organic fuels, such as hydrocarbons, resins, etc., have a considerable sensitivity to friction.

Binary mixtures of chlorates with metal powders (magnesium or aluminum) are sensitive to friction and very dangerous to handle. [Note: In view of the great danger of explosion, in certain countries it is prohibited to make a mixture of $KClO_3$ + Al.]

Among the binary mixtures containing the nitrates of metals, only mixtures with powders of highly active metals, such as magnesium, aluminum, zirconium, and others, have a significant sensitivity to shock and friction, if we do not consider the very sensitive mixtures with phosphorus and sesquisulfide. The sensitivity of mixtures of nitrates with organic fuels is not great in most cases.

The self-ignition point of aluminum is considerably higher than that of magnesium and therefore the self-ignition temperatures of aluminum compositions are considerably higher than the same quantity for magnesium compositions. Because of this, for ignition of compositions containing aluminum we must use igniting compositions having higher combustion temperatures to set off aluminum compositions.

As a rule, it is easier to ignite compositions having a lower flash point with the flame tongue of a Bickford fuse.

Black powder, binary mixtures of potassium nitrate with coal or iditol and the majority of chlorate compositions are readily ignited by a flame.

As a rule, compositions in pressed form are much less sensitive to a flame than in powder form.

The sensitivity of the compositions to a flame increases with decreasing grain size of the components, so that the compositions presenting the greatest fire hazard are those prepared from components ground to a dust.

On the basis of available factual data, it may be concluded that the most sensitive compositions are those in which the components are either in stoichiometric proportions, or there is a slight deficiency of the fuel; as a rule, compositions with a large excess of fuel are less sensitive to initial impulses.

Among physical factors, we should note the comminution of the components. When this question is examined from a theoretical point of view, it should be pointed out that a finer comminution of the components entering into the composition should promote a more intimate contact between them, and hence, facilitate the conditions leading to the combustion reaction.

Indeed, as was indicated above, the sensitivity of compositions to a flame increases with decreasing particle size of the components; less affected by the degree of comminution of the components are the self-ignition temperature and sensitivity to impact to many compositions. [Note: In cases where the particle size of the fuel is very small (of the order of a micron or less), a very large increase in the surface of the substance takes place that leads to a significant lowering of the self-ignition point of the fuel.

For example, sublimed magnesium deposited on the walls of a container in the form of fine crystals ignites spontaneously upon contact with air even at room temperature.] The effect of comminution of the components on the sensitivity of the compositions to friction has been insufficiently studied. It is known that in some cases, the presence of coarse grains of solid oxidizer may slightly increase the sensitivity of a composition to friction.

Increasing the density of a composition reduces its sensitivity to initial impulses (except for the self-ignition temperature), since when the density is higher, the same amount of energy is expended on a larger amount (mass) of the composition, and the effectiveness of the action on individual particles of the composition is lower.

As the initial temperature rises, the sensitivity of pyrotechnic compositions to mechanical effects and to a flame increases, and therefore, suitable precautions must be taken when drying pyrotechnic compositions.

The introduction of various impurities into pyrotechnic compositions frequently has a major effect on their sensitivity to mechanical and thermal impulses.

The addition of deterrents, i.e., soft plastic substances such as paraffin, vaseline, stearin, or various oils, to pyrotechnic compositions reduces their sensitivity to friction and to a flame.

The sensitivity to friction decreases as a result of the distribution of the friction force over a large mass of the composition.

The sensitivity to a flame decreases because the deterrent coats the individual particles of the composition with a film and thus hinders the access of heat to these particles. The deterrent itself in most cases is a substance of low activity igniting with difficulty.

The introduction of organic deterrents has an insignificant effect on the sensitivity to impact. Occasionally, the introduction of paraffin or similar substances into the composition increases the sensitivity to impact, as for example in the case of a mixture of potassium chlorate and magnesium.

Effective deterrents which decrease the sensitivity of compositions to all types of initial impulse when introduced into the composition in large amounts are inert substances which do not take any active part in the combustion process. Such substances include magnesium oxide, magnesium fluoride, barium fluoride, etc., and, if free metals are absent from the compositions, also carbonates, oxalates, etc.

For example, in the red light composition which contains potassium chlorate - strontium carbonate - iditol, the role of deterrent is played by strontium carbonate. Iditol can not be called a deterrent in this case, even though it is a soft plastic substance. In this composition, iditol is the chief fuel and takes an active part in the combustion process.

Admixtures of sand, glass, and other solid substances substantially increase the sensitivity of pyrotechnic compositions to mechanical effects and make them very dangerous to handle. In some cases, an increase in the sensitivity of the compositions to impact and friction takes place when solid metal powders (for example, ferrosilicon powder) or solid crystals of certain oxidizers are introduced.

For many substances, the value of hardness of the Mohs scale may be found in handbooks or determined experimentally.

When a very hard substance is added to a composition, the mechanical forces (impact, friction) concentrate on these particles. This local energy concentration causes the sensitivity of the composition to increase.

Available literature data on the sensitivity of pyrotechnic compositions are scarce and dispersed in the form of addenda to individual papers in periodic publications.

Among papers dealing mainly with the sensitivity of pyrotechnic compositions should be mentioned those of Blinov [114], Tomlinson and Audrieth [122], and Langhans [159]. Important information is also contained in the pyrotechnics of Ellern [9], who, while noting the particular danger of chlorate mixtures, also points out the high sensitivity of mixtures in which the oxidizers are ammonium and alkali metal perchlorates and certain oxygen compounds of lead, particularly PbO_2. He writes that the danger of accidental initiation by static electricity is high in the presence of fine powders (micron units) of such metals as Zr, Mg, and Al.

Chapter 10

MECHANISM OF COMBUSTION AND FACTORS AFFECTING THE RATE OF COMBUSTION OF PYROTECHNIC COMPOSITIONS

1. MECHANISM OF COMBUSTION OF COMPOSITIONS

The combustion process of pyrotechnic compositions may be divided into three separate stages: ignition, inflammation, and combustion.

Ignition of a composition is usually achieved by means of a thermal impulse which is imparted to a limited portion of the surface of the pyrotechnic composition.

Inflammation of the composition refers to the spreading of combustion over the entire surface of the composition.

Combustion proper refers to the propagation of the process into the interior.

The inflammation rate of pyrotechnic compositions is many times as fast as their combustion rate; a similar relationship is observed in the case of powders (both black and smokeless).

For the same composition, the inflammation rate depends on the following factors:

1. Degree of comminution of the components: the finer the components, the greater the total surface of the composition and the easier and faster the inflammation.
2. Density of the composition: the higher the density, the smaller the total surface of the composition (the number of pores in the composition decreases), and the more difficult and slow its inflammation.
3. Initial temperature of the composition: the higher this temperature, the easier and faster the inflammation process.
4. External pressure: as the external pressure rises, the inflammation rate increases markedly; this may be explained by the fact that compressed gases transmit more heat per unit time to the surface of the condensed system which they ignite than do the same gases at atmospheric pressure.
5. Composition of the gas phase and particularly its content of oxygen, which actively participates in the inflammation processes of many compositions (cf. [111], p. 154).

The highest inflammation rate is observed in the case of slightly compacted photomixtures and black powder; the inflammation rate of a black powder tract in an open space is 3.5 m/sec. It should be noted that the inflammation processes of pyrotechnic compositions as well as powders and explosives have been insufficiently studied, and an objective of the immediate future will be to accumulate data and use them for the erection of a complete theory of inflammation.

The combustion process of pyrotechnic compositions is extremely complex. Comparatively little attention has been given to its study until recently. At the same time, a large number of papers on the problems of the mechanism of combustion of explosives and powders have been published in the Soviet literature in the last 10 to 15 years.

The process of combustion of pyrotechnic compositions is of course characterized by a whole series of specific features which differentiate it from the combustion of explosives or powders, but many relationships are similar in the two cases, as has been demonstrated by a whole series of experiments.

For this reason, the treatment below includes certain cases in which analogies are drawn between, on the one hand, the relationships governing the combustion of pyrotechnic compositions, and, on the other hand, the relationships involved in the combustion of explosives and powders.

The process of combustion of pyrotechnic compositions is made up of many exo- and endothermic chemical processes and includes the physical process of heat transfer.

If as a rough approximation the combustion process is divided into two stages, it should be pointed out that it begins in the condensed phase and ends in the gaseous phase (in a flame).

The processes occurring in the condensed phase must generally be evaluated as endothermic or slightly exothermic ones; the degree of exothermicity of a reaction in the condensed phase depends on the formula of the composition and on the conditions of its combustion (external pressure). As a whole, processes taking place in the gaseous phase (in the flame) are always exothermic. In many cases, processes taking place in the condensed phase can occur only at the expense of the heat entering this phase from the gaseous phase (from the flame). This is confirmed by the observation that all the pyrotechnic compositions studied from this point of view up to the present time lose their capacity to burn at low pressure (i.e., a residual pressure of 1 Hg).

The question of the proportion of heat evolved in the condensed phase and gas phase should be considered individually for each type of composition. Quantitative data on this subject are still lacking.

According to Andreyev, during the combustion of certain explosives, an exothermic reaction prevails in the condensed phase at low pressures, and a flame reaction prevails at high pressures. Obviously, this assumption also applies to the combustion process of many pyrotechnic compositions.

An indispensable condition for achieving a normal, uniform combustion is the equality of the heat inflow and outflow in all reaction zones. When this equality is disturbed, the combustion either attenuates or becomes nonuniform, i.e., a pulsation takes place. [Note: For more detail on pulsating combustion, see Andreyev's papers [110].]

However, the division of the process into only two stages is a first and very rough approximation of the actual process.

In analyzing the combustion mechanism of pyrocompositions in more detail, we should note that the considerable difference between explosives on the one hand and pyrocompositions on the other lies in the fact that pyrocompositions are heterogeneous systems.

In order to achieve a fast reaction in a heterogeneous system, it is first necessary to bring the reactants into close molecular contact. This can not be achieved in the solid state, and therefore the reactions between solids are known to occur very slowly, even at very high temperatures; the fact that many low gas compositions burn fairly rapidly does not contradict this statement, since it is very probable that the main oxidation reaction of the fuel in these compositions takes place in the liquid or gaseous phase existing at the combustion temperature.

A rapid interaction between two components in a pyrotechnic composition can begin only when at least one of them is in the liquid or gaseous state.

If the processes are distinguished according to the state of aggregation of the reactants, the occurrence of two types of processes, liquid + solid and liquid + liquid, is possible in the condensed phase.

The components of compositions frequently have markedly different melting points as well as boiling points or thermal decomposition temperatures. Taking this into consideration, we can understand why in many cases the reactions between the components (or between their decomposition products) take place at the interface between the condensed and the gaseous phase; the gas + solid and gas + liquid variants are possible in this case.

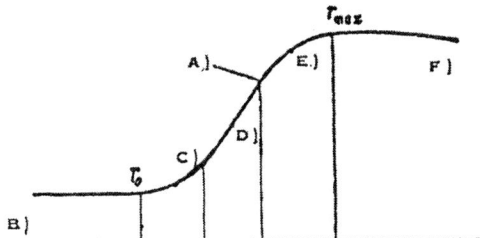

Fig. 10.1. Diagram of combustion of a pyrotechnic composition.
a) Reaction at interface, b) Pyrotechnic composition, c) Warmup zone,
d) Zone of reaction in condensed phase, e) Zone of reaction in gaseous phase, f) Reaction products.

The possibility of occurrence of chemical reactions will be largely determined by the rate of the physical processes: gaseous (and liquid) diffusion and the possibility of rapid removal of the combustion products from the reaction zone.

Finally, in the zone of highest temperature, all the reactants will be in the gaseous (vapor) state, and the reaction here will take place in the gas + gas system, i. e., in a homogeneous gaseous system.

In this last stage of combustion of pyrocompositions, an important part is played by atmospheric oxygen in many cases. It should be noted that the temperature in the flame of many pyrotechnic compositions is much higher than in the flame of a powder or nitro compound burning at atmospheric pressure.

All that has been stated above concerning the combustion mechanism of pyrocompositions can be clearly illustrated by a scheme which has many similarities with the powder combustion scheme proposed by Ya. B. Zel'dovich (Fig. 10.1).

A temperature rise in the reaction zone in the condensed phase may be due to the heat transferred from the gaseous phase or from the interface, and sometimes to the heat of the reaction taking place in the condensed phase itself.

The answer to the problem, which processes, in the condensed or in the gaseous phase, predominate during the combustion of any given pyrocomposition, can be obtained by studying the dependence of the combustion rate on the pressure of the ambient medium.

The more the combustion rate depends on the pressure, the higher will be the specific weight of the reactions taking place in the gaseous phase or at the interface between the gaseous and the condensed phase.

However, the above-described scheme of combustion of a pyrotechnic composition is simplified and does not fully correspond to the reality.

It has now been demonstrated experimentally that the combustion of pyrotechnic compositions at atmospheric pressure, as in the case of combustion of involatile explosives [111, p. 171], involves the dispersion of particles of the unreacted components of the composition, mainly powder particles of metals such as magnesium or aluminum, and also droplets of melt of the oxidizer (nitrate).

This ejection of particles of unreacted substances into the gaseous phase is due to the nonuniformity of the course of the reaction in the condensed phase layer close to the surface and to the fact that the gaseous products formed in the condensed phase detach solid (and liquid) particles of the components of the composition and carry them into the flame. Hence, the heterogeneity of the system is also preserved in the gaseous phase in many cases, and the flame zone closest to the condensed phase should be called the smoke-gas zone of the flame. During their further motion in the flame, these particles of unreacted substances disappear by reacting with the gaseous medium surrounding them.

In individual cases, the explanation of the combustion mechanism of pyrocompositions is facilitated by considering the properties of the components included in the composition.

As an example, we can cite an analysis of the mechanism of combustion of a composition prepared from potassium chlorate and magnesium.

We will indicate the melting point of the oxidizer and the melting and boiling points of the fuel. Potassium chlorate melts with comparatively little decomposition at 360°C; magnesium melts at a high temperature, 650°C, and boils at 1100°C at atmospheric pressure. [Note: In the presence of catalytic admixtures (MnO_2, Co_2O_3, etc.) in potassium chlorate, its decomposition takes place at 200-250°C.]

On this basis, the process of interaction of these two substances may be represented:

1. In the condensed phase:

   a) $KClO_3$ liquid + Mg solid →

   b) $KClO_3$ liquid → KCl + oxygen →

2. At the interface between the condensed and the liquid phase and in the smoke-gas zone of the flame:

   a) Mg solid + oxygen →

   b) Mg liquid + oxygen →

3. In the gaseous phase (in the flame):

   a) Mg vapor + oxygen →

   b) Mg vapor + atmospheric oxygen →

2. FACTORS AFFECTING THE COMBUSTION RATE OF COMPOSITIONS

There are two ways of quantitatively expressing the combustion rate: we can write the linear combustion rate u in mm/sec, but we can also calculate the mass rate of combustion u_m, expressed in units of g/cm² sec; u_m defines the amount of composition burning off in 1 sec from a unit of burning surface; knowing the linear combustion rate and composition density, we can calculate the mass rate of combustion from the formula $u_m = 0.1 \times u \times d$, where d is the cubic density of the composition in g/cm³.

As already noted, the combustion process takes place uniformly only when the composition has been sufficiently compacted. However, in order to evaluate the degree of compaction of a composition, it is necessary not only to know its cubic density d, but also to calculate the compaction coefficient K, which is the quotient of the practically attained density d by the maximum density of the composition d_{max} found by calculation on the basis of the density of the components included in the composition:

$$K = \frac{d}{d_{max}} \; ; \; d_{max} = \frac{100}{\frac{a}{d_1} + \frac{b}{d_2} + \cdots \frac{n}{d_n}}$$

where $d_1, d_2, \ldots d_n$ is the density of the components of the composition; a, b, ... n is the content of these components in the composition in percent.

For the majority of pressed compositions, the compaction coefficient varies in the range 0.7-0.9. [Note: The porosity of a composition will be characterized by the value of (1-K). Hence, the porosity of pressed compositions lies in the range 0.3-0.1.]

For powdered composition, the so-called bulk density is equal to 40-60% of d_{max}. For pyrotechnic compositions of different formulations, the linear rate of their combustion fluctuates considerably from tenths of mm/sec (for smoke compositions) to 20-30 mm/sec (for rapidly burning illumination compositions).

What factors determine the combustion rate of compositions?

The rate of the complex physicochemical process, combustion, is determined by the rate of the individual (elementary) chemical reactions and by the conditions of heat transfer from one reaction zone to another.

The conditions of heat transfer at a constant external pressure and initial temperature and also in the presence of constant density of the composition and constant comminution of its components are substantially determined by the difference of temperatures in the different reaction zones.

It has been found experimentally that as a rule, compositions having the highest flame temperature are also the fastest-burning ones. However, the presence of a significant number of deviations and exceptions to this rule shows that a high temperature in the flame is only one of the factors determining the combustion rate of the compositions. The combustion rate strongly depends on the absence or presence of low-melting or volatile components in the composition.

The heat which under other conditions would cause a sharp temperature rise in the reaction zone and hence, a sharp acceleration of a series of chemical processes, is expended on the transition of these substances from one state of aggregation to another in the presence of low-melting or volatile substances in the composition. This in our view explains the fact that such low-melting organic substances as resins, paraffin, stearin, etc., when introduced into binary mixtures (oxidizers-metal), sharply reduce their combustion rate. An even more decisive influence on the rate of the combustion process is that of the rate of the individual (elementary) chemical reactions.

The main reactions in the combustion process will be highly exothermic (flame) reactions. However, the rate of a complex multistage process will on the whole be determined by the minimum rate at which the most difficult and slowest stage of the process takes place. The most difficult and slowest are endothermic chemical processes.

For this reason, the combustion rate of compositions is determined in many cases precisely by the rate of decomposition of the oxidizer.

An objective indicator characterizing the ease of decomposition of the oxidizer is the partial pressure of the oxygen above it at various temperatures.

The simplest decomposition mechanism appears to be that of metal oxides, which are used in thermite ignition compositions.

It is well known that the rate constant of a chemical reaction K increases very markedly with rising temperature.

$$K = B \times e^{-\frac{E}{RT}}$$

where B is the proportionality coefficient (preexponential factor);
E is the activation energy in kcal/g-mole;
R is the gas constant.

However, even if the maximum temperature and activation energy of the process are known, the rate can not be calculated beforehand, since "all the combustion phenomena are related to the rate of the chemical reaction taking place under nonisothermal conditions" [137].

It is very important of course to know all the intermediate stages of a combustion process. However, in order to establish the composition of the intermediate reaction products, it is necessary in most cases to carry out a very complex experimental study; at the present time, these data are unfortunately lacking for most combustion reactions of pyrotechnic compositions.

Considering the factual data accumulated thus far, we should point out that the combustion rate of pyrotechnic compositions is determined by both the formula of the composition (chemical factor) and the combustion conditions of the compositions (physical factors).

An examination of data on the combustion rate of strongly compacted compositions (compaction coefficient K ≥0.85) at atmospheric pressure and room temperature (~20°C) suggests that some of the fastest-burning compositions are binary mixtures of alkali (or alkaline earth) metal nitrates with magnesium, containing 50-60% magnesium. Compositions containing zirconium and titanium also burn very rapidly.

Compositions with aluminum burn much slower than those with magnesium, provided that the comminution of the metal is the same. Compositions containing beryllium, boron, or silicon as the fuel burn slowly. The higher the ignition temperature of the fuel, the slower the combustion rate of the composition for a given fuel, other things being equal. It is possible that a correlation also exists between the combustion rate of the composition and the Pilling and Bedworth number for the metal (as well as B and Si) present in the compositions.

For fast-burning metals Mg and Zr, these numbers are respectively equal to 0.81 and 1.45; for Be, Si, and B, these numbers are greater, equal to 1.75, 2.04, and 4.08, respectively (see also Table 5.4).

The combustion rate of oxidizer-metal binary mixtures increases rapidly with rising content of metallic fuel in the composition (obviously, up to a certain limit); for magnesium, this limit is 60-70%; such an increase in combustion rate is to some extent related to the increase in the thermal conductivity of the composition as its metal content rises.

For the same metal content, binary mixtures of alkali metal nitrates and magnesium burn faster than mixtures of alkali metal chlorates and magnesium. This may be due to the exothermic reaction of the nitrate melts with magnesium in the condensed phase; for more detail on the oxidation of magnesium powder at elevated temperatures, see Makolkin and Vernidub [61].

Among mixtures containing no metal fuels, many chlorate mixtures and black powder burn rapidly. Belyayev's recently published paper discuss the effect of sulfur on the combustion rate of black powder [127, 128]l; there is also the paper of Benture et al. [140] dealing with the effect of various organic additives on the combustion rate.

Mixtures of potassium nitrate with coal and iditol have an intermediate combustion rate.

Data on the combustion rate of various type of multicomponent pyrotechnic compositions are listed in Table 10.1.

Table 10.1

Use of Composition	Rate of Combustion at atmos. pressure & K≥0.85mm/sec.
Illumination	1-10
Tracer	2-10
Signal light	1-3
Thermite ignition	1-3
Smoke forming	0.5-2

Nitrate-containing compositions which do not include metallic fuels burn slowly and with little intensity in most cases. Comparative data on the combustion rate of chlorate and nitrate binary mixtures containing no metallic fuels are shown in Table 10.2.

The introduction of drying oil or colophony into multicomponent compositions markedly slows down their combustion process.

As is evident from Table 10.2, mixtures of chlorate with sulfur burn rather vigorously; however, mixtures of sodium or potassium nitrate with sulfur at room temperature ignite with great difficulty, and their combustion is rather unstable. This appears to be due to the fact that the amount of heat evolved by the combustion of sulfur is insufficient to cause the decomposition of the nitrate and in addition, to warm up of the neighboring layers of the composition.

Table 10.2 LINEAR COMBUSTION RATE OF BINARY MIXTURES in mm/sec.
(Oxidizer & fuel taken in stoichiometric amounts; K=0.5 to 0.6)

Fuel	Oxidizer			
	Potassium chlorate	Potassium nitrate	Sodium nitrate	Barium nitrate
Sulfur	2	Does not burn	Does not burn	--
Charcoaol	6	2	1	0.3
Sugar	2.5	1	0.5	0.1
Shellac	1	1	1	0.8
Drying oil	0.5	--	--	--
The compositions were burned in cardboard tubes 16mm dia.				

At temperatures up to 300-350°C, the decomposition reaction of nitrates takes place with the formation of nitrites, and the proposed reaction between sodium nitrate and sulfur may be expressed by the equation

$$2NaNO_3 + S = 2NaNO_2 + SO_2 + 21 \text{ kcal}$$

At the same time, only 0.1 kcal of heat is evolved per 1 g of composition, this being insufficient for the spontaneous propagation of the reaction. However, in the presence of a strong initial impulse, combustion is possible, since when the reduction of nitrate is complete, the reaction is associated with a much more significant evolution of heat:

$$2NaNO_3 + 2S = NaSO_4 + SO_2 + N_2 + 173 \text{ kcal}$$

which corresponds to 0.7 kcal/g of composition.

In 1941, Ye. P. Molchanov was able to ignite a composition of 68% $NaNO_3$ and 32% S at room temperature and p=1 at in a cardboard tube 16mm in diameter; the combustion rate of the composition was ~0.4mm/sec.

It is easier to ignite compositions containing excess sulfur and a small percentage of sodium nitrate (17%); most of the sulfur in such compositions is burned by atmospheric oxygen; such compositions, developed by A. I. Sidorov, have found applications in the national economy (see Ch. 22).

Among the physical factors affecting the combustion rate, the following should be indicated.

Increasing the composition density markedly decreases the combustion rate of most compositions. Particularly strong is the influence of density on the character of combustion of photomixtures: 1 kg of photomixture in the powdered state burns up in a few tenths of a second, while the combustion time of the same amount of photomixture pressed under a pressure of the order of 1000 kg/cm^2 is expressed in several tens of seconds.

The dependence of the combustion rate of one of the illumination compositions on the specific pressure used for the pressing is represented by the following figures:

Pressure of compaction	1000	2000	3000	4000	kg/cm^2
Combustion rate of composition	5.0	4.2	3.8	3.6	mm/sec

Increasing the compaction pressure above 3000 kg/cm^2 has comparatively little effect on the density of the composition, and hence, on the rate of its combustion.

The effect of composition density on the combustion rate is explained by the fact that as the composition density increases, the possibility of penetration of burning gases into the composition decreases, and thus, the process of warmup and inflammation of deeper layers slows down.

In this connection, it would be of interest to study the gas permeability of compacted pyrotechnic compositions. This study should be conducted at elevated temperatures, care being taken to reconstruct the real combustion conditions of the compositions as fully as possible.

For compositions that do not contain a metallic fuel and have a low flame temperature, the combustion rate changes little with changing composition density. [Note: For the dependence of the combustion rate of explosives on density, see Andreyev [110].]

It should be noted that there also exist low-gas compositions whose increase in density promotes heat transfer in the condensed phase, and therefore their combustion rate increases only slightly with increasing density. The determination of the dependence of the combustion rate on the density of compositions may help considerably in providing an answer in each specific case to the question of how much greater is the role played during the combustion of a composition by processes taking place in the condensed phase.

Comminution of the components also affects the combustion rate of compositions to a significant extent. The smaller the size of the grains entering into the composition of the components, the faster the combustion rate of the composition prepared from them [125]. A particularly strong influence is exerted on the combustion rate by the degree of comminution and particle shape of the aluminum entering into the compositions. Compositions prepared from aluminum powder burn several times more slowly than those prepared with aluminum dust.

Two important factors affecting the combustion rate of pyrotechnic compositions are the temperature and pressure. As the initial temperature of the compositions rises, their combustion rate increases. This becomes understandable if one considers that at a higher initial temperature of the composition, the combustion reaction requires the supply of a smaller amount of heat from the flame zone to the condensed phase. [Note: As the initial temperature of the composition rises, the flame temperature increases, but not in a well defined manner.]

According to Andreyev [119, p. 406], the temperature coefficient of the combustion rate is smaller the higher the flame temperature (combustion temperature), and the higher the pressure at which the combustion takes place. The temperature coefficient for the same composition at a constant pressure is not constant, but increases with rising initial temperature of the composition [226, p. 95].

A low value of the temperature coefficient is displayed by rocket mixture powders (see Ch. 20), which have a high surface temperature of the condensed phase [223, p. 269].

Apparently, substances or mixtures having a high self-ignition temperature have a lower temperature coefficient [223, p. 269].

To calculate the temperature coefficient α of the combustion rate, the formula is used [128].

$$\alpha = \frac{1}{u} \times \frac{du}{dT}$$

where u is linear combustion rate
T is temperature.

For individual explosives, α usually ranges from 3×10^{-3} to 8×10^{-3}; this means that for a temperature change of 1°C, the combustion rate changes by 0.3-0.8%. For JPN nitroglycerin colloidal powder in the range 20-60°C, $\alpha = 4\text{-}5 \times 10^{-3}$.

The temperature coefficient of the combustion rate of pyrotechnic compositions is smaller than that of explosives or colloidal powders.

It can be calculated from Baum's data [113] that in the interval from -30°C to +44°C, the temperature coefficient for black powder $\alpha = 0.65 \times 10^{-3}$. According to our measurements, the temperature coefficient of black powder in the interval from 0 to 100°C at P=1 atm is

$$\alpha = (1.5 \pm 0.4)\, 10^{-3}$$

The temperature coefficient for all the pyrotechnic compositions studied from this point of view does not exceed 3×10^{-3} (at a maximum initial temperature not above 100°C). Andreyev [110, p. 166] gives the temperature coefficient as the ratio of $u_{100°}$ to $u_{0°}$. On this basis, he cites values of the temperature coefficient for a series of explosives and powders, including 2.9 for nitroglycerin powders and 1.15 for black powder.

The physical meaning of the effect of pressure on the combustion rate according to Andreyev is that at high pressure, the reaction rate in the gaseous phase (in the flame) is higher, and the high temperature zone moves closer to the surface of the condensed phase. The amount of heat transmitted per unit time from the gaseous to the condensed phase increases correspondingly.

Therefore, as the external pressure rises, the combustion rate of pyrotechnic compositions increases. Conversely, in a rarefied space, pyrotechnic compositions burn more slowly than at atmospheric pressure, and when the rarefaction is considerable (a few mm Hg), they lose entirely their capacity for propagating the combustion.

As was noted by Andreyev [124], "the existence of a minimum pressure below which the combustion does not propagate should be due to the fact that as the pressure decreases, the reaction rate in the gaseous phase slows down, and consequently, the amount of heat supplied to the condensed phase by the gases is reduced." Since the rate of heat transfer in the condensed phase is independent of pressure, as the latter decreases, this rate becomes faster than the rate of heat inflow, and the combustion attenuates.

Compositions containing a large excess of fuel over the stoichiometric proportions or compositions with fuels difficult to oxidize lose their capacity to burn faster than others as the pressure is decreased.

For powders and explosives, the relationship between the combustion rate and pressure is frequently expressed by the formula

$$u = A + BP^\gamma \qquad (10.1)$$

where A, B, and γ are constants. For a series of explosives, on the basis of experimental data, Andreyev [119, p. 408] assumes $\gamma=1$; if in the formula $u=A + BP$, the rate u is given in units of g/cm^2 sec, and the pressure P in kg/cm^2, the baric coefficient B of the combustion rate ranges for high explosives from 0.0071 (TNT) to 0.031 (gelatinized nitroglycerin 97:3).

For modern nitroglycerin powders, the exponent γ is close to 0.7 [111, p. 190]. Belyayev [127] states that for most powders and explosives, γ ranges from 0.5 to 1.0.

If the dependence of the combustion rate of pyrotechnic compositions is described by formula (10.1), then for all compositions, the exponent γ will be considerably smaller than unity (in contrast to explosives). However, examples are known among both explosives and powders where the dependence $u=f(P)$ changes substantially with changing pressure (for example, for potassium picrate [111, p. 172] or for black powder [127].)

On this basis, an attempt to describe $u=f(P)$ by using the same values of B and γ over a wide pressure range will not always lead to correct results. The slight dependence of the combustion rate on the pressure shows that exothermic reactions take place in the condensed phase to a significant degree.

According to Baum's data [113], the exponent γ in formula (10.1) for tubular powders is smaller the smaller the volume of the gaseous combustion products v_0 per unit weight of powder.

The dependence of the combustion rate of black powder on pressure in the range from 2 to 30 atm is expressed according to Andreyev [110, 123] by the formula $u=1.21 \cdot P^{0.24}$, where p is the pressure in atm and u is the combustion rate in cm/sec. Andreyev found that black powder still burns steadily at a pressure of 0.1 kg/cm^2.

Belyayev [127] proposes that $u=f(p)$ for black powder be described by two formulas: for P< 5atm, $u=0.88 \times P^{0.5}$; and for P>5 atm, $u=1.5 + 0.2 \times P^{0.47}$.

It would seem that for gasless compositions, the combustion rate should not depend on the external pressure at all. Experiments of Belyayev and Komkova [126] established that the combustion rate of aluminum and magnesium thermites increases, though slightly, with rising pressure.

Thus, for the thermites Fe_2O_3+Al, MnO_2+Al, and Cr_2O_3+Mg, as the pressure rises from 1 to 150 atm, the combustion rate increases by a factor of 3-4. Thus it becomes certain that the combustion of these compositions involves reactions (possibly, oxidation of aluminum by oxygen formed by the decomposition of ferric oxide) taking place in the gaseous phase or at the interface. In addition, Belyayev's experiment showed that there also exist compositions for which the combustion rate is independent of pressure: thus, thermite Cr_2O_3+Al burns at a uniform rate equal to 2.4 mm/sec, at both 1 at and 100 at. This seems to suggest that in this case, the combustion reaction takes place in the condensed phase from start to finish.

The combustion of compositions forming a large quantity of gases in a closed or semienclosed space produces a high pressure. The combustion rate increases substantially, and the combustion may change to an explosion. Such gradual acceleration of combustion ending in an explosion is most frequently observed in the case of chlorate compositions containing a large amount (89-90%) of chlorate, for example, in the combustion of the green light composition $Ba(ClO_3)_2$ + resin (for more detail on the combustion-to-explosion transition, see Ch. 11).

The rate of heat exchange with the ambient medium also effects the combustion rates of the compositions. Thus, the combustion rate of compositions in narrow tubes (small diameter of the article) should be several times slower, but in narrow tubes the outflow of gases is hindered, and excess pressure is created, particularly in the case of rapidly burning compositions, and for this reason a decrease of the combustion rate is not always observed in practice. At very small diameters, the heat loss to the surroundings becomes so large that the composition located in a tube of small diameter completely loses its capacity for propagating the combustion.

The magnitude of the maximum tube diameter at which a stable combustion of the composition is still possible depends on a whole series of factors: formula of the composition, initial temperature, pressure, material and thickness of the tube walls, and composition density. As a rule, the more heat is evolved by the combustion of a composition per unit time, i.e., the faster the combustion, the smaller its minimum diameter.

In connection with the heat loss to the surroundings, we should also discuss the question of the minimum possible combustion rate of compositions. To achieve a combustion process of very slow rate (for example, 0.001 mm/sec) at normal temperature and pressure is impossible because, as a result of the slight heat inflow per unit time and relatively high heat loss to the surroundings, it is impossible to create that significant difference between the temperature in the gaseous and condensed phases which is one of the most characteristic features of the combustion process.

One of the most slowly burning mixtures consists of 96% NH_4NO_3 and 4% wood charcoal, burning at P=1 atm and t=20°C (d=0.94 g/cm³) at a rate u=0.008 mm/sec. Assuming the volume of gases formed by the combustion of the mixture to be approximately v_0=700 cm³/g and a combustion temperature T=900°K, we obtain the rate of flow of the gases in the flame, and hence, the rate of their combustion:

$$u' = 0.008 \times 700 \times \frac{900}{293} \approx 17 \text{ cm/sec.}$$

This figure is close to the combustion rate of the slowest burning gaseous mixtures. Zel'dovich [130] points out that the minimum possible combustion rate of the gaseous mixture $CO+O_2$ at standard conditions should be about 2 cm/sec.

Thus far, the problem of catalysis in the combustion of pyrotechnic compositions has not been adequately solved. In practice, catalysts are used in solid rocket fuel containing ammonium nitrate oxidizer (see Ch. 20); the additives used in this case are potassium or ammonium chromates and dichromates or Prussian blue.

Concerning the influence of catalytic additives on the dependence u=f(P) for colloidal powders see [111, p. 191]. According to the data of Volodina, the introduction of catalysts into chlorate compositions increases the combustion rate of the binary mixtures $KClO_3$ + organic fuel (at P=1 atm); chlorate compositions with a low percentage of fuel (2-4%) which do not burn at P+1 atm acquire this capacity when a catalyst (for example, MnO_2) is added under these conditions.

Chapter 11

EXPLOSIVE PROPERTIES OF PYROTECHNIC COMPOSITIONS

[Note: Concerning the mechanism of generation and propagation of an explosion and the methods of studying explosive processes, the corresponding chapters can be recommended in textbooks on explosives [111, 1, 154, 156, 148, 152A].]

The majority of pyrotechnic compositions are designed for uniform combustion, and it is therefore desirable that such compositions possess either minimal explosive properties or none at all.

The production of compositions having a high sensitivity to impact and friction and at the same time possessing a significant explosion rate and force is a very dangerous process.

The technological conditions of production of pyrotechnic compositions are substantially determined by the explosive properties of the composition. The latter also determine what amount of composition can be present in the plant, what distances should be established between the individual shops, the type of plant buildings and production facilities (wall thickness, presence of knockout-type windows and ceiling, size of cabins, necessity of constructing enclosing yards, etc.). The choice of equipment (type, size, and loading of mixers, type of presses, possibility of mass pressing of articles, etc.) is also largely determined by the explosive properties of the composition.

In addition, it is necessary to have a clear idea of the conditions in which the initiated combustion process of the pyrocomposition may change to an explosion.

To elucidate the explosive properties of pyrotechnic compositions, laboratory tests are conducted that differ considerably from tests of explosives.

Tests of explosives are aimed at characterizing the power of the explosion and comparing the explosives in the effectiveness of their action.

The chief purpose of tests conducted with pyrotechnic compositions is to establish the absence or presence of their capacity to generate an explosion and propagate it in a stable manner. [Note: The necessary density of the pyrotechnic composition must be provided for in the tests.] Only in cases where it has been established that the composition is capable of a stable propagation of the explosion is it useful to conduct further tests to determine the propagation rate of the explosion in the composition and to determine the power of the explosion.

In order to protect oneself from an unexpected manifestation of previously unknown explosive properties of a composition under industrial conditions, it is necessary to create the strictest conditions during tests for its capacity to produce an explosion under laboratory conditions.

This is quite indispensable, since in most cases, an explosion in pyrotechnic compositions, except chlorate mixtures and solid rocket fuel with the oxidizer NH_4ClO_4, is much more difficult to induce than in ordinary secondary explosives.

The strictness of the testing conditions amounts to the following:

1) Use of charges of tested compositions of as large a weight as possible (not below 50-100 g) in powdered form;
2) Use of a powerful initial impulse (additional tetryl detonator weighing at least 8 g);
3) Placing the charge being tested in a strong casing (iron tube or even better, in a thick-walled lead casing).

In practice, the test can be conducted in a Trauzl block of lead by increasing its channel to 40 mm. An indication that the composition (a 50 g sample in powder form) has the capacity to produce an explosion is a significant (≥ 100 cm^3) expansion of the channel of the block of lead after the test.

The test of the capacity of the composition for a stable propagation of the explosion should consist in blasting elongated charges of pyrocompositions not less than 40 mm in diameter and not less than 15-20 cm long. To create strict conditions, it is also necessary to use an additional detonator and to enclose the composition in a strong casing.

A suitable indicator of whether the explosion will reach the end of the charge or will die out along the way may be a lead column (40 mm in diameter) used in the Hess test, or a plate of two-millimeter sheet iron.

For compositions that have shown the capacity for stable propagation of an explosion, the propagation rate of an explosion (detonation velocity) can be determined by the Dautriche method. The detonation velocity is determined more accurately by using a photo-recorder with a mirror scan [111]. When conducting all these tests, it is necessary to keep in mind that the capacity for generating and sustaining a steady explosion propagation decreases sharply in most compositions as their density increases. In pressed pyrotechnic compositions, an explosive decomposition is very difficult to induce, and, once induced, dies out easily and rapidly.

For this reason, the most dangerous operations involved in the preparation of pyrotechnic compositions are those where the composition is still in the unpressed form.

It must be emphasized once again that tests for the explosive properties of pyrotechnic compositions according to the tests adopted for explosives, i.e., the brisance test, the Trauzl block test (in its usual version), and the determination of the detonation velocity are not very representative of pyrotechnic compositions, and should be carried out only when there is reason to believe that a given pyrotechnic composition can sustain a stable propagation of explosive decomposition. Tests have shown that explosive decomposition is easily excited and propagates reliably only in chlorate pyrotechnic compositions (containing no less than 60% chlorates), in solid rocket fuel with NH_4ClO_4 oxidizer (see Ch. 20), or in compositions containing explosive additives.

Mixtures of chlorates with magnesium or aluminum develop a very high temperature when exploding, but yield a comparatively small amount of gases. This may explain the fact that mixtures of chlorates with magnesium or aluminum have lesser explosive properties than mixtures of the same oxidizers with organic substances. The explosive properties of certain chlorate mixtures are listed in Tables 11.1, 11.2, 11.3.

For a mixture consisting of 69% potassium chlorate and 31% aluminum, tested for brisance by Hess' test, the reduction of the lead columns is found to be 7 mm.

The introduction of a large amount of inert admixtures to chlorate compositions markedly decreases their explosive properties. Thus, compositions of masking and color signal smokes con-

gaining large amounts of low-activity substances such as NH₄Cl, organic dyes, and a comparatively small amount of KClO₃ (not more than 40-45%), show almost no tendency toward explosive decomposition. A marked deterrent effect in chlorate compositions of color signal lights is also exerted by alkali or alkaline earth metal oxalates or carbonates in amounts of 20-30%.

Table 11.1 EXPLOSIVE PROPERTIES OF GREEN LIGHT COMPOSITION AND OF EXPLOSIVES CHLORATITE-3 & TNT

COMPOSITION IN %	EXPANSION IN TRAUZL BLOCK IN cm³, 10 g SAMPLE	DETONATION VELOCITY m/sec.
Barium chlorate 81; resin 19	155	1600-2000
Chloratite-3: Potassium chlorate 91 Kerosene 9	255	3600
TNT	285	6700

Table 11.2 EXPLOSIVE PROPERTIES OF MIXTURES OF $KClO_3$ WITH VARIOUS CHEMICALS

FUEL	CONTENT OF FUEL IN COMPOSITION, %	DETONATION VELOCITY, m/sec.	AMOUNT OF MIXTURE, g	EXPANSION cm³
			TEST IN TRAUZL BLOCK	
Wood meal	25	2600	10	220
Sugar powder	10	--	13	105
Carbon black	10	--	13	160
Aluminum	25	1500*	10	160

* Initial impulse, Black powder

Table 11.3 DETONATION VELOCITY OF MIXTURES OF $KClO_3$ WITH VARIOUS FUELS

FUEL	CONTENT OF FUEL IN %	DETONATION VELOCITY TEST	
		DENSITY OF COMPOSITION	DETONATION VELOCITY m/sec.
Charcoal	13	1.27	1620
Graphite	13	1.44	500
Sulfur	28	1.36	1600

A necessary but by itself insufficient condition for generating an explosion is the formation of a considerable amount of gaseous products by the reaction. For this reason, gasless compositions or those with very little gas will either have no explosive properties or will have them to a minimum degree.

Iron-aluminum thermite forms almost no gaseous phase on combustion, and therefore possess no explosive properties. [Note: We are referring to the gaseous phase, not at room temperature, but at the temperature of a given reaction.] Scattering of individual particles during the combustion of thermite is due solely to the air present therin, which, as the temperature rises, expands and causes sparking of thermite.

The combustion of ignition compositions which in addition to thermite contain other components such as barium or potassium nitrate as well as organic binder forms a certain though slight amount of gas, and therefore such compositions may have explosive properties. However, it is very difficult to produce an explosive decomposition in compositions containing more than 50-60% of iron-aluminum thermite; this requires the presence of a strong shell and very powerful initial impulse (a significant amount of some explosive), and even then the results of the experiment will be dubious.

A second indispensable but by itself insufficient condition for a system to have explosive properties should be considered a high exothermicity of the reaction. Since the rate of a chemical reaction depends to a very large extent on the temperature the reaction of explosive decomposition may take place only when the temperature produced is not below 500-600°C.

This condition (T≥500-600°C) is satisfied by nearly all pyrotechnic compositions, the only exception being certain compositions of masking smokes containing 40-50% of a substance of low activity such as NH_4Cl.

The third condition determining the capacity of a system to generate and develop process of explosive decomposition is the homogeneity of the system, i.e., a property possessed by pyrotechnic compositions only to a very relative degree.

As is usually stated in courses on the theory of explosives, solid explosive mixtures not containing detonation conductors which are individual explosives always have a low brisance, and the explosion process propagates with great difficulty in such mixtures.

[Note: The situation is different for mixed liquids which form solutions. In the case of pyrotechnics, a molecular degree of dispersity is present, and therefore the explosion waves can propagate unchecked.] An example of such a mixture is black powder. Pyrotechnic compositions are solid mixtures. They can have pronounced explosive properties only when they contain an explosion (detonation) conductor, a pure substance capable of a spontaneous exothermic decomposition reaction.

Such a conductor in chlorate pyrocompositions is potassium chlorate, whose decomposition does not require any inflow of heat from the outside, and in solid rocket fuel (mixture powders), ammonium perchlorate (or ammonium nitrate).

Potassium perchlorate decomposes with a very slight evolution of heat:

$$2 KClO_4 = 2KCl + 4O_2 + 1.2 \text{ kcal}$$

and therefore, an explosive decomposition in compositions with $KClO_4$ should take place and propagate with greater difficulty than in chlorate compositions.

Nitrates (except NH_4NO_3) require a very large inflow of heat from the outside for their decomposition:

$$2KNO_3 = K_2O + N_2 + 2.5O_2 - 151 \text{ kcal}$$

and therefore the process of explosive decomposition in nitrate compositions (except those with NH_4NO_3) is excited with great difficulty, and conversely, dies out easily and rapidly.

Nitrate mixtures possess much weaker explosive properties than chlorate mixtures.

A powerful initial impulse is required to excite an explosive decomposition in illumination compositions containing barium nitrate as the oxidizer and aluminum powder or dust as the main fuel.

Table 11.4 lists comparative data on the explosive properties of mixtures of potassium perchlorate or barium nitrate with aluminum dust.

Table 11.4 EXPLOSIVE PROPERTIES OF BINARY MIXTURES OF POTASSIUM PERCHLORATE AND OF BARIUM NITRATE WITH ALUMINUM

OXIDIZER	OXIDIZER CONTENT, %	EXPANSION IN TRAUZL BLOCK cm^3	DENSITY OF COMPOSITION	DETONATION VELOCITY, m/sec.
				VELOCITY DETONATION TEST
KClO$_4$	66	172	1.2	760
Ba(NO$_3$)$_2$	73	34	1.4	Misfire

Note: The detonation velocity was determined in iron tubes 30mm dia and 250mm long; the initial impulse was provided by #8 caps + tetryl squib (10g).

The process of explosive decomposition in nitrate illumination compositions containing magnesium is generated somewhat more easily. However, the rate of explosive decomposition of nitrate illumination compositions does not exceed 1000 m/sec in most cases; the rate of explosive decomposition of black powder, which also may be treated as a nitrate pyrotechnic mixture, does not exceed 400 m/sec.

The explosive decomposition of uncompacted binary mixtures of nitrates with magnesium or AM alloy, if in any significant amount (more than 50-100 g), is readily excited not only by the action of an explosive impulse (detonating cap), but also by the action of a fire impulse (black powder, quick match). Such mixtures should be considered some of the most dangerous pyrotechnic compositions and must be handled with great care. It should be noted that a high combustion temperature favors a combustion-to-explosion transition, both in pyrotechnic compositions and in explosives.

Andreyev states that hot gases penetrating into the pores of the charge ignite the explosive particles more readily, the higher their temperature (temperature of the gases). Moreover, an increase in the surface of combustion and the associated pressure rise lead to the formation of a detonation wave. The pronounced tendency of PETN and hexogen to change from combustion to an explosion in comparison with other secondary explosives is due precisely to their high combustion temperature [151, 119, p. 411]. The combustion of ammonium perchlorate does not change into an explosion under the same conditions as when this transition readily occurs in PETN. However, it is sufficient to add a slight amount of finely dispersed aluminum to ammonium perchlorate to make the combustion of the mixture in the presence of rising pressure change into an explosion just as readily as in the case of PETN or hexogen [151]. The influence of aluminum consists in raising the temperature of the combustion products: by penetrating into the porous powder, the hotter gases ignite its particles more easily.

We will not consider the behavior of compositions acted upon by various types of initial impulses. An impact (or friction) corresponding to an individual portion of the surface of a pyrotechnic composition in the absence of conditions promoting an increase in the pressure during combustion usually causes only a local explosion of the composition. The remaining mass of the composition burns normally, as in the case of the action of an ordinary thermal impulse.

The penetration into pyrotechnic articles of a bullet that has retained a significant velocity may produce ignition in many cases, and when the composition is enclosed in a strong shell, sometimes an explosion of the composition as well. In a closed space (for example, in the presence of a flash in the course of pressing in a steel die) and in all cases where the composition is enclosed in a very strong metal shell, increasing pressure causes an increase in combustion rate, and in many cases, the process ends in an explosion even when ordinary fire devices are used for the initial impulse.

The same kind of pressure rise causing the combustion to change to an explosive decomposition arises in some cases involving simultaneous combustion of a large amount (10 kg or more) of powdered, rapidly burning compositions.

A very simple device for determining the possibilities of a transition from the combustion of an explosive in a closed space to an explosion was proposed by Andreyev [153, p. 59]. This device consists of a strong iron tube (200 mm long with an inner diameter of 40 mm) closed on all sides, partly filled with an explosive (50 g); the explosives are ignited by an ignition composition charge. The combustion process in the ignition composition is excited by a fine wire heated by an electric current (Fig. 11.1).

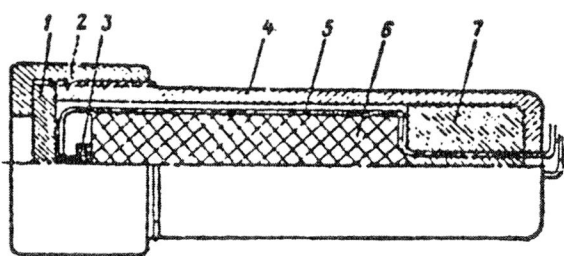

Fig. 11.1. K. K. Andreyev's tube for determining the possibility of the combustion-to-explosion transition.
1 - steel disc, 2 - screw-on head, 3 - ignition charge, 4 - tube housing, 5 - electrical ignition wire, 6 - composition tested, 7 - plaster.

Fragmentation of the tube into a large number of pieces (5-6 or more) indicates that the combustion changes to an explosion (Fig. 11.2).

In the same manner, one can determine the capacity of pyrotechnic compositions for a combustion-to-explosion transition in a closed volume; if desired, this test may be supplemented by weighing the fragments and calculating the average weight of a fragment.

Recently, Andreyev and co-workers [149] developed a test for the combustion-to-explosion transition capacity by burning in a manometric bomb minute charges of explosives pressed into plexiglas cups (the initiator is black powder; the pressure change is recorded on an oscillograph film).

From the shape of the manometric curve, this method makes it possible to follow the entire course of the process, whereas with Andreyev's tube, only the final test results are recorded.

In our view, the drawbacks of the method include a small sample (a few grams) of the tested substance and the fact that the charge burns only on the end. This sets the testing conditions apart to some extent from the conditions in which the combustion of explosives (or pyrotechnic compositions) may take place in practice.

In some cases, it is of interest to follow the behavior of a composition during its combustion in a semienclosed volume, i.e., when the outflow of the gases is severely limited.

Andreyev's iron tubes may be used for this purpose, provided that a hole of suitable diameter (from 2 to 12 mm) is drilled through the steel disc closing off one of the ends of the tube.

The same purpose of determining the behavior of a pyrocomposition during combustion in a semienclosed volume is served by a Trauzl block test using a small charge of black powder as the initial impulse instead of a detonating cap.

As is evident from Table 11.5, when a firing initial impulse is used, the expansion in the Trauzl block is found to be approximately 50% of the normal expansion, and for compositions of low sensitivity (composition 3), the combustion does not change to an explosion at all, and no expansion of the channel is obtained.

Table 11.5 DEPENDENCE OF EXPANSION IN TRAUZL BLOCK ON THE CHARACTER OF INITIAL IMPULSE

TEST NO.	COMPOSITION (UNPRESSED) %	EXPANSION IN cm^3 20g SAMPLE, INITIAL IMPULSE	
		BICKFORD FUSE	NO. 8 CAP
1	Potassium perchlorate 85 Charcoal 15	198	318
2	Potassium perchlorate 59 Magnesium 41	49	88
3	Barium nitrate 89 Iditol 11	0	120
4	Pressed TNT (for reference)	---	718

It should be noted in conclusion that in some cases, the danger of an explosion is not eliminated even in handling the individual components used for the preparation of the compositions.

Thus, for example, an explosion may result from the presence in air of large concentrations of dust, aluminum dust, fine magnesium powder, AM alloy powder, zirconium and titanium powders, and dust of other fuels, for example, carbohydrates, resin powders, etc.

For more detail, see the book of Godshello [152], and [111, p. 152]. It was found, in particular, that the lower explosive limit of aluminum powder (particle size 0.1-0.3μ) corresponds to its content of 40 mg per liter of air, and that of magnesium powder, to 25 mg per liter of air.

Moistened powders of many metals possess explosive properties under certain conditions. The reaction of magnesium or aluminum with water is associated with the evolution of a very large amount of heat and a significant quantity of gases:

$$H_2O + Mg = MgO + H_2 + 76 \text{ kcal}$$

This is equivalent to 1.82 kcal/g of mixture of the reactants; the amount of gas is 530 cm^3/g. Thus, the necessary conditions for the formation of an explosion exist. As was shown in [155], this explosion can be realized by using a sufficiently strong initial impulse. However, because of the absence of molecular contact between the particles of the oxidizer (water) and fuel, i.e., because of the inhomogeneity of the mixture, the system, while having the capacity for the formation of an explosion, does not have the capacity for its stable propagation, as was shown experimentally [160]. Concerning the possibility of the oxidizing action of water on nonmetals (B, Si, and P) and lower oxides of elements, see [119, p. 540].

The explosion of an individual component - potassium chlorate, which has a large destructive force, may take place only when it contains significant admixtures of fuels (over 1%).

The amount of heat evolved by the decomposition of the pure salt (without impurities) at room temperature is low (0.08 kcal/g); therefore, if an explosive decomposition of $KClO_3$ does take place in the presence of a very powerful initial impulse, the destructive action of the explosion will still be slight. Concerning the explosive properties of the oxidizers ammonium perchlorate and ammonium nitrate, see Ch. 20.

Chapter 12

PHYSICAL AND CHEMICAL STABILITY OF PYROTECHNIC COMPOSITIONS

The term "stability of pyrotechnic compositions" is taken to mean their ability to retain their properties in storage for a long time.

1. CHANGES OCCURRING IN COMPOSITIONS IN STORAGE

Physical and chemical changes take place in pyrotechnic compositions when pyrotechnic articles are stored in warehouses. In some cases, these changes are so substantial that the products become unsuited for use, and sometimes even dangerous to handle. Therefore, in each specific case it is necessary to determine the nature of the changes in the composition and also the influence exerted by various factors on the rate of the decomposition processes in the composition.

Studies are made to establish the conditions of storage in which the decomposition processes are minimized, as well as to determine the permissible storage periods of individual types of pyrotechnic articles.

The physical changes taking place in a composition in storage are mostly due to moistening of the composition. This is associated with a partial dissolution of the components of the composition and a change in the density and shape of the pressed charge. A less frequent cause of deformation of a charge (flare or pellet) are mechanical actions or changes in the temperature of the ambient medium. Particular mention should be made of changes in the formula of the composition, caused by sublimation of the volatile components of the composition (for example, sublimation of naphthalene, hexachlorethane, etc.).

The hygroscopicity of a composition depends chiefly on the hygroscopicity of its components, and also on the density and state of the surface of the composition, exposed to the action of moist air. As a rule, compositions in the pressed state attract moisture from air to a considerably lesser extent than those taken in the form of a powder. Other things being equal, the hygroscopicity of a composition increases with increasing degree of comminution of its components.

To prevent a composition from absorbing moisture from air in cases where complete sealing of the product is impossible, the particles of the composition or of the individual components are coated with a protective film of some plastic organic substance. Most frequently used for this purpose are mineral oils, stearin, and stearates of various metals, paraffin, and also lacquers based on various artificial and natural resins, drying oil, etc.

The protective action of resins is particularly effective when they are introduced into a composition in the form of lacquers; in dry form (or powder), resins are much less effective in protecting a composition from the penetration of moisture. The chemical changes occurring in compositions may be very different, owing to the very large assortment of the components used for their production. Nevertheless, certain general statements may be made.

COMPOSITIONS CONTAINING METAL POWDERS AND INORGANIC OXIDIZERS

Magnesium or aluminum powder enters into many pyrotechnic compositions as a fuel. The decomposition of these compositions begins with the corrosion of these powders in the presence of moisture:

$$Mg + 2H_2O = Mg(OH)_2 + H_2.$$

$$Al + 3H_2O = Al(OH)_2 + 1.5H_2.$$

The reaction of magnesium with water is much more vigorous than that of aluminum with water. The magnesium oxide film is porous, loose, and inadequately protects the metal from further corrosion. Aluminum powder (or dust) usually oxidizes only on the surface, and the dense oxide film forming on its surface prevents any further deeper oxidation.

Figure 12.1 shows a graph indicating the change in temperature as a function of time, resulting from the addition of 10% of water to magnesium or aluminum powder.

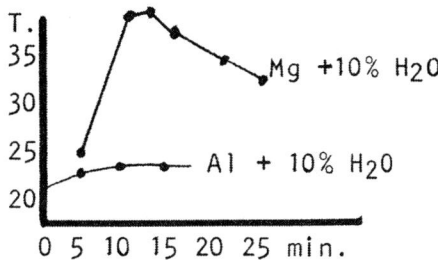

Fig. 12.1. Increase in the temperature of metallic powders upon the addition of water.
1 - Mg + 10% H_2O; 2 - Al + 10% H_2O.

Since a considerable amount of heat is evolved by the reaction of magnesium or aluminum with water, substantial moistening of binary mixtures of oxidizers (for example, nitrates) containing powders of these metals may cause the composition to heat up; if the mass of the composition is large (kilograms) or the composition is chemically unstable (see below), self-ignition may take place.

The corrosion of metal powders follows the same pattern as the corrosion of monolithic metals, but takes place much more intensively because of the very large specific surface of the powders.

It is generally known that the corrosion of magnesium takes place intensively in an acid medium, but that magnesium is stable in alkaline solutions. This is due to the properties of the magnesium oxide film, which dissolves readily in acids and is insoluble in alkalis.

The aluminum oxide film dissolves in both acids and alkalis. A characteristic feature of aluminum is its intensive corrosion in a strongly alkaline medium.

If upon being moistened, the oxidizers present in a composition produce an acid or alkaline medium, the presence of the oxidizer will obviously be reflected in the corrosion rate of the metal powder.

In the case of alkali metal or alkaline earth nitrates, the hydrogen formed by the corrosion of the metals reduces the nitrates to ammonia.

The decomposition reaction of moistened mixture $Ba(NO_3)_2$ + Mg may be written as follows:

$$8Mg + 16H_2O = 8Mg(OH)_2 + 16H(8H_2)$$
$$+\quad Ba(NO_3)_2 + 16H(8H_2) = Ba(OH)_2 + 4H_2O + 2NH_3$$
$$Ba(NO_3)_2 + 8Mg + 12H_2O = Ba(OH)_2 + 2NH_3 + 8Mg(OH)_2$$

In this case, the decomposition products, including $Ba(OH)_2$, produce an alkaline medium in the moistened composition, and this inhibits further corrosion of magnesium.

Many years experience has shown that compositions containing magnesium and alkaline earth or alkali metal nitrates are stable in storage, particularly if they contain a few percent of an organic binder.

The decomposition reaction of the moistened mixture $NaNO_3 + Al$ may be expressed by the equations

$$3NaNO_3 + 8Al + 8H_2O = 3NaOH + 8Al(OH)_3 + 3NH_3$$
$$+\quad 3NaOH + 3Al(OH)_3 = 3NaAlO_2 + 6H_2O$$
$$3NaNO_3 + 8Al + 12H_2O = 3NaAlO_2 + 5Al(OH)_3 + 3NH_3$$

In this case, the sodium hydroxide solution formed by the decomposition of $NaNO_3$ significantly accelerates the corrosion of aluminum powder. The $NaNO_3 + Al$ mixture has a low chemical stability and can be used only if it is effectively protected from atmospheric moisture by greasing additives or by sealing the product.

The stability of nitrate compositions containing metal powders strongly depends on the presence of moisture in the composition, and hence, on the hygroscopicity of the oxidizer. Hence it may be assumed that under the same storage conditions, the slightest chemical changes will take place in compositions containing the least hygroscopic salt, barium nitrate, as the oxidizer.

With the exception of mixtures with alkali metal nitrates, compositions containing aluminum are unquestionably more stable chemically than the corresponding compositions with magnesium powder. This is true even when a coarse magnesium powder and a finely divided aluminum powder are used.

As was stated previously, the presence of an organic binder considerably hinders the access of moisture to the particles of the composition. Therefore, in the absence of effective sealing of the product, illuminating and tracer compositions with organic binders will be chemically more stable than the corresponding photomixture-type binary compositions.

The addition of sulfur to a mixture of barium nitrate and aluminum powder or dust does not decrease the chemical stability of these mixtures. According to the literature data, aluminum does not react with sulfur even when sulfur is present in molten form. Mixtures consisting of barium nitrate, aluminum and sulfur are among the most stable pyrotechnic compositions during storage.

Mixtures of sulfur and magnesium are chemically unstable, as was noted long ago by the Russian pyrotechnists of the 19th century, F. F. Matyukeviches and A. Ordvnskiy.

D. I. Mendeleyev [165, Ch. 20] notes the effect of high pressure on the capacity of solids for combining with one another. He states: "At a pressure of 6000 atm, sulfur can combine with many metals at room temperature"; see also [164]. When a mixture of magnesium and sulfur which already contains MgS is moistened, the following reaction takes place:

$$MgS + 2H_2O = Mg(OH)_2 + H_2S$$

Mixtures containing magnesium and aluminum powders simultaneously are more sensitive to the action of moisture than similar compositions with one of these metals. Thus, when it is moistened with water, a composition containing powders of both metals heats up much more than a composition with magnesium alone. This is due to the fact that microscopic galvanic cells (metal - salt solution - other metal) whose presence considerably accelerates the corrosion of magnesium are formed in the moistened composition.

Compositions containing powders of AlMg alloys are much more stable chemically than compositions containing a mixture of the powders of these two metals. This is due to the fact that the particles of the alloy are coated with a dense protective film of aluminum oxide.

Mixtures of nitrates of heavy metals $Pb(NO_3)_2$ or $Cu(NO_3)_2$ with magnesium are not chemically stable upon moistening owing to the occurrence of exchange reactions, for example:

$$Pb(NO_3)_2 + Mg = Pb + Mg(NO_3)_2$$

Concerning the mixture $Pb(NO_3)_2 + Al$, there are indications that in this case aluminum is coated by a layer of basic lead nitrate, which protects the aluminum from further action of the solutions.

Chlorate mixtures with metal powders. Mixtures of potassium chlorate with magnesium powder under high humidity conditions show substantial chemical changes and can not therefore be recommended for long-term storage.

It is possible that the rapid oxidation of magnesium in a mixture with $KClO_3$ is largely due to the catalytic effect of the oxidizer. In this connection, we should recall the low hygroscopicity of potassium chlorate and note that this salt can not be a good transmitter of moisture from air to magnesium powder.

An aluminum - potassium chlorate mixture which produces a sound effect on burning is chemically much more stable than a magnesium - potassium chlorate mixture.

According to the literature data, mixtures of metal powders with potassium permanganate have a low chemical stability.

Mixtures of barium peroxide with metal powders. When anhydrous barium peroxide comes in contact with water, it hydrates, forming $BaO_2 \cdot 8H_2O$. Taken separately, barium peroxide at room temperature is decomposed by water only to a slight degree; at 0°C, the equilibrium constant of the reaction in solution:

$$BaO_2 + 2H_2O = Ba^{++} + 2OH^- + H_2O_2$$

is $K = 18 \cdot 10^{-12}$

When moist barium peroxide is mixed with aluminum powder, OH^- ions are consumed by the reaction with the oxide film on aluminum, the decomposition reaction equilibrium shifts to the right, and the decomposition is very vigorous.

Experiments have shown that when a $BaO_2 + Al$ mixture is moistened, its temperature rises by 60-100°C depending on the grade of aluminum used.

Iron-aluminum thermite and thermite compositions are very stable chemically. The oxidizer which they contain does not decompose at low temperatures. To prepare thermite compositions, very coarse aluminum powder is used which at room temperature has absolutely no tendency to react with moisture.

Effect of ammonium salts. The addition of magnesium-containing ammonium salts, particularly ammonium chloride, to compositions is undesirable, since ammonium salts are highly hygroscopic, and the aqueous solutions of ammonium salts of strong acids formed by hydrolysis have an acid reaction (for example, a 10% NH_4Cl solution has pH ≈5):

$$NH_4Cl + HOH = HCl + NH_4OH$$

The strong acids formed by hydrolysis of the salts react with the oxide film coating the metal powder particles, and magnesium from which the oxide film has been removed corrodes rapidly in acid solution.

It is possible that in NH_4NO_3 solution, the corrosion of magnesium will be slower than in NH_4Cl solution, since Cl^- ion stimulates corrosion.

Corrosion of aluminum powder is accelerated only slightly when ammonium salts are introduced into the compositions (see [167A]). It was noted in [162] that in experiments with aluminum shavings immersed in concentrated NH_4NO_3 solution, the evolution of gas was observed at room temperature, although it was very slow; at 60°C, it was much faster and produced more than 1 cm^3/hr; the gas in these experiments contained 6-13% hydrogen and 45-60% nitrous oxide.

One should also avoid introducing into compositions containing metal powders, particularly magnesium, any salts whose aqueous solutions have an acid reaction due to hydrolysis.

In aqueous solution of potassium chloride, which as a neutral reaction, magnesium corrosion proceeds somewhat faster than in water, but many times slower than in solutions of ammonium salts.

Corrosion of aluminum powder is markedly accelerated in the presence of copper and particularly mercury, and also in the presence of soluble salts of these metals, and therefore their addition to the compositions is entirely inadmissible.

In pressed compositions whose components are in more intimate contact, the chemical changes in certain cases take place more rapidly than in powdered compositions, despite the fact that compositions in powder form are more hygroscopic.

Concerning the corrosion of metals (including metal powders), the monographs of Evans [170], Akimov [163] and Tomashov [168] can be recommended.

Compositions containing iron metal powder are chemically unstable in most cases, since the oxidation of finely ground iron in the presence of moisture is very fast. In firework compositions containing iron filings, to prevent their oxidation, the latter are subjected to the burnishing operation, i.e., treatment with hot flaxseed oil; for more detail, see the book of Solodovnikov [6].

COMPOSITIONS CONTAINING NO METAL POWDERS

When such compositions are moistened, no significant chemical changes take place in most cases. An exception are mixtures containing two soluble salts which in solution are capable of undergoing a double displacement reaction with the formation of a precipitate.

Exchange reactions. An example of such an undesirable combination of salts is the yellow light composition, which includes a mixture of barium nitrate with sodium sulfate, carbonate, or oxalate. When such mixtures are moistened, exchange reactions occur with the formation of insoluble barium salts, for example:

$$Ba(NO_3)_2 + Na_2CO_3 \rightarrow BaCO_3 + 2NaNO_3$$

The formation of the hygroscopic salt $NaNO_3$ by such reactions causes further moistening of the composition, so that in some cases, the reaction in the solution formed can proceed almost to completion.

Compositions of signal lights containing no metal powders or hygroscopic salts (or combination of salts capable of undergoing double displacement) do not usually undergo any appreciable chemical or physical changes in storage. An example of a composition stable in storage is that of red light: potassium chlorate - strontium carbonate - resin.

Signal smoke compositions containing potassium chlorate, sugar, or starch and some organic dye are fairly hygroscopic, but do not undergo any appreciable chemical changes in storage. This is also frequently observed in the case of masking smoke compositions containing no metal powders.

<u>Formation of an unstable salt, ammonium chlorate.</u> Upon moistening of masking smoke compositions containing the smoke-forming substance ammonium chloride in addition to potassium chlorate, the following exchange reaction is theoretically possible:

$$KClO_3 + NH_4Cl = KCl + NH_4ClO_3$$

with the formation of ammonium chlorate, which is capable of spontaneous decomposition and even explosion when the temperature is raised slightly (30-60°C). Nevertheless, practice shows that masking smoke compositions which in addition to potassium chlorate and ammonium chloride contain a large amount of naphthalene or anthracene (for example, Yershov's mixture) are fairly stable in storage [211]; cases of spontaneous ignition of smoke mixtures similar to Yershov's mixture have not been observed thus far.

Because of the danger that an exchange reaction will occur with the formation of ammonium chlorate, the introduction of ammonium nitrate into chlorate compositions is inadmissible. The combined use of ammonium nitrate and chlorates is also forbidden in explosive mixtures. According to the literature, a mixture of potassium chlorate with a stoichiometric amount of ammonium nitrate explodes even at 120°C.

Concerning water-induced spontaneous ignition of chlorate compositions containing NH_4NO_3, copper salts and metal powders, see [7, p. 133].

<u>Unstable mixtures of chlorates and sulfur.</u> These mixtures are never used for loading articles to be kept in long-term storage. The possibility of self-ignition of $KClO_3 + S$ mixtures is particularly strong when the sulfur contains traces of sulfuric acid. The introduction into such mixtures of carbonates (for example, chalk - $CaCO_3$), which neutralize the acid traces, lowers the probability of self-ignition, but apparently does not eliminate it completely. Among mixtures undergoing rapid self-ignition is a mixture of glycerin with finely divided potassium permanganate, which self-ignites 10-20 sec. after the components are mixed together.

As previously noted, a mixture of potassium chlorate and red phosphorus is so sensitive that it explodes even when lightly rubbed with a rubber stopper. If this mixture is prepared very carefully, self-ignition may apparently follow after a short time without any external stimulation. The oxidation rate of pure red phosphorus increases with rising temperature of the ambient medium; the oxidation of red phosphorus is considerably decelerated when its particles are treated with iditol (or shellac) lacquer.

2. METHODS OF DETERMINATION OF THE HYGROSCOPITICITY AND CHEMICAL STABILITY OF COMPOSITIONS

Under ordinary conditions of storage of pyrotechnic products, i.e., at an air humidity of not more than 80% and a temperature corresponding to the climate of the given locality (we are referring to a dry, unheated warehouse), moistening of the compositions takes place very slowly, and in the majority of cases, their chemical decomposition is also slow under these conditions.

Appreciable chemical changes during storage of stable pyrotechnic compositions under such conditions take place only over the course of several years. In order to produce these changes artificially in the laboratory during a comparatively short period, the composition is stored for a certain period at a higher humidity and sometimes simultaneously at a higher temperature.

The term "test for the chemical stability of a pyrotechnic composition" is usually applied to the exposure of the composition to artificially created "strict" conditions and to the study of the changes which have taken place in the pyrotechnic composition (by weighing, chemical analysis, measuring the pressure of the gases above the composition, etc.).

It is very difficult to establish a definite pattern in the character and magnitude of the changes occurring in pyrotechnic compositions as a function of changes in storage conditions. It can not be indicated in advance what time interval in the storage of pyrotechnic products under ordinary warehousing conditions will be necessary to produce in the stored compositions the changes that the "chemical stability test" has shown to occur in the laboratory during a certain much shorter time interval.

Usually, the chemical changes detected in the composition after it has been subjected to the "test" are compared with the changes that occurred in the same "test" in a chemically stable composition. The composition studied is acknowledged as being chemically stable and is placed in production if data on the "test" do not exceed the changes of the chemically stable composition, i.e., the composition already checked in practice. However, a definitive evaluation of the chemical stability of compositions can and should be formulated only on the basis of observations in the changes occurring under actual conditions in long-term storage of products loaded with them.

The hygroscopicity of compositions is usually determined in hygrostats, i.e., instruments in which a definite and constant air humidity is maintained. The most frequently used type of hygrostat is the desiccator, the bottom part of which contains water or a saturated salt solution. The composition being tested, contained in weighing bottles or small crystallizing dishes, is placed on the desiccator plate.

A serious drawback of such a hygrostat consists in the fact that changes in the temperature of the ambient medium may cause appreciable changes in the absolute humidity inside the instrument, only the relative humidity actually remaining unchanged. This test will be more valid if the desiccator is placed in a thermostat.

To produce a given air humidity inside the desiccator, either water (testing at 100% relative humidity) or salt solutions are poured into its bottom part. At 20°C, a relative humidity of 92.5% is produced above a saturated potassium nitrate solutions, and a relative humidity of 77.5% is produced above a saturated sodium chloride solution at the same temperature.

If testing for hygroscopicity involves a powder composition, a certain weighed amount of the composition (for example, 10 g) is placed in an open beaker on the desiccator plate. Occasionally, pellets and even completely finished small pyrotechnic articles are tested in hygrostats.

In comparative testing of many compositions, it is necessary to maintain a constant surface of moistening (use of beakers of standard size) and to provide a small thickness of the layer of the composition being moistened; for this purpose, the diameter of the beakers should be no less than 50 mm for a 10 g sample.

During the testing, the composition is weighed periodically. The total duration of the hygroscopicity test of a composition when carried out at room temperature (20°C) is usually no less than 30 days.

When the experiment has been completed, a graph can be drawn (Fig 12.2) which shows the change in the weight of the composition as a function of its residence time in the desiccator. The residence time in the desiccator in days is plotted on the abscissa axis, and the relative increase in the weight of the composition in percent of the initial weight is plotted on the ordinate axis.

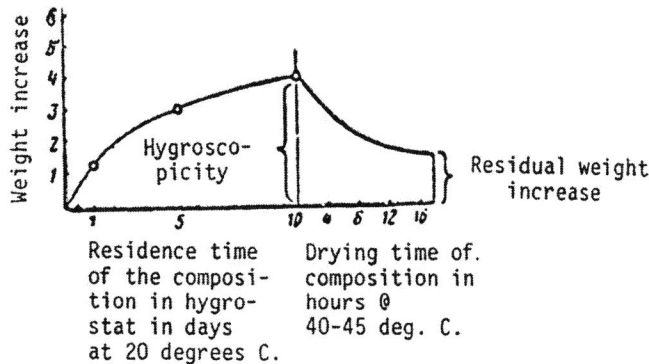

Fig. 12.2. Weight change of the composition during its exposure to a higher humidity followed by drying.

There are standards for the weight increase of a composition after ten days of testing in a hygrostat at t=20°C and 100% relative humidity: the result of the test is considered to be completely satisfactory if the weight increase of the composition does not exceed 2%.

For black powder, its weight increase after being kept at room temperature at 92.5% relative humidity for one day should not exceed 1.75%.

It should be noted that testing of compositions involving exposure to a higher relative humidity must not be referred to merely as testing for hygroscopicity of the composition, since several parameters characterizing the chemical stability of the composition are simultaneously determined while the compositions are exposed according to this method. The chemical stability of pyrotechnic compositions can be determined in different ways.

In the case of tests for chemical stability, the pyrotechnic compositions are frequently divided into three categories:

1) compositions containing metal powders and inorganic oxidizers;
2) compositions containing metal powders and organic oxidizers containing chlorine;
3) compositions containing no metal powders.

In most of the methods of testing for chemical stability employed at the present time, the action of moisture on the pyrotechnic compositions has been anticipated in one form or another.

The compositions are usually kept in desiccators at an increased relative humidity for at least 30 days. The composition is taken out (a sample is withdrawn) every 10, 20, and 30 days. If the compositions tested contain a metal (Mg or Al) powder, the change in the activity of the metal and the residual increase in weight (see below) are determined, and a graph of the dependence of these quantities on the exposure time is plotted. Since the weight of metal powders increases during oxidation due to their reaction with water, a composition subjected to the action of moisture, does not usually return to its initial weight after drying. In order to determine the stability of compositions (illuminating, etc.) containing metal powders, it is sufficient to know only the increase in weight (residual) following the test (see Fig. 12.2).

In addition, it should be noted that the residual increase in weight cannot be used as a measure of the chemical stability of pyrotechnic compositions when they contain volatile components.

The pressure developed by gaseous products of the decomposition reaction of a moistened composition is measured rather infrequently; this method of checking the chemical stability of a composition has been considered unreliable, since the hydrogen evolved by the decomposition of many compositions may be partly consumed by the reduction of the oxidizers (nitrates, etc.).

A complete analysis of the compositions during their testing for chemical stability is not usually performed, not only because it is very time-consuming, but also because in most cases, with the existing analytical methods, it is difficult to estimate the true changes that have taken place in the compositions on the basis of the data obtained. [Note: Concerning the methods of analysis of pyrotechnic compositions, see Danner and Goldenson [171]; the methods of analysis of certain components of pyrotechnic compositions are described in the first edition of our book (1943).]

3. DECREASES IN THE EFFECTIVENESS OF COMPOSITIONS IN STORAGE AND PERMISSIBLE PERIODS OF THEIR STORAGE

Moistening of compositions usually decreases the special effect. Moist compositions burn more slowly, develop a lower temperature during their combustion, and emit less light. A decrease in the "activity" of metals due to decomposition processes taking place in illuminating, tracer and other compositions leads to the same results (see Table 12.1)

Table 12.1 EFFECTIVENESS OF THE ACTION OF COMPOSITION: 64% $BA(NO_3)_2$; 24% MG; 12% IDITOL.

Activity of Metal, %	Light Intensity in Thousands of Candles	Combustion Time, Sec.
95	56	5.6
88	35	6.4

For this reason, for pyrotechnic products, it is necessary to establish the maximum permissible storage periods during which they should be used. Depending on the formula of the composition and degree of sealing of the product, these periods may range from one to two years to several dozen years.

A normal storage period of the products is considered to be ten years or more.

Signal-flare compositions containing no metal powder are considered to be among the most stable ones. Among illumination and ignition compositions, the most stable ones chemically are compositions containing only aluminum as the main fuel, and barium nitrate as the oxidizer.

The addition of magnesium to such compositions decreases their chemical stability and shortens the possible storage periods of the pyrotechnic products containing them. Since, other things being equal, compositions in the powdered state are more sensitive to the action of moisture than pressed ones, the possible storage periods for photomixtures must be assumed to be somewhat shorter than for other types of pyrotechnic compositions.

If the product is completely sealed, the permissible storage compositions of pyrotechnic products may be very significant in the majority of cases and be measured in dozens of years.

In individual cases, after the permissible storage periods have elapsed, pyrotechnic products may be subjected to tests for the effectiveness of action and chemical stability of the composition. If the test results are satisfactory, the storage period of the products may be extended.

PART TWO

Chapter 13

ILLUMINATION COMPOSITIONS

1. SPECIAL REQUIREMENTS IMPOSED ON ILLUMINATION COMPOSITIONS

[Note: Necessary information on optics and illumination may be found in the books of Landsberg [178], Meshkov [182], and Gershun [175].]

In addition to the general requirements placed on compositions indicated in Part One of the book, illumination compositions should also satisfy two special requirements.

1. The combustion of a unit weight of illumination composition should evolve the maximum quantity of luminous energy (measured in lumen-seconds). In addition to an adequate sharpness of observation, the emitted light should allow the observer to correctly identify the color. For this reason, the radiation of an illumination composition can not be monochromatic, i.e., it must contain luminous energy of all parts of the spectrum. For cases of visual observation, it would be desirable to have a spectral radiation intensity distribution close to that of the solar radiation, to which the human eye has become accustomed and adapted. The sensitivity of the human eye to light of different wavelengths is shown in Fig. 13.1. It should be noted however, that whereas until recently, the chief radiation detector has been the human eye, various physical detectors of radiant energy are now being used: photocells, photo-resistance, thermocouples, etc. Thus the spectral range of useful radiation has expanded from the previous values of 0.4-0.7μ to 0.4-5μ.

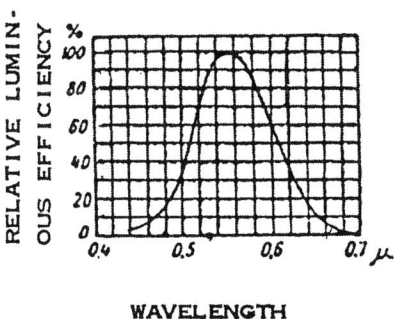

Fig 13.1. Dependence of relative luminous efficiency on wavelength.

2. The linear combustion rate of pressed illumination compositions should amount to units of millimeters per second. For large-sized illumination articles (flares of illumination bombs and shells), where a sufficiently long total combustion time is required, illumination compositions having a combustion rate of not more than 1-2 mm/sec in the compacted state are usually employed; for small-sized illumination articles (illumination pellets for pistol cartridges and rifle grenades), compositions having a combustion rate of 5-10 mm/sec in the pressed state are used.

To fulfill the first requirement, it is necessary to select the components of an illumination composition so that the composition evolves the maximum amount of heat during its combustion.

It has been shown in practice that a sufficiently large amount of radiant energy is obtained only from the combustion of compositions evolving no less than 1.5 kcal/g. This figure is used as one of the criteria for checking the validity of the components of an illumination composition.

The luminous efficiency of illumination compositions is a function of the quantity of thermal energy converted into luminous energy in the flame.

The numerical value of the luminous efficiency obtained during the combustion of illumination compositions is determined by many factors: the emissive power of the combustion products, spectral composition of the radiation, size and optical properties of the flame, combustion rate of the compositions, etc.

To obtain the highest values of luminous efficiency, it is necessary, by selecting a suitable composition formula and design of the article, to make sure that the flame formed during the combustion of the compositions has the maximum temperature, contains a sufficient quantity of solid and liquid particles which adequately radiate the light in the incandescent state, and has the largest radiation surface.

Efforts should be made to increase the surface of the flame, since the quantity of luminous energy radiated by the flame per unit time depends only on the mean brightness of the flame in stilbs and on the radiation surface in cm^2

$$F \text{ (lm)} = \pi B \text{ (sb)} \cdot S \text{ (cm}^2\text{)}.$$

[Note: This relationship applies only as a first approximation, since the emission of a flame is not planar, but three-dimensional; however, in many cases, when the flame size is sufficiently large and the absorptivity of the intrinsic radiation of the flame is high, illumination compositions should be treated as sources, not of three-dimensional, but of surface radiation, since a significant part of the radiation of the inner zones is absorbed by the outer layers of the flame.]

All of the above also applies to the power of a radiant flux in a spectral range beyond the spectral sensitivity of the human eye. In this case it would be necessary to speak, not of the luminous efficiency, but of the luminous efficiency of radiant radiation, the spectral radiation interval in each case being indicated as a function of the radiation detector.

The perception of radiant energy emitted by pyrotechnic flames is frequently hindered by the smokiness of the flame. Smoke is produced when finely dispersed particles of metal oxides emerge from the flame and are cooled. The presence of metal oxides (MgO, Al_2O_3, ZrO_2, etc.) in the flame is necessary, since they are the ones that cause the bright emission of the flame. It follows that the smokiness of an illumination composition may be decreased, but the smoke can not be eliminated completely.

In view of the fact that illumination terms will be used in the treatment below, here are the definitions of the basic illumination units: candle, lumen, lux, and stilb.

1. The principal unit of illumination is an arbitrary one, the new candle (c), equal to 1/60 of the light intensity received from 1 cm^2 of the surface of a blackbody in the direction of the normal at the solidification temperature of platinum (2046.6°K).

2. The unit of luminous flux is the lumen (lm), i.e., the flux emitted by a light source with an intensity of 1 c within a solid angle of 1 steradian. A source having a light intensity of 1 c in any direction emits a luminous flux

$$4\pi \text{ lm} = 12.5 \text{ lm}.$$

3. The unit of illumination used is the lux (lx), which is the illumination produced by a luminous flux of 1 lm uniformly distributed over an area of 1 m^2.

4. The unit of brightness used is the stilb (sb), which is the brightness of a uniformly luminous surface in the direction of the normal, emitting a light of 1 c intensity from 1 cm² in the same direction.

2. THERMAL AND LUMINESCENT RADIATION

Every substance whose temperature is above absolute zero continuously emits radiant energy.

The process of conversion of thermal energy into radiant energy is known as thermal radiation. Thermal (or thermic) radiation is caused by the motion of atoms and molecules of solids and liquids. Raising the temperature of substances increases their energy by increasing the velocity of the motion (vibrational or rotational) of the atoms or molecules. The thermal radiation of solids or liquids obeys the laws of radiation of ABB (Absolute Black Body) (see Sec. 2 in Ch. 8). At high temperatures (500°C or higher), thermal radiation increases, and a luminous radiation visible to the eye appears, i.e., the substances begin to emit light.

The presence of incandescent solid and liquid particles in the flame of an illumination composition is unquestionably necessary, since incandescent vapors and gases (without which there can be no flame at all) themselves emit a comparatively small amount of luminous energy. As already stated, the emission of light by the solid and liquid particles present in a flame obeys the laws of radiation of ABB.

Thus, taking the Stefan-Boltzmann law into consideration, we see that the energy radiated by a flame increases rapidly as its temperature rises.

Table 13.1 shows the temperature dependence of the brightness and luminous efficiency of an absolute blackbody.

The spectral distribution of the radiation energy of ABB (at 3000°K) and of the flame radiation of a typical illumination composition, and the curve of luminous efficiency of the eye are shown in Fig. 13.2.

Illumination compositions whose flame temperature is below 2000°C should not be used at all, since the quantity of luminous energy evolved by their combustion is very slight (the luminous efficiency factor of an absolute blackbody at this temperature is only 0.7%).

Bodies which on heating give a spectral radiation energy distribution curve analogous to the curve of an absolute blackbody at the same temperature but with smaller ordinates are known as gray radiators. Such radiators include carbon and certain black oxides. The total emissivity of graphite at 100-1500°C amounts to 52% of the radiation of an absolute blackbody heated to the same temperature.

Table 13.1 BRIGHTNESS & LUMINOUS EFFICIENCY OF AN ABSOLUTE BLACKBODY AT DIFFERENT TEMPERATURES

Temperature, °K	Brightness, sb	Luminous Efficiency lm/watt	Temperature, °K	Brightness, sb	Luminous Efficiency lm/watt
1600	2	0.2	2800	1552	13.9
2000	44	1.5	3000	2872	19.2
2200	136	3.2	3500	9432	34.7
2300	223	4.4	4000	$2.34 \cdot 10^4$	50.3
2400	350	6.6	5000	$8.41 \cdot 10^4$	74
2600	779	9.4	6000	$1.98 \cdot 10^5$	84

The radiation spectrum of such white oxides as MgO or Al_2O_3, whose radiation is used in illumination compositions, is continuous, as is the spectrum of an absolute blackbody; the radiated energy also increases rapidly with rising temperature. The emissivity of white oxides, both monochromatic and total, is lower than that of an absolute blackbody.

Some data on the emissivity of oxides at high temperatures are given in Table 13.2.

Magnesium and aluminum oxides at high temperature radiate with particular intensity in the shortwave range, where their radiation becomes almost equal to the radiation of an absolute blackbody in some cases.

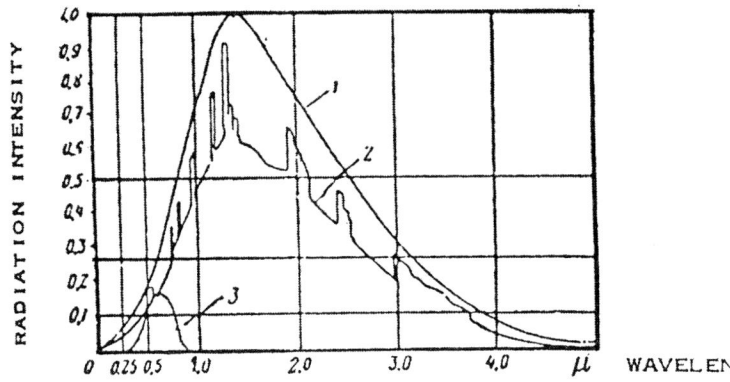

Fig. 13.2. Graph of spectral energy distribution. 1 - ABB at 3000°C, 2 - Pyrotechnic illumination composition, 3 - Curve of luminous efficiency of the human eye.

Table 13.2

Oxide	Formula	Temperature, °K	Emissivity of Oxides in % of Emissivity of ABB (wavelength in mμ)						
			750	700	650	600	550	500	450
Aluminum oxide	Al_2O_3	1600	24	25	31	40	53	81	90
Aluminum oxide	Al_2O_3	1900	31	33	38	50	65	89	99
Magnesium oxide	MgO	1500	--	23	--	35	45	65	--
Magnesium oxide	MgO	1900	37	41	--	53	61	65	83
Thorium dioxide	ThO_2	2000	--	47	--	48	49	49	50
Beryllium oxide	BeO	1700	06	08	--	19	30	45	--
Chromium oxide	Cr_2O_3	1500	--	78	--	78	64	43	--

In addition to thermal radiation, luminescent radiation is observed in the flame of illumination compositions in many cases.

A change in the energetic properties (decrease in stored energy) of electrons in atoms and molecules (excited by an explosion or by the formation of chemical bonds) is considered to be the cause of luminescent radiation.

The absence of presence of luminescent radiation in the flame of pyrotechnic compositions can be established in accordance with the Vavilov-Wiedemann criterion [183], by means of which luminescence is sharply distinguished from other radiation processes. Vavilov considers luminescence to be the excess over the thermal radiation of a body when this excess radiation has a finite duration that substantially exceeds the period of the vibrations of light.

The presence of individual lines and bands in the radiation spectra of flames of illumination compositions must not be regarded as an unconditional proof of the presence of luminescent radiation, since in this case, the lines and bands may also owe their origin to thermal excitation of atoms and molecules.

The phenomena of luminescence in flames of illumination compositions have been little studied thus far, and the influence of many "flame" additives on the luminous indices and character of flame radiation remains largely unexplained.

3. LUMINOUS CHARACTERISTICS OF COMPOSITIONS

In tests of illumination pellets or flares, the following quantities are found experimentally:

1) weight of the burned composition in grams, m;
2) diameter and height of the burned pellet or flare, d and h;
3) average light intensity in candles, I;
4) time of combustion of the pellet or flare in seconds, t;
5) occasionally, the flame shape is determined by being photographed, and the photograph is used to calculate the flame surface in cm^2.

It should be noted that neither the light intensity nor the combustion time can be used as the characteristics for illumination compositions; these quantities indicate only the quality of the products (flares, pellets) made from illumination compositions.

The illumination characteristics which refer directly to illumination compositions are:

1) specific light sum L_0 (c·s/g);
2) luminous efficiency C (lm/W);
3) luminous efficiency factor K (in %);
4) flame brightness B (in sb);
5) linear combustion rate u (in mm/sec), with an indispensable indication of the density d of the pressed composition.

The characteristics of greatest practical importance are the specific light sum and linear combustion rate.

Specific light sum. The amount of luminous energy evolved by the combustion of 1 g of illumination composition is obviously more correctly expressed in units of lumen-second/gram, abbreviated as lms/g, but, considering that the testing of illumination products involves not a light flux but the light intensity in candles, the specific light sum signifies the quantity candles per second obtainable from 1 g of burned composition:

$$L_0 = \frac{I \cdot t}{m} \text{ candle per s/g}$$

Example 1. Calculate the specific light sum for an illumination composition if a pressed pellet of the latter weighing 50 g burns for 9 sec while developing a light intensity equal to 100,000 c;

$$L_0 = \frac{100000 \times 9}{50} = 18,000 \text{ candles per s/g}$$

To change from lumens to candles and vice versa, the following formula is used:

$$F \text{ (lumen)} = 4\pi \cdot 1 \text{ candles}$$

Luminous efficiency. By luminous efficiency C is meant the ratio of the total light sum L to the total energy Q expended in producing a given quantity of light:

$$C = \frac{L}{Q} \text{ lm/Watt}$$

From the obtained experimental data on the light intensity of the composition, the luminous efficiency is calculated from the formula

$$C = \frac{l \cdot t}{m} \cdot \frac{4\pi}{q \cdot 4.18 \cdot 1000} = \frac{l \cdot t \cdot 3.003}{m \cdot q \cdot 1000} \text{ lm sec/Joule or lm/Watt}$$

where q is the amount of heat in kcal evolved by the combustion of 1 g of illumination composition; 4.18 is the coefficient of conversion from small calories to Joules.

Example 2. Calculate the luminous efficiency for an illumination composition if its specific light sum L_0=18,000 candles per s/g, and the heat of combustion q=1.65 kcal/g.

Solution:

$$C = \frac{1800 \cdot 3.003}{1.65 \cdot 1000} = 32.7 \text{ lm/W.}$$

Luminous efficiency factor. The luminous efficiency factor is the ratio of the luminous efficiency of a given light source to the maximum luminous efficiency obtained for a 100% conversion of the entire energy into luminous energy; for monochromatic light with a wavelength of 555 mμ, the latter quantity is equal to 621 lm/W, whence

$$K = \frac{C \cdot 100}{621}$$

For light sources with a complex radiation composition, the calculation of luminous efficiency from such a formula is not possible. In this case, it is necessary to take into consideration the curve of relative luminous efficiency of the human eye (see Fig. 13.2). N. F. Zhirov [5] points out that the luminous efficiency factor for white flames may be calculated with a certain approximation by using the formula

$$K = \frac{C \cdot 100}{248}$$

The mean flame brightness is expressed in stilbs. It is calculated by dividing the light sum in candles by the flame surface S in cm^2, calculated from a photograph:

$$B = \frac{L}{S} \text{ c/cm}^2$$

[Note: This calculation is very approximate, since it is impossible to determine the true flame surface; the flame has no regular geometrical shape, and its dimensions keep changing in the course of combustion.]

The linear combustion rate is expressed in mm/sec. It is calculated by dividing the height h in mm of the burned pellet or flare by the combustion time of the composition t in sec:

$$u = \frac{h}{t} \text{ mm/sec}$$

The linear combustion rate of a composition is not a constant quantity, but depends on the composition density d, and therefore the quantity u is always accompanied by the remark "at some density of the composition".

In order to be able to compare the illumination indices of pyrotechnic and electric light sources, Table 13.3 lists some characteristics of such sources. As is evident from the table, the best luminous efficiency is that of gas lamps producing luminescent radiation.

Table 13.3

No.	Lamp Type	Light Source	Temperature °K	Luminous Efficiency, lm/W	Brightness, sb
1	Incandescent	1000 W gas-filled incandescent lamps	2990	20	1,220
2	Incandescent	250 W " etc.	3350	26	--
3	Park	Carbon arc	4200	12	70,000
4	Park	Carbon arc w/high intensity carbons	5000	35	120,000
5	Gas-filled	Sodium lamps	--	50	Slight
6	Gas-filled	SP-800 superhigh pressure mercury w/water cooling	--	62	91,000

4. FORMULATION OF BINARY MIXTURES

The basis of every illumination composition is the fuel-oxidizer binary mixture.

The total quantity of the other components in an illumination composition seldom exceeds 10-15%, and therefore the qualities of a composition are determined mainly by the selection of the fuel-oxidizer binary mixture. For such a binary mixture of fuel and oxidizer, it is also necessary to establish their optimum proportions.

SELECTION OF FUEL

In discussing various fuels, one must take into account the quantity of heat evolved by their combustion, as well as other physicochemical properties of the fuel itself and of its oxidation products.

As the fuels for an illumination composition, elements are chosen for which the heat of formation of 1 g of oxide Q_2 is at least 2.0 kcal. It is evident from Table 3.1 that such fuels include the following elements:

$$Li, Be, Mg, Al, Ca, Ti, Zr, H, C, B, Si, P.$$

In selecting a fuel, it is necessary to consider that the greater or at least a considerable part of the combustion products of a pyrotechnic composition should melt at a very high temperature and should not vaporize at the temperature of the combustion reaction, so that a significant quantity of solid and liquid particles will be present in the flame. Thus, it is necessary to select for illumination compositions a fuel whose combustion products are high-melting and involatile substances.

Hydrogen, carbon, and phosphorus do not meet this requirement, and can not therefore be used as the main fuels in illumination compositions. It should be noted that the temperature developed by the combustion of phosphorus in air does not exceed 1500°C.

All light sources based on the use of combustion of organic substances have a very low luminous efficiency (not more than 1 lm/W [5]). Elemental carbon burning in an oxygen atmosphere produces a luminous efficiency of only 1.1 lm/W.

Beryllium and zirconium can scarcely be used as fuels in illumination compositions on a practical scale. According to the amount of energy evolved by the composition of these metals, only beryllium can be of interest.

Data on the quantities of heat evolved by the combustion of certain compositions in which the indicated metals are as fuels and barium nitrate is the oxidizer are listed in Table 13.4.

One could expect a large amount of luminous energy to be evolved by the combustion of composition 2 (see Table 13.4) only if the luminous efficiency for zirconium were much higher than for magnesium. However, experience shows that the luminous efficiency of zirconium, even when highly favorable conditions for its combustion are created, is only slightly higher than that for magnesium. Thus, Van Limpt states that the luminous efficiency of zirconium burning in an oxygen atmosphere amounts to 36 lm/W, whereas the luminous efficiency of magnesium burning under the same conditions is 28 lm/W.

Table 13.4

No.	Fuel, %	Barium Nitrate	Heat of Combustion kcal/g
1	Beryllium - 15	85	1.94
2	Zirconium - 46	54	1.12
3	Magnesium - 32	68	1.66

Experiments conducted by I. V. Bystrov on the photometry of illumination compositions with zirconium showed that the specific luminous energy for these compositions is even lower than could have been assumed.

For a composition consisting of 49% Zr, 51% Ba$(NO_3)_2$, C=20 lm/W, which corresponds to a specific light sum L_0=7400 candles per sec/g [2].

Beryllium has not thus far been tested as a fuel in illumination compositions.

Consequently, the remaining six elements Al, Mg, Ca, B, Si, and Ti are best suited for use as fuels in illumination compositions.

Some characteristics of these elements are listed in Table 13.5. As is evident from the table, the largest amount of heat is produced by the combustion of binary mixtures of barium nitrate with <u>magnesium</u> or <u>aluminum</u>. In addition, oxides of these metals possess a satisfactory emissivity. All this taken together forms a sufficient basis for the use mainly of aluminum or magnesium, but also their alloys or mixtures, as fuels in illumination compositions.

Table 13.5

Fuel	Barium Nitrate	Heat of Combustion kcal/g	Luminous Efficiency in lm/W for Burning in Oxygen
Aluminum - 26	74	1.57	26
Magnesium - 32	68	1.65	28
Calcium - 43	57	1.44	--
Boron - 12	88	1.37	--
Silicon - 21	79	1.28	--
Titanium - 31	69	1.18	25

The use of calcium or its alloys in illumination compositions does not appear possible in view of their high corrodibility.

In all probability, the use as fuels of any alloys containing a significant amount of boron or silicon will not yield illumination compositions with satisfactory luminous indices. Such alloys can only cause an inhibition of the combustion of illumination compositions, this being useful in certain cases.

The luminous efficiency of titanium burning in oxygen is found to be somewhat lower than that of magnesium and aluminum tested under the same conditions, (see Table 13.5). The quantity of heat evolved by the combustion of titanium is also less than in the case of magnesium or aluminum. Thus there is no basis for expecting satisfactory luminous indices from illumination compositions prepared by using metallic titanium or its alloys.

CHOICE OF OXIDIZER

In order to obtain an illumination composition of high calorific value, it is necessary to select for it an oxidizer requiring the minimum amount of heat for its decomposition. However, mixtures containing chlorates are usually highly sensitive to mechanical effects. Therefore, despite the fact that compositions prepared with these oxidizers produce the largest amount of heat on burning, they are not used in illumination compositions for practical purposes.

To a lesser extent, the same considerations apply to perchlorate oxidizers. Barium perchlorate, a highly hygroscopic substance, has not been used in pyrotechnics thus far.

In the last few years, indications have appeared in the foreign literature (in the paper of Tavernier [185] and in the book of Izzo [12]) to the effect that sodium perchlorate $NaClO_4$ can be used as the oxidizer in illumination compositions. Indeed, compositions based on this oxidizer can have high luminous indices; however, $NaClO_4$ is very hygroscopic (the hygroscopic point at room temperature is 69-73%), so that the compositions can be prepared without risk of their moistening only in an atmosphere of low humidity. Izzo [12] gives the following formula of an illumination composition for flares: $NaClO_4$ 65%, Al 30%, resin (gum arabic) 5%.

Nitrates are used most frequently as oxidizers for illumination compositions.

Mixtures of nitrates with magnesium or aluminum prepared in stoichiometric proportions evolved 1.5 to 2.0 kcal per 1 g of composition during their combustion.

Among nitrates, those used most frequently in illumination compositions are barium nitrate, a nonhygroscopic salt, and sodium nitrate, a hygroscopic salt, but having the advantage that when it is introduced into the composition, an intense radiation in the yellow portion of the spectrum is generated in the flame.

Mixtures of fuels with sulfates yield less heat during combustion than mixtures with nitrates of the same metals. For this reason, the use of sulfates as well as peroxides as oxidizers in ordinary illumination compositions does not appear desirable. Table 13.6 lists some comparative characteristics of barium compounds: nitrate, sulfate, and peroxide, and also the composition and heat of combustion of mixtures of these compounds with magnesium.

As a general rule in selecting oxidizers for illumination compositions, it may be stated that, other things being equal, salts of metals of low atomic weight should be used as oxidizers since these salts contain a large amount of oxygen, and compositions prepared from them evolve a large amount of heat on burning.

It has already been stated that the use of heavy metal salts (for example, $Pb(NO_3)_2$, etc.) as oxidizers is undesirable. It should also be noted that the use of potassium salts as oxidizers in illumination compositions is also undesirable because of the low luminous indices of such compositions (see Table 13.7).

Table 13.6 THERMOCHEMICAL CHARACTERISTICS OF BINARY MIXTURES

Oxidizer	Amount of Heat Expended in the Evolution of 1g of Oxygen by the Oxidizer, kcal	% Magnesium in the Mixture	Equation of Combustion Reaction	Heat of Combustion, kcal/g
$Ba(NO_3)_2$	-1.2	32	$Ba(NO_3)_2 + 5Mg = BaO + 5MgO + N_2$	1.7
$BaSO_4$	-3.7	23	$BaSO_4 + 4Mg = BaS + 4MgO$	1.2
BaO_2	-0.7	13	$BaO_2 + Mg = BaO + MgO$	0.7

Table 13.7 LUMINOUS INDICES OF BINARY MIXTURS WITH VARIOUS OXIDIZERS
(Pellet diameter 24mm, cardboard shell)

Composition	Density of Mixture, g/cm3	Compaction Coefficiency	Linear Combustion Rate, mm/sec	Specific Light Sum, C·s/g
$Ba(NO_3)_2$ 60 Mg 40	1.94	0.80	8.0	13000
$NaNO_3$ 60 Mg 40	1.71	0.85	11.0	15200
KNO_3 60 Mg 40	1.69	0.87	8.7	10600
NH_4NO_3 60 Mg 40	1.72	0.99	1.8	5600
$Ba(NO_3)_2$ 60 Al 40	2.70	0.90	4.9	15600
$NaNO_3$ 60 Al 40	2.17	0.89	2.6	15300
KNO_3 60 Al 40	2.18	0.94	0.8	1300
NH_4NO_3 60 Al 40	2.02	1.00	1.6	800

On the contrary, the introduction of sodium salts into illumination compositions increases the luminous indices of the compositions. Sodium nitrate is one of the best oxidizers, imparting high illumination indices to the compositions (see Table 13.7).

The values of the specific light sum of compositions in which the oxidizers are barium and strontium salts are close to one another and may be acknowledged as being sufficiently high. Barium salts impart a slightly greenish tinge to the flame of an illumination composition; strontium salts give a pale pink color to the flame.

Strontium nitrate is seldom used as an oxidizer in illumination compositions, since this salt is more hygroscopic than barium nitrate.

DETERMINATION OF THE PROPORTION OF THE OXIDIZER AND FUEL IN A BINARY MIXTURE

The oxidizer and fuel in binary mixtures for illumination compositions either are taken in stoichiometric proportions, or a certain excess of fuel is taken in such an amount that it can burn by combining with atmospheric oxygen. For example, the content of magnesium or AM alloy in binary mixtures for illumination compositions may reach 50-60% in some cases. However, it is not always possible to use binary mixtures with a large content of metallic powders, since the combustion rate increases markedly with rising content of metallic fuel.

The luminous characteristics of certain binary mixtures with magnesium powder are listed in Table 13.8 (the combustion was carried out in cardboard shells 24 mm in diameter).

Table 13.8

	Composition						
No.	$NaNO_3$	Mg	Density g/cm^3	Linear Combustion Rate, m/sec	Specific Light Sum C·s/g	Heat of Combustion, kcal/g	Luminous Efficiency, lm/Watt
1	70	30	1.9	4.7	9800	1.3	22.6
2	60	40	1.7	11.0	15200	2.0	25.0
3	50	50	1.7	14.3	20000	2.6	23.0

The combustion reaction of mixtures whose characteristics are listed in Table 13.8 may be approximately represented by the following equations:

1) $2NaNO_3 + 3Mg = Na_2O + N_2 + 3MgO + O_2$;
2) $2NaNO_3 + 4.7Mg = Na_2O + N_2 + 4.7MgO = 0.15O_2$;
3) $2NaNO_3 + 7Mg + O_2$ (atmospheric oxygen) $= Na_2O + N_2 + 7MgO$.

According to our calculations, the average flame brightness during the combustion of mixture 3 (see Table 13.8) amounts to about 600 sb.

The luminous efficiency of pyrotechnic compositions expressed in lm/W is of the same order as for electric incandescent lamps (cf. Tables 13.8 and 13.3).

Table 13.9 lists data showing the dependence of the illumination indices of binary mixtures on their aluminum content. Twenty-four-mm pellets in cardboard shells were burned.

Table 13.9

	Composition					
No.	$Ba(NO_2)_3$	Al	Density cm^3	Linear Combustion Rate, mm/sec	Luminous Intensity, I, thous. c.	Specific Light Sum C·s/g
1	74	26	2.7	3.0	51	13800
2	64	36	2.7	5.0	86	14500
3	61	39	2.6	5.5	87	13300
4	55	45	2.6	6.6	82	10700
5	49	51	2.6	5.9	60	8600

The combustion reaction of composition 1 (see Table 13.9) may be expressed by the equation

$$3Ba(NO_3)_2 + 10Al = 3BaO + 3N_2 + 5Al_2O_3.$$

which corresponds by calculation to the evolution of 1.6 kcal/g of heat.

Compositions 2-5 contain excess fuel which partly burns by combining with atmospheric oxygen and partly forms the nitride AlN. As is evident from Table 13.9, the combustion rate increases with rising aluminum content of the composition, but only up to a certain limit (45%); better luminous indices are exhibited by compositions moderately overloaded with the fuel (aluminum content, 36-39%).

5. MULTICOMPONENT ILLUMINATION COMPOSITIONS

As was stated previously, the chief indices of an illumination composition are the specific light sum and linear combustion rate.

The actual formulation of an illumination composition is made on the basis of a specified linear combustion rate, care being taken to obtain a specific light sum of at least 20,000-25,000 candles·sec/g. In order to slow down the combustion of a composition, give it strength in pressed form, and increase its chemical stability, various organic substances, i.e., resins, mineral oils, drying oil, paraffin, stearin, etc. are added to the above-described binary oxidizer-metal powder mixtures.

Multicomponent compositions formulated in this manner usually have a much lower linear combustion rate than the corresponding binary mixtures, but also a significantly lower luminous intensity.

The specific light sum of compositions with magnesium powder is decreased to a much lesser degree by the introduction of organic substances than in the case of compositions containing aluminum dust or powder as the fuel.

Ellern [9] gives the following two formulas of illumination compositions for parachute flares (in % by weight):

	I	II
Magnesium	52	48
Sodium nitrate	39	42
Polyester resin	9	8
Polyvinyl chloride	--	2

He indicates that a polymerization catalyst is added to the polyester resin: composition I burns faster than composition II, and produces a greater luminous intensity from 1 cm² of burning surface.

As an example of an illumination composition containing sodium nitrate, the following formula may also be given:

Magnesium	45%
Sodium nitrate	48%
Binders	7%

Pellets of this composition weighing 36-38 g burn for 10 sec and produce a luminous intensity of about 120 thousand candles. Hence, the specific light sum obtained is

$$L_0 \approx 30{,}000 \text{ candles per s/g.}$$

Ellern [9] gives formulas of illumination compositions with barium nitrate oxidizers (with a slight admixture of strontium nitrate):

	III	IV
Magnesium	52	17
Aluminum	--	15
Barium nitrate	38	55
Strontium nitrate	7	5
Flaxseed oil	3	3
Asphaltite	--	5

Composition IV has almost the same combustion rate as composition II, but a lower luminous intensity.

As an example of a ternary composition, we can cite one consisting of 30% magnesium, 66% barium nitrate, and 4% iditol.

Aluminum-containing compositions, when coarsely ground aluminum is used or when organic binders are introduced in large amounts, frequently spark vigorously during their combustion.

The sparking phenomenon is due to the fact that particles of unburned metal and incandescent cinders are ejected from the flame by the gases formed by the combustion of the composition. Careful inspection through blue-colored eyeglasses may reveal that the incandescent metal particles have a high velocity and markedly differ in color from the dark-red particles of the unburned cinder.

The ejection from the flame of unburned metal particles is known as "forced" sparking, in contrast to another type of sparking, cinder sparking.

The decrease in the luminous characteristics of aluminum-containing compositions upon introduction of organic binders is most probably due to the fact that in the presence of forced sparking the combustion of the metal is incomplete, so that the combustion temperature of the composition decreases significantly.

Upon introduction of an appreciable quantity of organic substances, the oxygen balance of illumination compositions usually becomes sharply negative.

Thus it should be assumed that the introduction of organic binders and organic substances of similar properties in amounts exceeding 5-6% into illumination compositions is undesirable in the majority of cases. This should be kept in mind all the more that other ways of slowing down the combustion of compositions have been known for a long time, particularly applicable to illumination compositions, namely: 1) change in the degree of dispersity of metal powders; 2) addition of sulfur, a low-melting and low-activity substance, to aluminum compositions.

During combustion, the oxidation of aluminum in illumination compositions containing sulfur takes place partly as a result of combination with sulfur, aluminum sulfide Al_2S_3 being formed. However, aluminum sulfide is only an intermediate reaction product, since it is oxidized by atmospheric oxygen in the outer zone of the flame. The fusion temperature of Al_2S_3 is 1100°C.

The reaction of aluminum with sulfur

$$2Al + 3S = Al_2S_3 + 140 \text{ kcal}$$

takes place readily at high temperatures, which explains the fact that illumination compositions containing sulfur spark comparatively little during combustion even when coarse-grained aluminum is taken for their preparation.

The introduction of over 10% sulfur into aluminum compositions decreases their luminous indices.

Practice has shown that the addition of sulfur to aluminum illumination compositions is very useful.

A composition from World War I is given below:

Barium nitrate	76%
Aluminum dust	10%
Aluminum powder	8%
Sulfur	4%
Castor oil	2%

For a 10.5 cm diameter of the flare, the indices of this composition are as follows: density 2.3 g/cm^3, u=1.7 mm/sec, L_0=11,600 candles per s/g.

To increase the luminous indices of illumination compositions, a few percent of so-called "flame additives" are often introduced.

Most frequently used for this purpose are nonhygroscopic poorly soluble sodium salts such as sodium fluoride, cryolite, etc.

The increase in the luminous indices of a composition upon introduction of such additives seldom exceeds 15-20%. These additives (introduced into the compositions in the amount of a few percent) have little effect on the combustion rate of illumination compositions.

In order to decrease the formation of dust by compositions containing finely ground components (for example, aluminum dust), various greasing substances are added to them in certain cases. Various oils are most frequently used as such processing additives. The introduction of greasing substances also promotes an increase in the stability of the compositions in storage.

To protect metal powders from corrosion and to slow down the combustion process, such substances as stearic acid or stearates of certain metals are added to illumination compositions.

To obtain Mg and Al powders stable to water and aqueous salt solutions during storage, U.S. Patent 2894864 (1959) proposes the treatment of the powders with a 10% solution of stearic acid in a 1:1 benzene-alcohol solution.

Various explosives (nitro compounds) may also be used in addition to metal nitrates as additional oxidizers in illumination compositions.

Thus, Norwegian patent 99194 (1961) indicates the following formula of an illumination composition:

Alkali metal nitrates	50-60%
Dinitrotoluene	13-25%
Magnesium	20-28%
Cellulose	4-8%

During World War II, the following gypsum illumination compositions were used by the German army [9]:

	V	VI
Mg-Al alloy	41	--
Magnesium	--	40
$NaNO_3$	11	13
$CaSO_4 \cdot 0.5H_2O$	32	40
Water	1	7
$CaCO_3$	5	--

Gypsum with the water it contains is used in this case as both the binder and the oxidizer, partially substituting for sodium nitrate.

An effective illumination composition which burns under water [9] has the following formula:

Magnesium	16%
Aluminum	13%
$BaSO_4$	40%
$Ba(NO_3)_2$	32%

Eight parts of flaxseed oil and one part of MnO_2 are added as the binder. This is an experimental formula dating from World War II.

The following slightly exothermic secondary reactions may take place during the combustion of such a composition in water:

$$MgO + H_2O = Mg(OH)_2 + 9 \text{ kcal},$$

$$BaS + 2H_2O = Ba(OH)_2 + H_2S + 5 \text{ kcal}.$$

6. PRINCIPLES OF FLAME PHOTOMETRY AND RADIOMETRY OF PYROTECHNIC COMPOSITIONS

The radiant energy of pyrotechnic flames is used in practice to act on the human eye, on photographic materials, photoelectric cells, and other electronic devices.

As is evident from Fig. 13.2, the radiation energy of illumination compositions is distributed over the spectrum in such a way that its bulk corresponds to the infrared region (IR) of the spectrum, and the radiation maximum corresponds to wavelengths of 1.2-1.8μ.

Thanks to the development of infrared technology, this portion of radiant energy, which earlier had been useless, proved useful for nighttime photography, for seeing in the dark, etc. [177, 181].

Pyrotechnic radiators have certain advantages over electrical sources of radiant energy: when designed into articles such as illuminating or signaling cartridges, they are of small size, require no electric power, and are easily tossed behind the adversary's lines; a disadvantage is the short duration of their action.

Concerning sources of IR radiation and their uses in military engineering, see the books of Ango [172], Deribere [176], Yu A. Ivanov and B. V. Tyapkin [177], and M. L. Margolin and N. P. Rumyantsev [181].

For the action of radiant energy on the human eye, the already familiar quantities, candle, stilb, and lux (see Sec. 1 of this Chapter) are used for quantitative characterization of a light source and for estimates of the illumination intensity.

The terms radiant intensity, energy brightness, and irradiance are used for the action of radiation on various photoelectronic devices. The interrelationship and dimensions of these quantities are described in courses in optics [178].

Practical flame photometry and radiometry are based on the principle of measurement of illumination of irradiance as a measure of the effect on a suitable detector.

From the illumination or irradiance, the luminous (or radiant) intensity is calculated by using the equation $I = E \times R^2$.

Radiant energy detectors most frequently used for the visible region of the spectrum are photoelectric cells. For IR radiation, in addition to photocells and photoresistances, use is made of thermocouples, bolometers, and opticoacoustic detectors.

The main characteristics of any radiation detector are its spectral and integral sensitivity, and also their stability with time.

Spectral sensitivity is the sensitivity of a detector to radiation of different wavelengths; it is determined by the nature of the substance comprising the light sensitive layer in the instrument, and may change over wide limits from 0.3 to 5μ (Fig. 13.3).

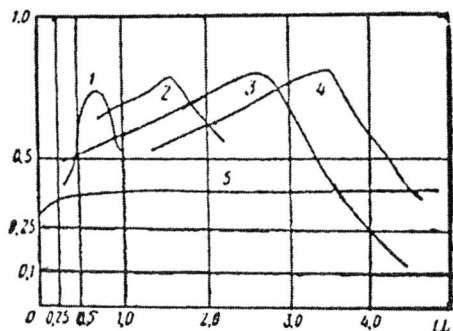

Fig. 13.3. Spectral sensitivity curves of photoelectric cells.
1 - selenium, 2 - sulfur-bismuth, 3 - sulfur-lead, 4 - selenium-lead, 5 - thermocouple.

Such radiant energy detectors as thermocouples, bolometers, and opticoacoustic detectors have no selectivity for radiations in different regions of the spectrum.

In addition, it is frequently necessary in practice to measure the radiant intensity, not over a wide spectral range sensed by a detector, but over a narrow interval corresponding to the sensitivity of the human eye, photographic material, or optical system. This problem is solved by using filters in the form of plates transmitting only the radiation of a definite spectral composition. Colored glasses are used for the visible region of the spectrum, and special glasses, mica, lithium floride, rock salt, sylvite, potassium bromide, etc. are used for IR filters.

Photocells directly convert radiant energy into electrical energy; this phenomenon is known as the photoelectric effect. The magnitude of the photoelectric effect is characterized by two laws:

1. Stoletov's law: the photoelectric current is directly proportional to the radiant flux striking the photocell.

2. Einstein's law: the maximum energy of photoelectrons increases linearly with increasing frequency of the incident light regardless of its intensity. By measuring the photocurrent formed in the photocell, one can determine the illumination (irradiance) produced by the radiation source.

At the present time, the luminous intensity of illumination devices is measured by means of various objective photoelectric instruments (luxmeters). The photoelectric luxmeter is made up of two main parts: a light detector consisting of one or several photocells provided with suitable light filters, and an electric measuring instrument to measure the photocurrents.

Selenium photocells are widely applied in the photometry of pyrotechnic flames [179].

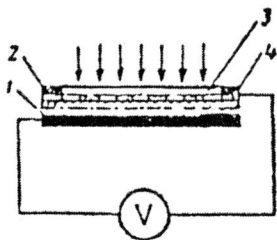

Fig. 13.4. Selenium photocells.
1 - iron plate, 2 - selenium layer, 3 - semitransparent gold layer, 4 - contact metal ring.

A photocell (Fig. 13.4) consists of a round iron plate 1 with a thickness of 1-2 mm and a diameter of 30-60 mm on which is deposited a thin layer 2 of selenium (or other light-sensitive material). A semitransparent gold layer 3 is deposited on the upper selenium layer by cathodic evaporation; metal ring 4, which serves as a conductor, is applied on this layer. Between the selenium and the semitransparent layer there is formed a barrier layer which does not allow the electrons to pass in the direction from the semitransparent layer through selenium toward the iron plate. The barrier layer does not interfere with the opposite movement of electrons (from the iron plate through selenium to the semitransparent layer).

If light coming from the semitransparent layer strikes the photocell, upon reaching the selenium, it liberates electrons which can move only from selenium to the semitransparent layer; at the same time, the latter becomes negatively charged, and the iron plate becomes positively charged.

By connecting ring 4 with the iron plate by means of a wire conductor, we obtain an electric current in the circuit; the intensity of this current remains for a comparatively long time proportional to the amount of light incident on the photocell.

The intensity of the current obtained is determined by two main characteristics: 1) the total or integral sensitivity and 2) the spectral sensitivity of the photocell.

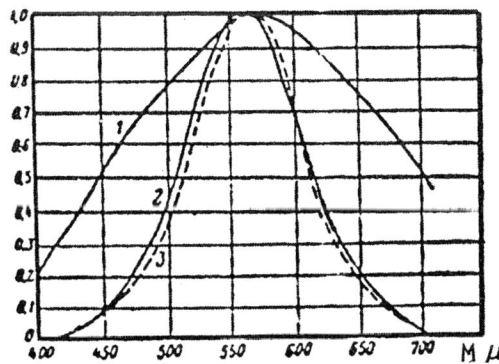

Fig. 13.5. Spectral sensitivity curves of the selenium photocell (1),
human eye (2), and selenium photocell with a compensation filter (3).

The integral sensitivity is the ratio of the current intensity obtained in the photocell circuit to the luminous power producing it. For modern selenium photocells, it amounts to 400-500 µA/lm (microamperes per lumen) for a photocell area ≈10 cm².

As is evident from graph (13.5), the spectral sensitivity of the selenium photocell (Curve 1) is similar to the spectral sensitivity curve of the average human eye (Curve 2).

Yellow-green compensation filters (Curve 3) are used for accurately reducing the sensitivity of the photocell to that of the human eye. Compensation filters are selected individually for each photocell so that a significant attenuation of luminous intensity corresponds to the region of wavelengths of 400-530 and 580-700 mμ. For convenience in handling, the photocells with the light filters are mounted in a frame having an optical sighting device that permits an accurate adjustment of the photocell, contact screws for connecting the wires, and a threaded opening for mounting the photocell on a tripod.

Electric measuring instruments used in photometry (radiometry) are built in the form of a pointer-type galvanometer or loop oscillograph [174]. The main part of the oscillograph is the loop or dipole (Fig. 13.7). As is evident from the figure, loop 1 made of fine wire is placed in the magnetic field of a permanent magnet. A small mirror 2 of square or round shape with an area of about 1 mm² is attached to the loop. When the electric current from the photocell passes through the wire, under the influence of the couple produced by the interaction of the current with the magnetic field, the loop together with the mirror rotates about the vertical axis. If the loop mirror is illuminated, the rotation of the reflected beam will indicate the angle of rotation of the mirror.

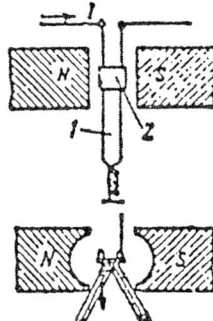

Fig. 13.7. Diagram of loop oscillograph.
 1 - loop,
 2 - mirror.

A diagram of the device for measuring the luminous intensity is shown in Fig. 13.8.

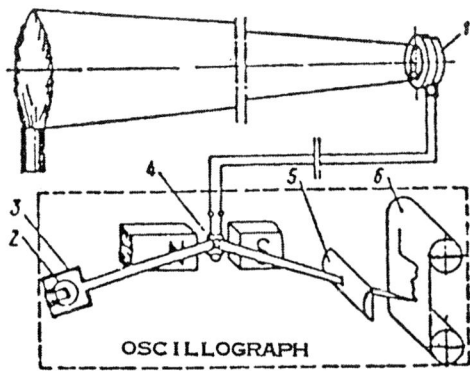

Fig. 13.8. Diagram of measurement of luminous intensity.
1 - photocell, 2 - illuminator, 3 - diaphragm, 4 - loop, 5 - lens, 6 - photographic paper.

The beam of light coming from illuminator 2 passes through diaphragm 3 and strikes loop mirror 4. The reflected beam is focused by lens 5 and srikes moving photographic paper or photographic film 6. If the movement of the photographic material is uniform, and the photocurrent changes as a result of the change in the illumination of the photocell during the combustion of the flare, an oscillogram of the combustion process is recorded on the photographic material. The following requirements are placed on electric measuring instruments used for measuring photocurrents:

a) the current sensitivity should be no less than $1 \cdot 10^{-6}$ A per scale division or per 1 mm of oscillogram width;
b) the minimum internal resistance should not exceed 100 ohm;
c) the temperature coefficient, i.e., the change in readings caused by a temperature variation should be no greater than 0.2% per 1°C;
d) the number of scale divisions for the galvanometer should be no less than 50, and for the oscillograph, the height of the ordinate should be no less than 35 mm;
e) the minimum period of damped oscillations should be 3-5 sec for pointer-type instruments, and no less than 25 cps for loop oscillographs;
f) the presence of a corrector for adjusting the pointer or light spot.

The instruments manufactured industrially which meet these requirements are the POB-14 (N-700) and the M-82 millivolt-milliammeter.

In order to switch from electric quantities to luminous (or energetic) ones, it is necessary to calibrate the detector with the electric measuring instrument by using a standard radiation source. Thus, the calibration consists in defining the scale value of the galvanometer or the value of 1 mm of the height of the oscillogram in lux or W/mm. Photometric lamps are used as standard sources of visible radiation, and various models of an absolute blackbody are used as standards for infrared radiation.

In the first case, the calibration consists in creating various illuminations on the light detector with the aid of a photometric lamp by changing the distance between the luminous element of the photometric lamp and the light-sensitive layer of the light detector. The calibration is made for 6-10 illumination values, the following conditions being observed:

a) the illumination on the surface of the light detector should not exceed 2000 lx;
b) the instrument readings should range from 10 to 90% of the width of the oscillographic paper or galvanometer scale;
c) the photometric lamp should be warmed up, and the light detector should be exposed to the light for 3-5 min.

The calibration of luxmeters and radiometers is carried out on the so-called photometer bench, consisting of a guiding rail on which is mounted the tripod for the lamp, the carriage for mounting the light detector, and a number of auxiliary devices (distance meter, screens, illuminators, etc.), as shown in Fig. 13.9.

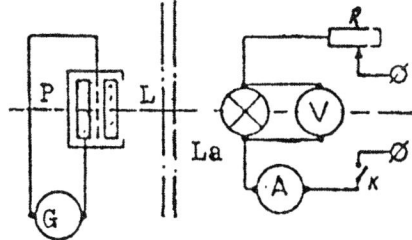

Fig. 13.9. Diagram of calibration of a luxmeter. L - light filter, P - photocell, G - galvanometer, La - lamp, A - ammeter, R - rheostat, V - voltmeter, K - key.

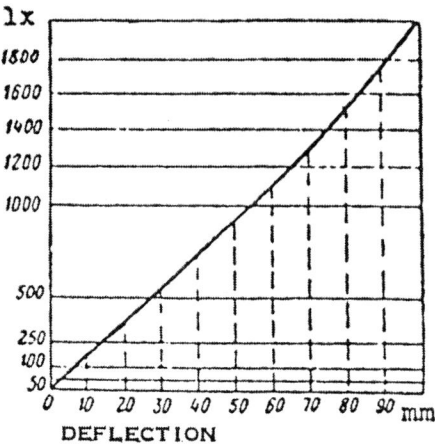

Fig. 13.10. Calibration graph of a luxmeter.

The following literature can be recommended on this subject: Meshkov [181], Bur'yanov [174], Bel'skiy and Bessekerskiy [173], Luk'yanov [179], Margolin and Rumyantsev [181], Shkurin [184], Landsberg [178], Zhirov [5].

7. EFFECT OF COMBUSTION CONDITIONS ON THE RADIANT INTENSITY OF A FLAME AND DEVICES FOR STATIONARY TESTING

The radiant force of a flame is determined by the temperature and spectral composition of the flame radiation, and also by the size and shape of the flame. The first two factors depend chiefly on the formula of the composition, and the size and shape of the flame depend not only on the diameter of the flare, but also on the conditions of combustion of the composition.

Depending on the conditions created during the combustion of the composition, the flame of a pyrotechnic product may assume different sizes and shapes. A particularly marked influence on the radiant intensity is exerted by blowing air on the flame at a high velocity (above 100 m/sec), by burning the compositions under vacuum or under pressure, and by absorption of the radiation by the cooled combustion products of the compositions (smoke).

Less influence is exerted on flame radiation by its reflection from the surrounding objects, by air transparency and atmospheric temperature and humidity. Under certain conditions, these factors may alter the radiant intensity over wide limits, introducing appreciable errors into its measurements.

The determination of the effectiveness of pyrotechnic flame compositions should be carried out under field conditions; however, in addition to the cost of such experiments and large investment of time, it is practically impossible to ensure the stability or reproducibility of the measurement results.

Comparatively simple installations for measuring the luminous intensity and combustion time of pyrotechnic products are photometric chambers. These installations stabilize the influence of external conditions to some extent, provide for the removal of smoke in the course of combustion of the product, and make it possible to conduct the tests in the daytime.

A typical large-model photometric chamber consists of an assembly which includes a hearth for combustion, and corridor (as the photometering base) and an exhaust ventilation system (Fig. 13.11).

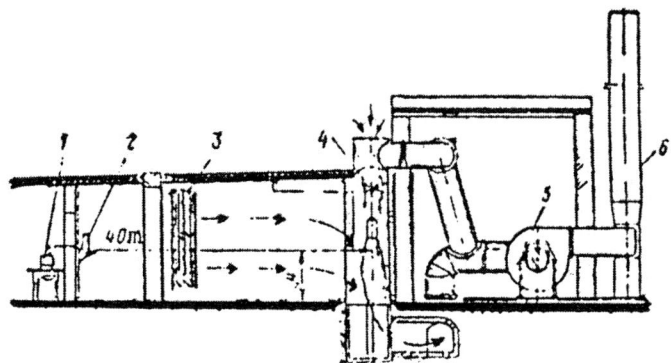

Fig. 13.11. Diagram of photometric chamber.
1 - oscillograph, 2 - light detector, 3 - louvers, 4 - lift, 5 - fan, 6 - smokestack.

The photometric chamber has the form of a corridor of rectangular cross section 2-3 m wide, 2.5-4 m high and 40-50 m long, one end of which has facilities for the equipment and personnel, and the other, the hearth for the combustion and the ventilation system.

The combustion hearth is a vertical shaft of circular or rectangular shape, whose interior is lined with refractory brick and provided with air ducts which supply air and remove the combustion products.

The ventilation system, which removes the combustion products and produces the required air speed around the burning product usually consists of two strong fans (VRS No. 8 or No. 10), air ducts and smokestack.

An important drawback of the measurement of the radiant intensity of a flame consists in the fact that during the operation of the photometric chamber, the walls of the hearth become coated with white cinders that multiply reflect the luminous flux coming from the flame, and on the radiant energy detector produce an additional irradiance which in some cases may exceed the main irradiance. The reflectivity of the hearth walls does not remain constant, and, depending on the operating and weather conditions, ranges from 40 to 150%, causing appreciable errors in the measurement results.

Recently, a method has been adopted for determining the reflectivity (multiple reflection coefficient η) of the hearth in the course of testing: a sensitive photometer is used to measure the illumination produced by an incandescent lamp of at least 5 kW. The lamp is mounted in the combustion hearth in place of the product undergoing combustion. Then the light detector is struck by the direct light of the lamp as well as the light multiply reflected from the hearth walls.

The multiple reflection coefficient is calculated from the formula

$$\eta = \frac{I_{tot} - I_i}{I_{tot}}$$

where I_{tot} is the total luminous intensity (of the lamp + reflected light);
I_i is the luminous intensity of the lamp.

The multiple reflection coefficient obtained is used to introduce a correction into the measured luminous intensity.

Depending on the use of a given pyrotechnic product, the latter is installed in suitable fashion in the airstream in the chamber (Fig. 13.12.)

The arrangement of pyrotechnic products during their photometric analysis is important, since, like any other real light source, a pyrotechnic flame emits different amounts of light in different directions [2, p. 192].

During the combustion of pyrotechnic products, not only the luminous intensity, but also the combustion time and character of the combustion (uniformity, pulsation, etc.) are recorded.

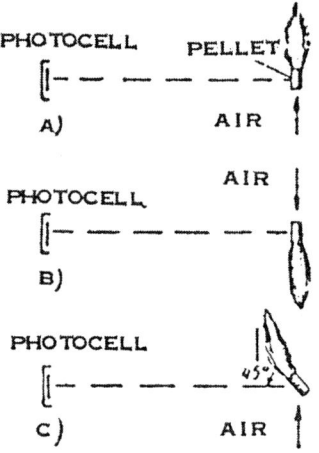

Fig. 13.12. Arrangement of illumination products during photo-metric measurements.
a - combustion of illumination pellets,
b - combustion of parachute illumination products, c - combustion of artillery illumination products.

Before the start of photometric measurements, the distance between the photocell and the product is calculated from the formula

$$L = \sqrt{\frac{I}{a \cdot n}}$$

where L is the distance in m;
I is the assumed luminous intensity in c;
a is the value of a single division (of the galvanometer or oscillograph) in lx;
n is the desired deflection of the galvanometer pointer or oscillogram height.

The following sequence of operations is observed in photometric measurements:

a) the product is placed at the combustion site (as indicated above);
b) the photocell is mounted at the required distance and is aimed at the product in such a way that the intersection of the cross hairs of the optical sight coincides with the flame height;
c) the action of the luxmeter is checked; for this purpose, the photocell is illuminated, and at the same time it is noted whether the pointer of the galvanometer or light spot of the oscillograph is deflected to the correct side;
d) the fan is turned on;
e) the product is ignited;
f) the start of combustion is noted by starting a stopwatch;
g) the readings of the galvanometer are recorded, and the character of the combustion is observed;
h) the end of the combustion is noted by stopping the stopwatch.

The galvanometer readings are taken every 3 to 5 sec; after the same time intervals, the ordinates of the oscillogram are measured (see below).

Using a law according to which the illumination is inversely proportional to the square of the distance from the light source, we obtain the formula

$$E = \frac{I}{R^2}$$

whence

$$I = E \cdot R^2$$

where E is the illumination in lx;
I is the luminous intensity in c;
R is the distance in m.

The average illumination is defined as the average of all the recorded deflections of the galvanometer or all the ordinates of the oscillogram, multiplied by the scale value:

$$E_{av} = \frac{n_1 + n_2 + n_3 + \ldots + n_i}{i} a$$

where n are numerical values of individual galvanometer readings or ordinates of the oscillogram;
i is the total number of readings;
a is the scale value.

Substituting the value of E_{av} into the formula of the illumination law, we obtain the average luminous intensity

$$I_{av} = \frac{n_1 + n_2 + n_3 + \ldots + n_i}{i} aR^2$$

Chapter 14

PHOTOMIXTURES

1. USE OF PHOTOMIXTURES. NIGHTTIME AERIAL PHOTOGRAPHY. PHOTOGRAPHIC MATERIALS

Photomixtures or photoillumination compositions are used to obtain light pulses ("flashes") of very short duration, usually lasting not more than a few tenths of a second.

Photoflashes are used in ordinary photography as an artificial light source at night in cases where no sufficiently powerful electric illumination devices, photographic lamps, etc., are available. During World War II, photomixtures used for loading aerial photoflash bombs found extensive applications in nighttime aerial photography.

Nighttime aerial photography requires the presence of the following elements:

1. A light source, i.e., a bomb loaded with a photomixture (photoflash bombs - FOTAS).
2. A nighttime aerial camera.
3. Photographic material for nighttime aerial photography.

Photomixtures used for ordinary ground photography at night must meet the requirement of minimum flash duration in order to obtain clear pictures of moving objects. This requirement applies even more to photomixtures used in nighttime aerial photography.

This requirement is due to the fact that an airplane carrying an aerial camera moves at a high velocity in relation to the portion of the earth's surface being photographed. During the exposure, all the points of the image of the object being photographed are displaced by a certain distance as a result of the motion of the aerial camera in air. [Note: In addition to the translational motion of the airplane, the displacement of the image is due to longitudinal and transverse vibrations of the airplane.] The displacement of the image during exposure is a characteristic of aerial photography that makes it difficult to obtain a clear picture. When the displacement is appreciable, the picture becomes blurred, and is difficult or altogether impossible to identify.

In ground photography, the camera is stationary during exposure; the object being photographed is either stationary or moves at a low velocity. Therefore, photography using photoflashes usually involves the use of a camera with a steadily open shutter. The duration of the exposure in this case is determined by the duration of the flash itself. Since the duration of the photoflash emission ranges from hundredths to a few tenths of a second, even pictures obtained of moving objects are usually sufficiently clear.

In nighttime aerial photography, cameras without a shutter were also used initially, i.e., the exposure time was equal to the duration of the flash emission.

However, aerial photographs obtained in this manner had a large image displacement because of the relatively long exposure time and were unsuitable for detailed analysis.

In addition, in the case of highly sensitive films, the film frame was fogged by various extraneous ground-based light sources (fires, various lights, shell explosions, etc.) that left longitudinal tracks on the film that considerably complicated the analysis of the film frame.

After numerous studies, these difficulties were overcome by using a nighttime camera with a shutter triggered automatically by a photoelectric relay or by the flash of a photographic bomb.

The intensity of radiation during the explosion of a photoflash bomb changes with time: at the start of the flash, it increases rapidly from zero to some maximum value, then falls comparatively slowly from the maximum back to zero. For this reason, in photographing with an automatic shutter, it is extremely important that the lens of the camera be opened precisely at the instant of highest flash intensity. This is achieved by carefully adjusting the release mechanism and coordinating the graph of the shutter operation with the "light-time" graph of the photographic bomb flash.

Figure 14.1 shows combined graphs of the operation of shutters and change in the radiation intensity of a photographic bomb flash according to Safronov [188]. In practice, only the unshaded area is used. It should be noted also that an excessive reduction of the flash time t is impractical, since this may lead to an incomplete utilization of the luminous energy (reduction of the effective light sum of the photoflash) as a result of the time lag of the shutter-opening mechanism, amounting to about 14 msec.

The nomenclature of photographic materials [186, 187] for nighttime aerial photography as used at the present time is quite varied. The photographic materials are differentiated according to their light sensitivity, contrast, and spectral sensitivity curve.

The light sensitivity is characterized by a quantity that is inversely proportional to the amount of illumination producing a given photographic effect on a given photographic emulsion (the light sensitivity criterion). The light sensitivity criterion chosen is the photographic effect, which can be expressed quantitatively (some optical density, gradient, etc.).

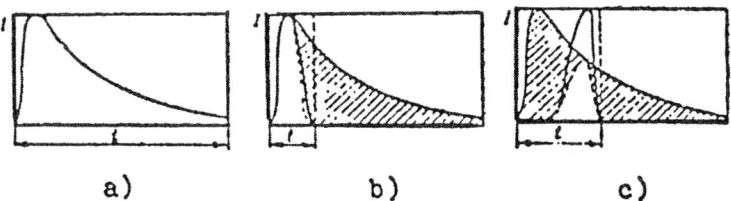

14.1. Combined graphs of the operation of shutters and flash radiation of a photographic bomb.
a - for a shutter of approximate synchronization (aperature open during the entire duration of the flash),
b - for a shutter operating by opening,   c - for a shutter operating by opening and closing.

According to GOST 2817-50, an optical density equal to $D_0 + 0.2$, where D_0 is the fog density, is taken as the light sensitivity criterion, and $H_0 = 1$ lx·sec is taken as the unit of illumination causing this effect under specified development conditions.

Films of high light sensitivity are used in nighttime aerial photographs. This is due to an increase in the altitude of photography in connection with the development of more powerful antiaircraft weapons.

The change in the optical density of a photographic material as a function of the change in the illuminations acting on it under various development conditions is determined by the contrast, which is expressed numerically as the contrast factor γ.

Photographic materials are not equally sensitive to all the rays of the spectrum. According to the spectral sensitivity, modern photographic materials are subdivided into the following groups.

1. Nonsensitized materials have a sensitivity in the blue-violet region of the spectrum with a limit up to 500 mµ, which corresponds to the natural sensitivity of silver halide.

2. Orthochromatic and isoorthochromatic photo materials are also sensitive in the green and yellow regions of the spectrum, sensitization limit up to 580 and 600 mµ, respectively.

3. Isochromatic photo materials sensitive to all visible rays and sensitized up to 650 mµ.

4. Panchromatic photographic material sensitive to rays of the entire visible spectrum up to 680-700 mµ, but having a slight decrease of sensitivity in the yellow-green (500-550 mµ).

5. Infrachromatic photographic materials sensitive to all visible rays and in addition, to the invisible infrared portion of the spectrum.

The spectral composition of the radiation of photomixture flames differs from the composition of natural daylight. The color temperature of the flame radiation of a photomixture approximately corresponds to 2500°K. Therefore, in order to make better use of the luminous energy of a photoflash bomb, it is necessary to use panchromatic materials with maximum sensitivity in the orange-red portion of the spectrum. Since the radiation maximum in the photoflash spectrum lies in the infrared region, the best photographic effect should be expected by using infrachromatic photomaterials materials. However, these materials have a relatively low sensitivity, and therefore are not used in photography.

2. PHOTOFLASH BOMBS

In nighttime aerial photography, the photomixtures are placed in the shells of photoillumination aerial bombs (abbreviated PHOTAB). A photoflash bomb must meet a number of special requirements. The chief requirement is that upon exploding, PHOTAB produce a flash with specified characteristics (luminous intensity, flash duration, spectral composition of the radiation), which must be coordinated with the operation of the shutter of the aerial camera and the general and spectral sensitivity of the photographic film employed. The structural and ballistic qualities of the bomb should provide for stability of the bomb during flight and its explosion at a given most convenient point of the trajectory.

Depending on the flight attitude and velocity, the qualities of the photographic film employed, relative aperature of the camera objective, object being photographed, etc., the position of the explosion point of the photoflash bomb may change. The structure of the bomb should allow for the possibility of controlling the explosion altitude as a function of the change in the above-enumerated conditions.

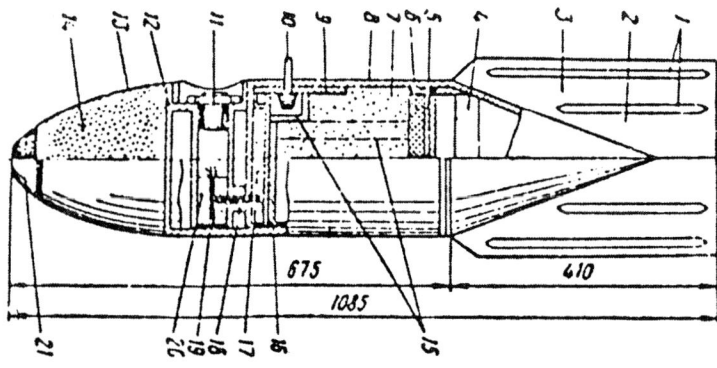

Fig. 14.3. The BLC-50 photoflash bomb.

1 - stiffness creases, 2 - stabilizer, 3 - openings for attaching sirens, 4 - intermediate ring for attaching the stabilizer, 5 - diaphragm of intermediate ring, 6 - felt gasket, 7 - pyrotechnics, 8 - bomb shell, 9 - molded stamped cylinder, 10 - lug, 11 - fuse, 12 - 3-mm diaphragm, 13 - stamped ogive, 14 - concrete, 15 - movable diaphragm with four angles, 16 - cardboard gasket, 17 - ejection charge, (Black Powder), 18 - 7mm diaphragm, 19 - quickmatch powder train, 20 - wooden boss, 21 - cast cap.

The uniformity of the photoflash bomb characteristics should be ensured, and the luminous intensity of the flash and its duration for an individual bomb must not deviate appreciably from the values established for photoflash bombs of this type. In addition, photoflash bombs also must be simple and safe to load and operate, must retain their basic characteristics during extended storage, and must meet a series of other special requirements. The armaments of the air forces of various countries include large numbers of various types of photoflash bombs. Figure 14.3 shows the BLC-50 photoflash bomb, used by the military air force of the former German army. The American patent 2,775,938 of 1 January 1957 [190] describes a photoflash bomb which differs substantially in design and operation from those described above.

Fuse 2 is placed in casing 1 (Fig. 14.4) Detonating cable 3 runs from the fuse to explosive charge 4 located at the center of the bomb. Magnesium powder 5 is spread around explosive charge 4. The patent indicates that finely divided aluminum may be used for the same purpose. A complete combustion of the magnesium (aluminum) powder scattered by the explosive charge can be achieved, according to the indications of the patent, if the powder particles are coated with a thin film of sodium. This is achieved in an inert gas atmosphere at t = 200°C; sodium is used in amounts of 5 to 25% of the weight of the metal powder.

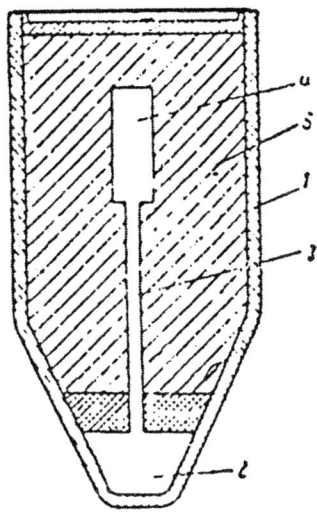

Fig. 14.4. Structural principle of a photoflash bomb (American Patent 2,775,938).

3. METHODS OF DETERMINING THE CHARACTERISTICS OF PHOTOFLASHES

The quantitative characteristics of photoflashes include: I_{max} - maximum luminous intensity of a flash in candles; t_{tot} - duration of the entire flash in seconds; t_{max} - time from the start of the flash to the onset of the maximum.

In many cases, the comparison of different photomixtures also involves the calculation of the light sum L of the flash, represented by the integral

$$\int_0^t I \cdot dt$$

and expressed in second·candles, the specific light sum of the flash L_0, which is the ratio of the total light sum of the flash L to the weight of the burning composition, and the effective light sum L_{ef}, i.e., the light sum received by the camera for a given shutter exposure time, for example, $\tau = 1/25$ sec.

Complete qualitative characteristic of flash radiation is provided by:

a) the curve of spectral distribution of the flash energy;
b) the color temperature of the flash;
c) the change in the color temperature and spectral distribution of energy in the course of the entire time of the flash.

The determination of the quantitative and qualitative characteristics of photoflashes involves considerable difficulties in view of the extremely short duration of the luminous phenomena accompanying the explosion or combustion of photomixture charges. An exact measurement of the quantitative radiation characteristics of photoflashes became possible only after the development of the above-described (Ch. 13, Sec. 6) recording photoelectric luxmeters which make it possible to obtain a recording of the change in emission intensity during a flash.

The total duration of a flash and other time characteristics are calculated by dividing the length of the oscillogram by the linear displacement rate of the photographic material.

In order to calculate the maximum luminous intensity of a flash from the height of the largest ordinate of the oscillogram, the luxmeter is graduated with the aid of a photometric lamp of known luminous intensity. Then the illumination value corresponding to the largest height of the oscillogram is found from the graph. I_{max} is determined by multiplying this illumination by the square of the distance between the photocell and the flash studied.

Spectrographs of various systems are used for determining the qualitative characteristics of the radiation of photoflashes. Pictures of the emission spectra of the photoflashes show an intense continuous spectrum against the background of which lines and bands are clearly visible.

The continuous spectrum is due to the emission of incandescent particles of solid and liquid combustion products of the photomixture. The intensity of the continuous spectrum and the location of the maximum radiation energy depend on the flame temperature of the photoflash and on the nature of the radiating products. The appearance of lines and bands in the spectrum of photoflashes is due to the atomic and molecular emissions of the combustion products which are in the gaseous or vapor state. The system of bands and lines in the spectrum changes substantially with changing composition of the photomixture.

A suitable processing of the spectrograms makes it possible to plot a graph of the energy distribution in the radiation spectrum of a photoflash. Figure 14.6 shows a typical spectral distribution curve of the radiation energy emitted by the flash of a photoflash bomb. It is obvious that the flash radiation maximum lies in the infrared portion of the spectrum.

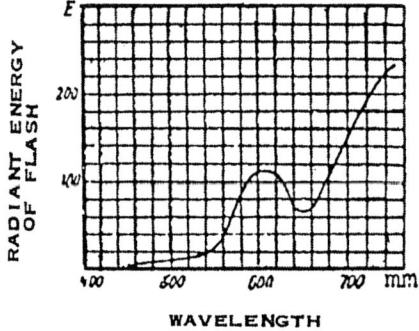

Fig. 14.6. Curve of spectral distribution of the total radiation energy of a flash by a photoflash bomb.

It should be noted that the above spectrograms and the spectral energy distribution curves plotted from them characterize the total action of the entire radiation of a flash from the start of combustion of the photomixture until the emission dies out completely.

It is sometimes important to know the spectral composition of the radiation at different instants of the flash. In that case, the spectrogram is recorded, not on a stationary photographic plate, but on a photographic material moving at a known rate in the focal plane of the spectrograph.

Analysis of time-swept spectrograms showed that the spectral composition of the radiation changes continuously over the entire duration of the flash. The longest radiation is that in the red region of the spectrum, and the shortest, in the violet region. The highest color temperature of the flash, characterized by the emission of all wavelengths of the visible spectrum, takes place at the instant corresponding to the instant when the maximum luminous intensity is attained. It follows that, other things being equal, the radiation intensity and effectiveness of photoflashes are determined mainly by the flame temperature.

4. PHOTOMIXTURES. REQUIREMENTS PLACED ON THEM. FACTORS AFFECTING THE OPTICAL CHARACTERISTICS OF FLASHES, AND PROPERTIES OF PHOTOMIXTURES.

Among the first photoflash compositions were powders or thin strips of magnesium and certain other easily burning metals, which burned in air. However, the combustion of magnesium and other metal powders in air takes place comparatively slowly, and the radiation intensity is relatively low. Some of the best results were obtained by blowing metal powders into the flame of a gas-air burner.

In oxygen, the combustion of magnesium, aluminum, their alloys, zirconium, and some other metals in the form of a thin wire or foil takes place at high rates and with a high luminous efficiency.

Photomixtures are most frequently prepared by mixing powders of magnesium or other metals with various salts containing a sufficient quantity of oxygen.

The following special requirements are imposed on photomixtures:

1. Minimum flash duration.
2. Maximum luminous intensity.
3. Highest conformity of the spectral composition of the flash radiation to the spectral sensitivity of the aerial photographic film.

The flash duration, i.e., the time during which the emission of the flame is observed and the time during which the photomixture burns are not identical and are determined by two entirely different processes. Whereas the combustion time is determined by the rate of the chemical reaction between the fuel and oxidizer, the duration of the flame emission is determined by the occurrence of physical processes of heating and cooling of the combustion products involving the heat produced by the combustion reaction. However, the combustion reaction rate is a decisive factor determining the duration of the flash.

The combustion rate of a photomixture in turn depends on:

1) the formulation of the composition, i.e., the nature of the components used and their proportions in the composition;
2) degree of size reduction of the compounds;
3) density of the photomixture;
4) nature and intensity of the initial impulse;
5) amount of simultaneously burned photomixture and its location in space.

Photomixtures with aluminum powder burn more slowly than mixtures with magnesium.

The combustion rate of photomixtures largely depends on the nature of the oxidizers employed. Mixtures with potassium perchlorate and other similar high activity oxidizers burn faster than mixtures with metal nitrates, oxides, or sulfates. For example, according to the data of Katz [7, p. 170], in standard American photoflash bombs M-46, the maximum luminous intensity (500 million candles) is achieved in the case of a composition with potassium perchlorate approximately 0.013 sec from the start of the flash, and when compositions containing barium nitrate are used, the maximum flash intensity is reached in approximately 0.025 sec.

D. Hart [10] indicates that a photoflash bomb containing 10 kg of composition produces a luminous intensity of 800 million candles for a total combustion time of 0.04 sec.

It follows from the paper of Yutsevich [189] that "a PHOTAB flash at a flight altitude of the order of 8-10 km takes place at a distance of 6-8 km from the airplane". He also reports that at a flight altitude of 10 km and speed of 800 km/hr, the average illumination of the terrain will be of the order of 200 lx. If the explosion height of a PHOTAB above the terrain is assumed to be 5 km, a luminous intensity of 5×10^9 c will be required to create such illumination.

Table 14.1 lists data on the flash duration for certain magnesium photomixtures. As a rule, the shortest flashes are produced by mixtures in which the components are taken in stoichiometric proportions. Mixtures with an excess fuel or excess oxidizer yield longer flashes.

Table 14.1

Mixture of 1g Magnesium Powder	Flash Duration	Mixture of 1g Magnesium Powder	Flash Duration
w/0.75g $KMnO_4$	0.03	w/1.00g $Sr(NO_3)_2$	0.10
w/1.00g KNO_3	0.07	w/1.00g $Th(NO_3)_4$	0.22
w/1.00g $Ba(NO_3)_2$	0.07	w/0.50g $Th(NO_3)_4$	0.24

The finer the size reduction of the components of a photomixture, the faster its combustion. Of great importance is the degree of size reduction of the metal fuel.

Photomixtures are usually employed in the powdered state. In this state they burn at high rates measured in hundreds, sometimes thousands of meters per second. As long as the powdered state is preserved, the powder packing density does not appreciably affect the combustion rate of the composition. However, when pressed under high pressures into briquets, photomixtures burn as do fast-burning illumination compositions, at a constant rate of 10-15 mm/sec.

The combustion rate of a photomixture substantially depends on the character and intensity of the initial impulse, and also on the location of the mixture in the charge. To decrease the total combustion time of a photomixture, not a thermal impulse (electric primer, black powder charge, Bickford fuse, etc.), but an explosive impulse (detonating capsule, charge of some explosive, etc.) is employed.

Also of great importance is the amount of simultaneously burning photomixture. As the amount of simultaneously burning photomixture increases, the combustion rate accelerates. When a photomixture is burned in amounts exceeding several tens of grams, combustion changes into explosion. The charge of a photomixture in the form of a compact mass burns faster than the same charge scattered in the form of a long train.

However, although the combustion rate of a mixture increases when the simultaneously burning photomixture charge is increased, the photoflash duration does not decrease, but increases. This is due to an increase in total combustion time as the amount of mixture increases, and also to an increase in the cooling time of the combustion products of the mixture.

Table 14.2 FLASH DURATION FOR CHARGES OF DIFFERENT WEIGHTS

Amount of Composition, g	Total Flash Duration, sec	Time from Start of Flash to Maximum Emission, sec.
50	0.028	0.011
100	0.040	0.013
500	0.074	0.017
1000	.080	0.026
1400	0.120	0.030

The luminous intensity of a photoflash is determined by the following factors:

1) the heat of combustion of the mixture and the flame temperature dependent on it;
2) the presence in the flame of solid and liquid combustion particles of high emissivity;
3) the chemical composition of the photomixture, which determines factors 1 and 2, and also the spectral composition of the flash radiation;
4) size of the photomixture charge;
5) size of the flash flame;
6) strength of the casing in which the photomixture charge is placed.

To produce intense flashes, the photomixtures should have the maximum thermal effect (over 2 kcal/g). For this reason, metal powders of the highest calorific power are used as fuels in photomixtures: magnesium, zirconium, titanium, magnesium alloys, aluminum. In selecting a metal fuel and oxidizer for photomixtures, one is guided by the same considerations as in selecting metallic fuels for slowly burning illumination compositions.

In some cases, special substances coloring the flame and other additives are added to the main binary oxidizer-fuel mixture to obtain the required combustion time, flame color, etc.

At the present time, a large number of various photoflash compositions have been patented, including those using rare earth elements in pure form and in the form of compounds. Many have not found any wide practical application because of the scarcity of the compounds employed.

The most effective and accessible for general practical use have proven to be mixtures of barium nitrate or strontium nitrate with powders of magnesium and other metals. For example, a panchromatic mixture of 60% strontium nitrate and 40% magnesium powder produces intense radiation in the red portion of the spectrum, due to the emission of strontium oxide, and can therefore be successfully used for nighttime photography with panchromatic photo materials.

Photomixtures with sodium nitrate and nitrite are not used because of the great hygroscopicity of these salts.

It is well known that flashes of highest luminous intensity are produced by photomixtures with a certain overload of the metal. In this case, the excess amount of fuel burns in atmospheric oxygen. The total heat effect is correspondingly increased, and the flame size also increases.

The simultaneous increase in flame temperature and radiating surface of the flame results in an increase of the photoflash intensity during the combustion of a photomixture overloaded with metal.

As the amount of a simultaneously burning photomixture increases, the flash intensity increases. However, this increase in flash luminous intensity is not proportional to the increase in the amount of mixture. It is evident from Table 14.3 that the specific luminous intensity per gram of composition, characterizing the luminous efficiency of the flash, decreases sharply with increasing amounts of burned composition.

Table 14.3

Amount of Photomixture, g	Maximum Luminous Intensity in Millions of Candles	Specific Luminous Intensity in Thousands of Candles per g of Mixture	Area of Projection of flame, m²
50	8.5	170	0.36
100	15.3	153	0.75
200	22.6	113	2.35
500	43.7	88	3.60
1000	50.2	50	6.50
1400	52.9	38	7.30

As is evident from Table 14.3, the flame size increases up to certain limits almost in proportion to the amount of photomixture, and as the photomixture charge increases further, the increase in flame size, like the increase in luminous intensity, gradually slows down, and lags increasingly behind the increase in the amount of simultaneously burning photomixture.

A certain influence on the luminous efficiency of a flash and particularly on its duration is exerted by the strength of the shell containing the composition. The photomixture burns faster in strong metal shells than in cardboard ones.

Of great importance is the sensitivity of photomixtures to mechanical and thermal effects. This property determines the degree of danger of the technological process during the production of the photomixture and during loading of the latter into photoflash bombs and also the possibility of safe use of these bombs.

Most sensitive to impact and friction are mixtures containing chlorates and permanganates as oxidizers.

Tomlinson and Audrieth [122] indicate that chlorates mixed with powdered metals are sensitive to impact and friction. Cases of explosions of such mixtures have been recorded. For this reason, it is recommended that the more stable potassium perchlorate be used whenever possible. These authors also note that mixtures containing permanganates and powdered metals are very sensitive to impact and friction.

In their view, one of the most commonly used photomixtures, consisting of finely ground powders of barium nitrate and aluminum-magnesium alloy, is not only easily ignited, but also extremely sensitive to friction and impact.

Photomixtures become particularly sensitive to impact when they contain even slight amounts of organic substances.

As the standard photomixture (F mixture) used by the U. S. Army, Ellern [9] cites a composition of the formula:

Aluminum powder	40%
Barium nitrate .	30%
Potassium perchlorate	30%

He also cites the luminous characteristics of a PHOTAB loaded with this F mixture (see Table 14.4).

It is evident from Table 14.4 that as the weight of the photomixture charge increases, the luminous intensity increases, but the light sum (sec c) decreases (cf. Table 14.3). It is evident from a comparison of the luminous intensity of two bombs with charge weights of 32-36 kg and 104 kg that the creation of large-sized PHOTAB with a concentrated effect is irrational.

The German "dust bombs" contain no oxidizer, but only aluminum dust or fine aluminum powder surrounding a central explosive charge. However, as was indicated in [9], AM alloy, which is currently used abroad, should be given preference over aluminum powder.

Table 14.4 LUMINOUS CHARACTERISTICS OF PHOTAB OF FOREIGN ARMIES

Type of Equipment	Formulation	Weight kg	Maximum Luminous Intensity, Millions of c	Light Sum, Millions sec-candles	Light Sum, sec-candles per 1g Metal
Photocartridge	F. Mixture	0.20-0.23	120	1.5	16700
Photocartridge	F. Mixture	0.7	400	5.0	15700
PHOTAB	F. Mixture	32-36	3200	76-90	5500
PHOTAB	F. Mixture	104	4500	111	1100
German dust bomb	Aluminum dust only	15	450	63	4200
German dust bomb	Pulverized aluminum only	30	800	--	--

In addition to their main use, for nighttime aerial photography, photocartridges and PHOTAB can be used as simulation devices, namely, for simulating explosions of shells, flashes during firing of guns and atomic explosions.

Chapter 15

TRACER COMPOSITIONS

1. BRIEF INFORMATION ON THE DESIGN OF TRACERS. ILLUMINATION CALCULATIONS

In order to render visible, at night and in the daytime, the flight trajectory of missiles, bullets or other objects moving at high speeds, special devices known as tracers are used [191].

A tracer consists of a metal casing into which a tracer composition has been pressed at a high pressure (from 3 to 8 T/cm^2). [Note: As a rule, small-caliber shells are loaded directly by pressing the tracer composition into the shell casing.]

Tracers are used for observation of the flight of:

1) jet aircraft;
2) jet antitank projectiles;
3) gun projectiles;
4) tracer bullets.

Tracers of the first and second groups are actuated by electric ignition devices supplied by an on-board power supply system.

The firing of shell tracers can be achieved either directly with powder gases (tracers with "radial firing") or by fuses ("Bofors" type). Tracer bullets usually have "radial firing".

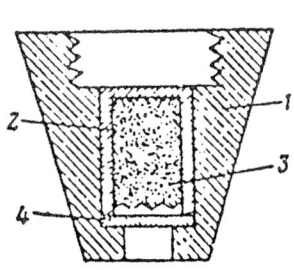

Fig. 15.1. Tracer for MD-5 fuse.
1 - knot, 2 - cup, 3 - tracer composition, 4 - celluloid disc.

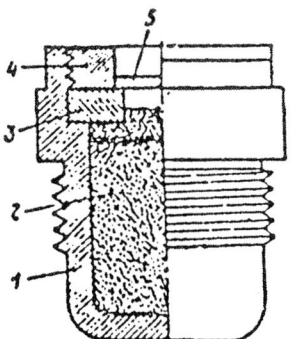

Fig. 15.2. Tracer
1 - tracer housing (steel), 2 - tracer composition, 3 - ring, 4 - knot, 5 - celluloid disc.

Thus, for example, in armor-piercing shells, cup 2 with composition 3 pressed into it is mounted into nut 1, which is then fastened to a fuse (Fig. 15.1).

In some cases, the tracer consists of a massive metallic shell (plug) filled with the tracer composition (Fig. 15.2).

PRINCIPLES OF PYROTECHNICS

In small-caliber fragmentation shells of antiaircraft guns, the heat generated by the tracer composition during its torching is frequently used to explode the equipment for the purpose of self-destruction of the shell in flight (Fig. 15.3). On firing, ignition composition 1 and tracer composition 2, both pressed directly into the bottom part of shell body 3, ignite. The shell has an impact-actuated fuse 4 and explosive charge 5. If the shell misses, the tracer composition burns up, and the heat of its combustion is transferred to pyrotechnic self-destructor 6, consisting of a metal shell with a pressed powder charge, and the heat of combustion of the latter is transferred through metal rod 7 to detonator 8, which explodes explosive charge 5.

Designs of a tracer and an armor-piercing tracer bullet are shown in Figs. 15.4 and 15.5.

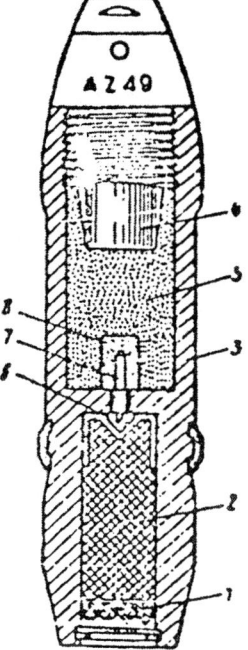
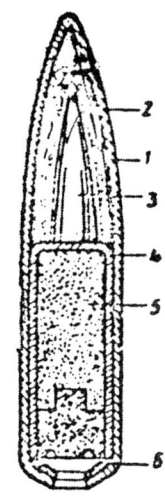
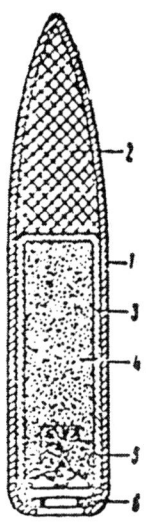

Fig. 15.3. Self-destroying fragmentation projectile, with tracer. 1) igniter; composition; 2) tracer composition; 3) casing; 4) fuse; 5) bursting - charge; 6) self-destruction charge; 7) rod; 8) detonator.

Fig. 15.4. BT-32 Armor-piercing tracer bullet. 1) bullet jacket; 2) lead jacket; 3) steel core; 4) cup; 5) tracer composition; 6) annulet.

Fig. 15.5. T-30 Tracer bullet. 1) bullet jacket; 2) lead core; 3) cup; 4) tracer composition; 5) igniting composition; 6) annulet.

Under daytime firing conditions, the use of smoke compositions in tracer devices is desirable if the trace is not longer than 1-2 km.

The literature of the period of World War I contains indications on the use of tracer bullets filled with yellow phosphorus as the smoke-forming substance. Attempts have also been made to produce smoke trails by using compositions close in formulation to signal smoke compositions.

At the present time, tracers are loaded primarily with light compositions. By increasing the luminous intensity of the flame, the visibility of the light trail is achieved in the daytime as well.

PRINCIPLES OF PYROTECHNICS

CALCULATION OF THE NECESSARY LUMINOUS INTENSITY OF FLAMES OF TRACER COMPOSITIONS

The perception by the eye of a luminous point located at a great distance depends primarily on the total illumination of the terrain and brightness of the background.

The background brightness may change over very wide limits depending on the time of day and weather: the brightness of the night sky at new moon is only 1×10^{-8} sb; the brightness of white clouds at noon is 0.5 to 3 sb.

For a luminous point to be perceived by the eye, it is necessary that the illumination it produces on the surface of the pupil exceed the threshold illumination corresponding to the given brightness of the background. [Note: The threshold illumination is the minimum illumination still perceived by the human eye.] Values of the threshold illumination are listed in Table 15.1.

Table 15.1

Color of Light	Threshold Illumination E_{th}, Lux		Color of Light	Threshold Illumination E_{th}, Lux	
	Night	Day		Night	Day
Red	$0.8 \cdot 10^{-6}$	$0.5 \cdot 10^{-3}$	Yellow	$2.0 \cdot 10^{-6}$	$1.0 \cdot 10^{-3}$
Green	$1.2 \cdot 10^{-6}$	$0.9 \cdot 10^{-3}$	White	$3.0 \cdot 10^{-6}$	$1.5 \cdot 10^{-3}$

As is evident from the table, the sensitivity of the eye decreases considerably in the daytime. For example, in order to observe a white trail at a distance of 2 km in the daytime, it is necessary to provide for a luminous intensity of no less than $I = E_{th} r^2 = 1.5 \times 10^{-3} (2 \times 10^3)^2 = 6000c$.

The absorption of light by the atmosphere makes it necessary to use as tracers light sources with an even higher luminous intensity. When the absorption of light by the atmosphere is considered, the minimum luminous intensity I required from a tracer is expressed by the formula

$$I = \frac{E_{th} \cdot r^2}{a^{\frac{r}{1000}}}$$

where a is the transparency coefficient of a layer of the atmosphere 1 km thick;
r is the distance from the luminous point to the eye in m.

The relation between the transparency coefficient and the state of the atmosphere is represented by the data shown in Table 15.2.

Table 15.2

Code Points	Transparency Coefficient A	Visual Estimate of the State of the Atmosphere
0	--	Very heavy fog
1	--	Heavy fog
2	0.0004	Medium fog
3	0.02	Light fog
4	0.14	Very thick haze
5	0.38	Thick haze
6	0.67	Light haze
7	0.82	Satisfactory visibility
8	0.92	Good visibility
9-10	0.97	Exceptionally good visibility

2. REQUIREMENTS PLACED ON TRACER COMPOSITIONS

First of all, tracer compositions must meet requirements similar to those imposed on illumination compositions, namely:

1) evolve the maximum quantity of luminous energy during combustion;
2) burn at a definite moderate rate measured in units of millimeters per second.

In addition, tracer compositions must also meet certain additional requirements determined by strict and very specific conditions in which the processes of their ignition and combustion take place. This specificity consists in the following:

1. In most cases (in particular, in the operation of the tracer designs illustrated in Fig. 15.1-15.5), tracer compositions are subjected to the direct action of the high pressures that develop during the combustion of powder in the bore of the weapon.
2. At the instant of firing, the entire shell as a whole and hence, the composition contained in the tracer are subjected to very large mechanical stresses.
3. Rotation of the shell in flight (the speed of rotation of small-caliber shells is measured in tens of thousands of revolutions per minute) causes a very large centrifugal force in the composition as well as in its combustion products.
4. During the free flight of the shell, rarefaction is frequently produced behind its bottom part, and eddy currents of air are generated.
5. The small flame size due to the small size of the tracers causes the latter to cool considerably.

Thus, in addition to everything else, tracer compositions should meet the following:

1. In pressed form, have a strength much greater than all the other types of pyrotechnic compositions.
2. Be in a highly compacted state, and be dependably ignited by suitable ignition compositions.
3. Leave the maximum quantity of cinders in the tracer shell after burning up.

Failure to meet the first requirement leads to a partial or complete combustion of the compositions in the bore, resulting in a "short" trail or the complete absence of a trail in flight and premature exhaustion of the weapon.

The second requirement makes it necessary to introduce easily ignitable fuels (for example, magnesium) into the composition.

The latter condition applies only when use is made of ammunition in which the weight of the tracer composition is appreciable in comparison with the total weight of the ammunition. This requirement is due to the necessity of obtaining a correct flight trajectory of the tracer bullet or shell. The more slag will remain in the article after the combustion of the composition, the smaller will be the weight loss of the shell and the smaller its deflection from the normal trajectory. Slag formation is particularly important for tracer bullets in which the weight of the composition is equal to approximately 10% of the total weight of the bullet.

In addition it should be noted that in practice, the latter requirement is not always fulfilled to a sufficient degree. Even though, as a rule, the weight of the solid reaction products amounts to no less than 60-80% of the weight of the composition, part of them are ejected by the flow of gases from the reaction zone into the atmosphere.

Experimental data [2, p. 109] show that the weight of the slag remaining in the tracer amounts to 35-45% of the initial weight of the composition.

In addition to the basic requirements enumerated above, it is desirable that the tracer composition have good pouring properties, since the dispensing of the composition dose necessary for pressing is usually done by volume with the aid of special pouring instruments.

3. FORMULAS OF TRACER COMPOSITIONS. TESTING METHODS.

The principles of formulation and the formulas of tracer compositions (particularly white trace compositions) are very similar to those of illumination compositions.

The main components of any tracer compositions are: 1) oxidizer, 2) fuel, 3) binder.

The action of the hot high-pressure gases of a powder charge on a tracer composition poses a great danger of an accelerated combustion. For this reason, Ellern [9] indicates that perchlorates and particularly chlorates must not be used as oxidizers in tracer compositions.

Fig. 15.6. Arrangement of tracer articles during photometric analysis.

The oxidizer most commonly used in tracer composition is barium nitrate, and in red compositions, strontium nitrate; sodium oxalate is added to yellow trail compositions.

The chief fuel in tracer compositions is almost exclusively magnesium. The use of aluminum as a fuel in tracer compositions involves major difficulties because of the difficult ignitability of such compositions.

Slowing of the combustion of tracer compositions may be achieved by introducing a small amount of organic binder, for example, calcium resinate, shellac, asphaltite, ethylcellulose, etc.

Some formulas of compositions of white, red, and yellow trails and their illumination characteristics are given in Table 15.3.

The binder in compositions 1 and 2 (see Table 15.3) is shellac, and in composition 4, calcium resinate.

In addition to the components indicated in Table 15.3, other compounds used in tracer compositions are strontium peroxide, carbonate, and fluoride, graphite, calcium silicide, etc.

The formulas and manufacturing technology of German tracer compositions used during World War II are described in [192].

The luminous intensity and combustion time of tracer compositions are sometimes determined by burning them in the stationary state (Fig. 15.6). However, in the majority of cases, tests of this kind can not provide an idea of what the luminous intensity and combustion time of a tracer composition should be in flight. In flight, the combustion rate of the compositions usually increases somewhat in comparison with the combustion rate in the stationary state, and the luminous intensity also changes to some extent.

It would be most desirable to carry out the laboratory tests of tracer compositions under conditions as similar as possible to those of their practical application, i.e., to provide for rotation of the tracer cup during the tests, and the tests should be conducted in a wind tunnel or other suitable apparatus with air blowing at a velocity close to the flight velocity of the shell.

PRINCIPLES OF PYROTECHNICS

In addition to these tests, it is very important to determine the density of tracer compositions in the pressed state. Occasionally, the quality of the pressing of tracer compositions is examined under the microscope to detect cracks. In some cases, the combustion rate of tracer compositions at reduced and increased pressures is also determined.

Table 15.3 FORMULAS & CHARACTERISTICS OF COMPOSITIONS OF WHITE (1-5), RED (6-7) & YELLOW (8) TRACERS. COMPOSITIONS 4, 7, & 8 WERE TAKEN FROM REF. 3

Components	Numbers of Compositions & Content in %							
	1	2	3	4	5	6	7	8
Magnesium	35	25	36	35	44	30	--	33
Al/Mg alloy	--	--	--	--	--	--	37	--
Barium nitrate	55	65	49	--	39	--	--	--
Strontium nitrate	--	--	--	32	--	60	56	40
Barium peroxide	--	--	--	31	3	--	--	--
Sodium oxalate	--	--	--	--	8	--	--	17
Polyvinyl chloride	--	--	--	--	--	--	7	--
Resin binder	10	10	15	2	6	10	--	10
Combustion Rate, mm/sec	4.7	3.1	4.0	--	--	3.1	--	--
Spec. Light Sum, c•s/g	5300	4400	6500	--	--	4400	--	--

(Note: The tests were conducted in a casing approx. 10mm dia.)

In estimating the effectiveness of pyrotechnic tracers, it is necessary to consider that:

a) the flame of a tracer is subjected to the action of an incident air flow of very high velocities;
b) the useful radiant flux penetrates through the smoke trail produced by the tracer;
c) in some cases, tracers operate at great altitudes in a rarefied atmosphere.

To reproduce these conditions, special stands are used consisting of an arrangement for mounting the tracer and blowing air on it, a compressor, and a control desk.

Chapter 16

SIGNAL-FLARE COMPOSITIONS

1. SIGNALING SYSTEMS. REQUIREMENTS PLACED ON COMPOSITIONS.

The compositions of signal flares are designed to give signals at night and in the daytime.

The most generally used system of signaling by means of pyrotechnic lighting compositions is three-color signaling involving the use of red, yellow, and green lights. If necessary, the yellow flare may be replaced by a white flare, obtained by burning illumination compositions.

The use of four- or five-color signaling, which includes red, yellow, green, blue, and white (moonlight white) colors can scarcely be acknowledged as completely rational, since the color discrimination in this case becomes insufficiently dependable at large distances.

The night signaling pyrotechnics used at the present time are signal cartridges, screw-type signal grenades, and powerful nighttime signals.

The compositions of signal-flares must meet the requirements common to all types of pyrotechnic compositions (see Chapter 1, Sec. 3) and in addition, the following special requirements:

1. The flame produced by the combustion of signal flare compositions should have a sufficiently characteristic color which permits an easy discrimination of the color of the signals. Quantitative estimates of the flame color show that the color "purity" (see Sec. 2 of this chapter) should be at least 70-75%.

2. The amount of luminous energy evolved by the combustion of the compositions should be maximum; the specific light sum of the compositions should be expressed by a value no smaller than several thousand c·s/g. In order to ensure their detection, the luminous intensity of signal-flares should be no less than several thousand candles under nighttime conditions. For example, in order to detect the green flare signal under nighttime conditions at a distance of 10 km (10^4 m) and a threshold illumination of 1.2×10^{-6} lx and an atmospheric transparency coefficient a = 0.8 (see Tables 15.1 and 15.2 in Ch. 15), the luminous intensity should be no less than

$$3. \quad I = \frac{1.2 \cdot 10^{-6} \cdot (10^4)^2}{0.8^{10}} = 1120 \text{ c.}$$

4. The combustion of the compositions should take place at a definite rate expressed in units of millimeters per second. Signal pellets usually burn at a rate of 3-6 mm/sec.

2. CHARACTER OF THE FLAME RADIATION

An ideal signal-flare composition would be one whose flame radiation would correspond entirely to a single portion of the spectrum. In this case, the flame radiation could be termed monochromatic, and the flame color purity of such a composition would be 100%. Actually, however, the flame of signal-flare compositions has a certain slight radiation in other parts of the spectrum as well.

The ratio of the intensity of monochromatic flame radiation E_λ to the intensity of the entire visible radiation E_{tot} is called the <u>color purity</u> of the flame p and is usually expressed in percent.

Thus, for example, the flame color purity of red flare compositions may be expressed:

$$p = \frac{E_\lambda \cdot 100}{E_{tot}} = \frac{E(620 - 760 m\mu) \cdot 100}{E(400 - 760 m\mu) \cdot}$$

In calculations, E_λ and E_{tot} may be expressed both in units of luminous flux, i.e., lumens, and in candles, which are proportional to lumens.

In some cases, the flame color purity is more conveniently determined with the expression:

$$p = \frac{E_\lambda \cdot 100}{E_\lambda + E_{wh}}$$

where E_{wh} is the intensity of the white color used to dilute the monochromatic rays.

A flame having a low color purity can not have a sharp color when observed with the eye, since the ordinary white color can be obtained by mixing 1/3 of red, 1/3 of green and 1/3 of blue flare fluxes.

Substances which are in the gaseous state and are brought to an excited state in some manner produce a discontinuous radiation spectrum, i.e., a line or band spectrum. Consequently, in contrast to the flame of illumination compositions, the radiation of the gaseous phase should predominate in the flame of signal-flare compositions.

A line radiation spectrum is produced only by monatomic vapors and gases, and it is therefore called an atomic spectrum. The lines in such a spectrum are arranged in regular characteristic groups called series.

In order to obtain a colored flame on the basis of atomic radiation, use is made of elements having bright spectral lines, only in a single portion of the spectrum. Foremost among such elements are sodium, lithium, thallium, and indium. They have characteristic bright lines: sodium - yellow 589 and 589.6 mμ, lithium - red 671 mμ and orange 610 mμ, thallium - green 535 mμ and indium - blue 451 mμ. The salts of these metals at high temperatures (1000°C and above) dissociate easily, so that the line spectrum of the metal vapor appears in the flame. In practice, the radiation of lithium, thallium and indium is not used in signal-flares because of the scarcity of the compounds of these metals.

The atomic radiation of sodium vapor is extensively used in yellow flare compositions.

A band spectrum consists of a single band or a series of bands with different widths corresponding to different portions of the spectrum. However, by using a powerful spectroscope producing a high dispersion, it is possible to observe that these bands consist of a large number of very close lines. A band spectrum is obtained from the radiation of molecules of substances in the gaseous state, and is therefore termed a molecular spectrum.

3. PRINCIPLES OF FORMULATION OF COMPOSITIONS AND BASIC REQUIREMENTS PLACED ON THEIR COMPONENTS

1. The amount of energy evolved by the combustion of a composition should be sufficient to excite or ionize the atoms or molecules present in the flame in the vapor state. In practice, a sufficiently powerful color radiation is obtained from the combustion of compositions whose head of combustion is no less than 0.8 kcal/g.

2. When molecular radiation is used, the flame temperature of the compositions must not exceed the limits at which dissociation of the radiator molecules is already observed. Thus, for example, the flame temperature of green light compositions must not exceed 2000°C.

3. The flame should contain only a small amount of solid reaction products. When metal (magnesium or aluminum) powders whose combustion forms involatile oxides are introduced in large amounts (above 15-20%) into signal light compositions, the flame brightness increases markedly, but the purity of this color is greatly decreased.

4. The components of the composition should be chosen in such a way that the undesirable radiation of the gaseous reaction products in other portions of the spectrum be minimal. This is achieved by a suitable choice of the oxidizer and other components.

5. The elements or compounds formed by the combustion of the composition and imparting color to the flame should be volatile at 1000-1200°C and pass completely to the gaseous state. For this reason, a colored flame is frequently obtained by using the radiation of the highly volatile alkaline earth chlorides.

The compositions of signal-flares usually contain the following components:

1) oxidizers;
2) organic fuels - binders;
3) salts coloring the flame.

In addition, they frequently contain:

1) substances that improve the flame color, i.e., chlorinated organic compounds;
2) inorganic fuels, i.e., magnesium or aluminum powders.

The oxidizers used in signal-flare compositions should be salts of a metal whose compounds provide the required flame color. For example, these are sodium salts in yellow flare compositions, strontium salts in red flare compositions, etc. If this is not possible for any reason (for example, hygroscopicity of the salt), and the oxidizer used is the salt of a different metal, the decomposition products of such an oxidizer should produce the minimum radiation in the flame so as not to appreciably alter the flame color. Potassium salts are frequently used as such oxidizers.

The organic binders used in signal-flare compositions must not alter the color of their flame during combustion. Ideal binders would be substances which on burning would produce an almost colorless flame, similar to the one produced by sulfur, hydrogen, or carbon monoxide burning in air. According to the data of Tideman and Sciborski [7, p. 197], when organic compounds burn in air, their flame will be nearly colorless if they contain at least 50% oxygen.

The yellow color of flames produced by the combustion in air of organic substances containing little oxygen is explained by the presence of unburned solid carbon particles in the flame. Since most signal-flare compositions are formulated with a negative oxygen balance, the conditions produced by their combustion can not be considered conducive to a complete combustion of the binders.

Hence, the organic substances used for the preparation of signal light compositions should contain the maximum possible amount of oxygen; this requirement applies particularly to green flare compositions. The ultimate composition of the organic substances most frequently used in signal-flare compositions is listed in Table 4.2 (Ch. 4). An inspection of this table shows that one of the best binders from this point of view is iditol, and some of the worst are colophony and its derivatives.

Starch and sugar are not binders, but because of their high oxygen content, their partial substitution for binders is entirely possible and sometimes even advantageous.

Metaldehyde $(C_2H_4O)_4$ used to be strongly recommended as a fuel in signal-flare compositions. It is the product of polymerization of acetaldehyde at low temperatures.

Patent disclosures indicate that it is useful to introduce a certain amount of urotropin into colored flare compositions. Urotropin $C_6H_{12}N_4$ (see Ch. 3) contains 51.4% carbon, 8.6% hydrogen, and 40.0% nitrogen; it has a negative heat of formation, and burns in air with a colorless, slightly bluish flame. Urotropin does not possess binding properties.

Polyvinyl chloride is a chlorinated organic compound most frequently used during World War II in red and green signal-flare compositions; its formula is $(CH_2=CHCl)_n$. This substance exists as a white powder or grains with a softening temperature of about 80°C, a density of 1.4 at room temperature, and a chlorine content of 56%. Polyvinyl chloride is soluble in dichloroethane and other chlorinated aliphatic hydrocarbons, and is plasticized by tricresyl phosphate or dibutyl phthalate. On heating to approximately 160°C, polyvinyl chloride begins to decompose with the evolution of hydrogen chloride [86].

4. YELLOW FLARE COMPOSITIONS

Only the atomic radiation of sodium is used to produce a yellow flame in pyrotechnic practice. The sodium salts entering into the compositions of yellow signal-flares should dissociate readily at high temperatures, have the highest possible sodium content, and be as nonhygroscopic as possible. The intensity of the sodium line in the flame is proportional to the amount of sodium introduced into the flame. In addition to the yellow sodium D line (589-590 mµ), other sodium lines (616, 568, 509, and 498 mµ) may also appear at high temperatures, but their intensity is comparatively low and they are of no practical importance.

The largest amount (up to 60-70%) of sodium salts can be introduced into the composition if the sodium salt is used as the oxidizer. Despite its hygroscopicity, sodium nitrate was used in yellow flare compositions during World War II (the physiocochemical properties of $NaNO_3$ are described in Ch. 2). Among other oxidizers in yellow flare compositions, potassium salts are used almost exclusively, since they produce a comparatively weak radiation in the flame.

The potassium compounds formed in the flame emit the atomic radiation of potassium, but the potassium lines are of low intensity with comparison with the sodium D line, and therefore, the presence of potassium vapor in the flame leave the flame color purity, equal to 80-85%, almost unaffected. Among potassium salts as oxidizers, use is made of potassium nitrate KNO_3 and also potassium chlorate and potassium perchlorate. In this case, sodium oxalate, sodium fluoride, cryolite and sodium fluosilicate are used more frequently than other sodium salts. Some physicochemical properties of these salts are listed in Table 16.1. Because of the slight solubility in water, these salts are slightly hygroscopic or even completely nonhygroscopic.

Table 16.1 PHYSICOCHEMICAL PROPERTIES OF SOME SODIUM SALTS

Substance	Molecular Weight	Sodium Content %	Density	Solubility g/100g solution. @20° C.	Remarks
Sodium oxalate	134	34	2.3	3.7	Decomposition w/formation of Na_2CO_2 begins at 200°C.
Sodium fluoride	42	55	2.7	4.1	MP 292°C BP 1695°C
Cryolite	210	33	2.9	Sparingly sol.	MP 920°C
Sodium fluosilicate	188	24	2.7	0.65	Dissociates at 1000°C or almost completely into $2NaF+SiF_4$

Old formulations of yellow flare compositions usually contain potassium chlorate as the oxidizer and resin as the fuel. One can cite, for example, a composition consisting of the following components: 60% $KClO_3$, 25% $Na_2C_2O_4$, 15% iditol. Its combustion products are KCl, Na_2CO_3, H_2O, and CO. Such compositions evolve little heat (less than 1.0 kcal/g) on burning, and their flame has a low luminous intensity.

Yellow flare compositions having a great flame brightness contain magnesium, and as the oxidizer, potassium or sodium nitrate. As an example, we will cite the following formula:

Potassium nitrate	37%
Sodium oxalate	30%
Magnesium	30%
Resin	3%

On combustion, such a composition evolves a significant amount of heat (>1.0 kcal/g), and its specific light sum amounts to about 4000 c s/g.

The following composition was used during World War II in Germany:

Sodium nitrate	56%
Magnesium	17%
Polyvinyl chloride	27%

The luminous intensity of yellow flare pellets 22 mm in diameter prepared from this composition turns out to be 11000 candles.

Typical of American signal-flare compositions is the use of potassium perchlorate as the oxidizer. An example [9] is a composition consisting of 19% Mg, 50% $KClO_4$, 15% $Na_2C_2O_4$, 7% C_6Cl_6, and 9% gilsonite (organic substance); a variant of this composition is one reproducing a flame of amber color: 25% Mg, 50% $KClO_4$, 13% $Na_2C_2O_4$, 2% C_6Cl_6, 10% sodium resinate.

The use of the chlorinated organic compounds polyvinyl chloride or hexachlorobenzene in yellow flare compositions apparently is not mandatory, since yellow flare has an atomic type of emission.

5. RED FLARE COMPOSITIONS

A red flame is obtained in practice exclusively by introducing strontium compounds into the composition. The emission of atomic strontium can not be used, since its radiation corresponds to the short wavelength portion of the spectrum (the 461 mμ) line.

Strontium oxide gives a broad, diffuse band in the orange-red portion of the spectrum with a radiation maximum of about 606 mμ. An identical spectrum is given by strontium nitrate.

The flame associated with the radiation of strontium oxide actually has a pink, not a red color, since as a result of the high volatilization temperature of strontium oxide (above 2500°C), it is difficult to produce a significant concentration of its vapors in the flame.

At high temperatures, strontium chloride dissociates, forming strontium monochloride and splitting off free chlorine:

$$2SrCl_2 = 2SrCl + Cl_2.$$

Strontium monochloride SrCl can be obtained by heating a mixture of $SrCl_2$ and Sr to 1000°C in an argon atmosphere.

Strontium chloride $SrCl_2$ has a melting point of 870°C; it has a significant vapor pressure above this temperature. Its boiling point, calculated by extrapolation, is 1250°C. On heating in an oxygen atmosphere, it gradually changes into SrO.

In a flame, strontium chlorate gives an emission spectrum identical to strontium chloride.

In the visible region of the spectrum, strontium fluoride has two band systems: from 678 to 628 mμ in the red-orange region and from 586 to 562 mμ in the yellow-green region. This eliminates the possibility of using strontium fluoride in red flare compositions.

In practice, red flare compositions are formulated only on the basis of the emission of strontium oxide or strontium monochloride, the emission of the latter being much more intense, and in addition, closer to the extreme red portion of the spectrum. This accounts for the effort to introduce chlorine into all the formulations of red flare compositions.

Strontium chlorate should not be used because of its hygroscopicity and the high sensitivity of compositions containing it to mechanical impulses.

Strontium chloride is also seldom used because of its hygroscopicity. The hygroscopic point of $SrCl_2 \cdot 6H_2O$ is 65.6%.

Chlorine is introduced into a composition most often as potassium chlorate or perchlorate or in chlorinated organic compounds.

Strontium carbonate, oxalate, nitrate, and more seldom sulfate are added as flame-coloring salts.

Table 16.2 lists the physicochemical properties of these salts, except for strontium nitrate, whose properties have already been described in Chapter 2. All the strontium salts listed in the table are very sparingly soluble in water.

Table 16.2 PHYSICOCHEMICAL PROPERTIES OF SOME STRONTIUM SALTS

Substance	Molecular Weight	Strontium Content %	Density	Remarks
Strontium carbonate	148	70	3.6	Dissociates completely into SrO & CO_2 at 1300°C
Strontium oxalate	196	53	--	Completely loses water at 150°C; decomposition of oxalate begins 200°C
Strontium sulfate	184	56	3.9	Decomposition begins at 1130°C; melting point 1605°C

The formula of certain red flare compositions are given in Table 16.3.

The flame color purity of compositions 1 & 2 ranges from 80 to 50%; the resin they contain is iditol. Composition 3 was taken from [9].

Pellets 22 mm in diameter prepared from composition 4 yield a luminous intensity of about 10000 c in combustion.

The hardening (polymerization) of the unsaturated hydrocarbon monostyrene present in composition 5 can be carried out by introducing the catalyst tin tetrachloride into it.

Red flare compositions are conveniently formulated with a negative oxygen balance, since the presence in the flame of a reducing atmosphere preventing the oxidation of SrCl to SrO promotes an increase in flame color purity. The presence of excess free chlorine in the flame shifts the equilibrium:

$$2SrCl + O_2 = 2SrO + Cl_2$$

to the left, and thus improves the flame color. Chlorinated organic compounds (for example, polyvinyl chloride, hexachlorobenzene, etc.) can be introduced into red flare compositions. In this case, chlorates may be completely replaced by other oxidizers, thereby reducing the sensitivity of the compositions to mechanical effects.

Table 16.3 FORMULAS FOR RED FLARE COMPOSITIONS

Component	Numbers of Compositions & % of Components					
	1	2	3	4	5	6
Potassium chlorate	60	60	--	--	--	--
Strontium nitrate	--	--	30	57	52	55
Strontium carbonate	25	--	--	--	--	--
Strontium oxlate	--	25	--	--	--	--
Magnesium	--	--	40	23	20	30
Resin	15	15	5	--	--	--
Hexachlorobenzene	--	--	5	--	--	--
Polyvinyl chloride	--	--	--	20	15	15
Monostyrene	--	--	--	--	13	--
Potassium perchlorate	--	--	20	--	--	--

Chlorinated organic compounds should contain the maximum quantity of chlorine (not less than 50%) and be nonhygroscopic and involatile.

With regard to calcium compounds, it should be noted that they impart an orange-red coloration of uncharacteristic color to flames, and can not be used in red flare compositions.

6. GREEN FLARE COMPOSITIONS

Practically the only way in which a green flare is obtained in pyrotechnics is by means of barium compounds. The emission of atomic barium can not be used, since atomic barium gives a whole series of lines in different portions of the spectrum (668, 611, 597, 554, 455, 429 mµ, etc.); the strongest are the yellow-green 554 mµ line and the blue 455 mµ line.

Barium oxide, a high-melting and involatile compound, gives broad diffuse bands in the yellow and green portions of the spectrum. Upon introduction of BaO into a flame, the latter acquires a dull yellow-green color. Barium nitrate produces an identical spectrum in a flame.

Barium chloride dissociates in a flame, forming barium monochloride and splitting off free chlorine:

$$2BaCl_2 = 2BaCl + Cl_2.$$

The emission spectrum of BaCl consists of numerous bands in the green portion.

The melting point of barium chloride $BaCl_2$ is 960°C; its boiling point, calculated by extrapolation, is 1835°C.

The following reaction of decomposition of barium monochloride with the formation of the corresponding amounts of barium oxide may take place in an oxidizing flame:

$$2BaCl + O_2 = 2BaO + Cl_2.$$

The presence of this reaction always leads to a substantial attenuation of green color in the flame.

Barium chlorate in a flame yields an emission spectrum identical to that of barium chloride.

On heating, barium fluoride dissociates with the detachment of fluorine, forming barium monofluoride BaF. The latter has a series of emission bands in the green (from 494 to 505 mμ) and red portion of the spectrum; this of course eliminates the possibility of using barium fluoride in green flare compositions.

The production of an adequate, pure green flame can be achieved in practice only by using the emission of barium monochloride. Hence, compounds containing chlorine must of necessity be introduced into green flare compositions.

Among chlorine-containing substances, green flare compositions include: 1) barium or potassium chlorates, or potassium perchlorate, 2) chlorinated organic compounds.

The use of barium perchlorate is not possible because of its extreme hygroscopicity.

The combustion reaction of the composition barium chlorate 89%, iditol 11%, may be expressed by the equation:

$$5Ba(ClO_3)_2 \cdot H_2O + C_{13}H_{12}O_2 = 5BaCl_2 + 13\ CO_2 + 11H_2O.$$

The flame color purity of the composition is 70-80%; in view of its high content of barium chlorate, this composition has a high sensitivity and marked explosive properties, and is not currently in use.

Compositions with potassium chlorate most frequently contain barium nitrate or carbonate as the flame-coloring salt. An example is a composition containing 27% potassium chlorate, 53% barium nitrate, and 20% shellac. The sensitivity of such compositions is comparatively low, and the flame color is only yellow-green.

The replacement of resin by sulfur in green flare compositions with potassium chlorate markedly improves the flame color, but the compositions become very sensitive, and, if the sulfur is not sufficiently pure, acquire a tendency to ignite spontaneously.

The introduction of chlorinated organic compounds into green flare compositions produces a high chlorine concentration in the flame and thus improves the flame color. As an example of compositions containing no chlorates and therefore safe in handling, we can cite compositions consisting of the following components:

1.
   - Barium nitrate 40%
   - Magnesium 28%
   - Hexachlorobenzene 30%
   - Linseed oil 2%

2.
   - Barium nitrate 59%
   - Magnesium 19%
   - Polyvinyl chloride 22%

The luminous intensity produced by the combustion of a pellet 22 mm in diameter prepared from composition 2 turns out to be 3500 c.

American green flare compositions frequently contain potassium perchlorate, for example [9]: Mg 26%, $Ba(NO_3)_2$ 45%, $KClO_4$ 16%, C_6Cl_6 7%, CuO 2%, oil (vegetable) 2%, gilsonite 2%.

A description of the formulas and technology of German signal-flare compositions used during World War II is given in [195].

Copper salts color the flame of a burner an intense green. However, no one has thus far been able to obtain a green flare composition of satisfactory quality on the basis of copper compounds. For military pyrotechnics, this problem is of no particular interest in view of the scarcity and hygroscopicity of most copper salts. The presence of boron compounds in the compositions imparts a green color to the flame, but the flame color purity obtained is inadequate.

7. BLUE AND WHITE FLARE COMPOSITIONS

Adequate blue flare compositions whose combustion produces a flame of sufficient brightness and of sharply defined blue color are as yet unknown. Blue flames are obtained mostly on the basis of cuprous chloride CuCl.

The presence of copper compounds in a flame imparts a green or blue color to it. The flame color in this case depends on the copper compound taken, temperature of the flame, and its reducing power. The blue emission of cuprous chloride can be obtained only in the reducing zone of the flame and at temperatures not in excess of 1000-1200°C.

Chlorine-containing compounds must be present in blue flare compositions. A typical blue flare composition has the formula

Potassium chlorate	63%
Mineral blue $2CuCO_3 \cdot Cu(OH)_2$	19%
Sulfur	20%

In the absence of sulfur, cuprous chloride is not formed in the flame. In this case, sulfur reacts with potassium chlorate to form free chlorine:

$$KClO_3 + S \rightarrow K_2SO_4 + SO_2 + Cl_2.$$

The combustion reaction of the composition may be approximately represented as follows:

$$KClO_3 + 2CuCO_3 \cdot Cu(OH)_2 + S \rightarrow CuCl_2 + KCl + K_2SO_4 + CO_2 + H_2O + SO_2.$$

The blue flare composition is sensitive to mechanical effects and is relatively unstable chemically. In addition to mineral blue, other copper salts such as malachite $CuCO_3 \cdot Cu(OH)_2$, cuprous sulfide, copper thiocyanate, as well as metallic copper can be used in blue flare compositions. The use of chlorinated organic compounds in blue flare compositions is also possible. However, the presence of sulfur in such compositions is not obligatory.

The flame color purity of the best compositions of blue (more accurately, sky-blue) flare does not usually exceed 25-30%.

The German white flare composition used during World War II, was prepared from the following formula:

Barium nitrate	56%
Potassium nitrate	11%
Barium fluoride	6%
Aluminum	19%
Sulfur	8%

Barium nitrate imparts a greenish tinge, and potassium nitrate, a pinkish one to the flame; when they are present together in the composition, the flame obtained is white and not very bright.

8. TESTING METHODS

Special tests of signal-flare compositions consist in determining the luminous intensity and color of their flames. The luminous intensity of the composition of signal-flares is determined with a photoelectric luxmeter by using the same procedure as for determining the luminous intensity of illumination compositions.

The determination of the flame chromaticity amounts to establishing the color tone and flame color purity. Differences expressed by the definitions "red", "green", "yellow", etc., are differences in hue. When the hue of a flame is said to be $\lambda = 520$ mμ, and the color purity of the flame p = 40%, this should be taken to mean that the color sensation of the human eye due to the perception of such a flame will be the same as the color sensation due to a luminous flux obtained by mixing 40% of a monochromatic radiation with wavelength $\lambda = 520$ mμ and 60% of the radiation of a white flare source.

The first attempt at a quantitative determination of the chromaticity of flames of pyrotechnic composition was made by the Russian scientist K. I. Konstantinov (1846). Essentially, the method he proposed, based on the use of several colored filters, is still being used.

At the present time, color, including the color of colored flames, is measured according to an international system (XYZ) [193], according to which the quality of a color is determined by its three-color coefficients X, Y, Z, i.e., by the proportion in which the three fundamental colors X (red), Y (green), Z (blue) must be mixed in order that the quality of the mixture be the same as the quality of a specified color. Then, by using a special diagram, the three-color coefficients XYZ can be converted to the color hue which determines the wavelength of the monochromatic radiation that must be added to white radiation in order to reproduce the color in question.

The same XYZ graph is used to determine a second quantity characterizing the chromaticity of radiation, namely, color purity.

In the 1930's, the coefficients X, Y, Z for colored flames were determined by means of the GOI (State Institute of Optics) three-color colorimeter of L. I. Demkina's system, whose operation consisted in the visual matching of two fields of view equal in hue and brightness. The short duration of the combustion of signal pellets substantially hindered this work, which was difficult even without this complicating factor.

Some data obtained with this instrument are listed in Table 16.4.

Table 16.4 DETERMINATION OF COLOR OF SIGNAL FLARE COMPOSITIONS ON THE THREE-COLOR COLORIMETER OF DEMKINA'S SYSTEM

Color Flame	Components	Coefficients			Hue λ mμ	Color Purity, p%
		X	Y	Z		
Red	Potassium chlorate; Strontium oxalate; Iditol	0.65	0.31	0.04	622	87
Yellow	Potassium chlorate; Barium nitrate; Sodium oxalate; Iditol	0.53	0.39	0.08	593	80
Green	Barium chlorate; Iditol	0.33	0.53	0.14	556	75

Note. The sum of X, Y, and Z is equal to 1 in all cases. To convert from the X, Y, Z system to values of p and λ, it is necessary to find a point with coordinates X and Y on graph 16.3. The value of p (color purity) is determined by the position of the point obtained on one of the arcs enveloping point B (white light source). The rays diverging from point B indicate the value of (color tone) in mμ.

In recent years, objective recording three-color photoelectric colorimeters have been designed which provide for a high degree of measurement accuracy. They include the VEI (All-Union Electrotechnical Institute) universal photoelectric colorimeter designed by Shklover and Ioffe [194].

The principle and most important part of the colorimeter is the colorimetric head, which consists of three selenium photocells covered with color filters. The filters are matched so that each of the three detectors reproduces the curve of the corresponding component in spectral sensitivity: red - the curve of component X, green - Y, and blue - Z.

The front opening in the colorimetric head is covered by a piece of glass absorbing UV radiation and in addition, protected from stray light by a removable tube. The main parts of the colorimeter, i.e., the colimetric head and standard lamps, are mounted on a metric optical bench.

The photocurrents generated in the photocells are measured with sensitive galvanometers (sensitivity 5×10^{-9} A/mm) or a multiloop oscillograph. The measurement of color on this instrument consists in reading the galvanometers and calculating the quantities X, Y, Z from calibration equations obtained previously. A special nomogram is available to simplify the calculation.

The colorimeter is calibrated by means of standard electric lamps with VNIIM (All-Union Scientific Research Institute of Metrology) rating plates. In measurement accuracy, this colorimeter surpasses the best visual colorimeters.

In addition to measuring the chromaticity, the same instrument can be used to measure the luminous intensity of pyrotechnic light sources, since detector Y (yellow-green) reproduces the relative luminous efficiency curve in spectral sensitivity.

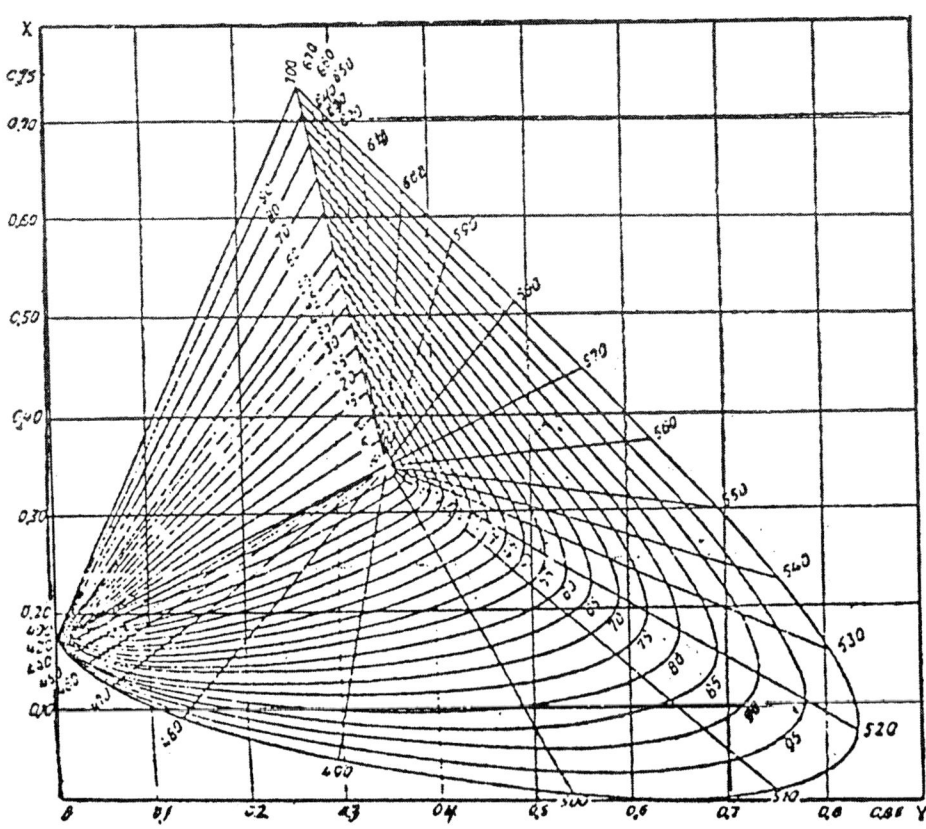

Fig. 16.3. Graph for conversion from three-color coefficients X, Y, Z to the hue and color purity coordinates.

Chapter 17

INCENDIARY COMPOSITIONS

The purpose of incendiary compositions is the destruction of all possible combustible materials. In addition, incendiary compositions (for example, thermite) disable metallic equipment. Particular mention should be made of incendiary compositions against live targets (flame-throwing compositions).

1. IGNITABLE MATERIALS

Incendiary compositions may cause the combustion of many substances and materials in air. However, the materials which are ignited most frequently are as follows:

a) wood in the form of construction materials, fuel, wooden structures, etc.;
b) grasses and cereals in dry form (fodder, grain elevators) or standing (crops);
c) liquid petroleum products in the form of light fuel (gasoline, ligroin, kerosene, etc.) or heavy fuel (petroleum, fuel oil, heavy oils, etc.).

Different ignition compositions and devices should be used for igniting materials which differ in case of inflammation. In order to achieve the kindling of various materials with the minimum investment of available means, it is necessary to know their properties. For this reason, some properties of wood and other fuel materials are cited below.

WOOD. The moisture content of wood may vary over very wide limits. Thus, moist standing wood, which is the most difficult to ignite, contains only 100% moisture (in relation to the weight of absolutely dry wood); air-dried wood contains from 15 to 35% moisture. The kindling temperature of wood ranges from 300 to 400°C.

The process of wood combustion consists in the evolution of volatile components during heating and their subsequent combustion. The processes taking place during the heating of wood may be described as follows:

At 110°C, wood dries up; volatile substances begin to be evolved.

At 110-150°C, the wood turns yellow, and the evolution of volatiles intensifies.

At 150-230°C, the wood turns a brown shade and carbonization begins.

At 230-300°C, carbonization takes place, and above 300°C the wood begins to burn. The purpose of an incendiary composition in kindling wood is to heat as large an area of a wooden object as possible to 300-400°C.

A major obstacle is the poor thermal conductivity of wood.

Further spreading of combustion following inflammation of wood takes place as a result of the heat evolved by the combustion. The calorific power of 1 kg of air-dried wood amounts to 4000 kcal. The combustion temperature of wood is approximately 800-1000°C. The combustion of wood is more intense the more air has access to the burning surfaces. Moist wood is very difficult to ignite, since most of the heat is consumed in evaporating the moisture.

Grass and cereals are much easier to ignite than wood. Dry hay is particularly easy to ignite.

LIQUID PETROLEUM PRODUCTS. The ignitability of liquid petroleum products is determined by their flash point. Gasoline and other liquids having a low flash point ignite immediately on contact with a flame. Such liquids as petroleum and kerosene for lighthouse lamps, etc., whose flash point (see Table 3.7 in Chapter 3) is much higher than room temperature, require preheating for ignition. What burns are not the liquids, but their vapor, and the ignition can therefore occur only when the minimum concentration of its vapors at which the flash of the vapor-air mixture can take place is produced above the surface of the fuel liquid.

Here are some minimum concentrations of vapors in mg per liter of air required for ignition:

Ethanol	73
Benzene	49
Ethyl ether	39
Gasoline	33

This will be illustrated by an example. At 0°C, the ethanol vapor concentration is 34 mg/l, and for the much more volatile ethyl ether, 780 mg/l. The concentration of ethanol vapor at 0°C is below the indicated minimum concentration, and therefore ethanol does not ignite at 0°C (its flash point is +12°C); the concentration of ether vapor at 0°C substantially exceeds this value, and therefore ether will unfailingly ignite not only at 0°C, but also at lower temperatures (its flash point is -20°C).

2. CLASSIFICATION OF INCENDIARY COMPOSITIONS AND REQUIREMENTS PLACED ON THEM

If treated from the standpoint of their atmospheric oxygen requirements during combustion, all existing incendiary compositions may be divided into two major groups.

I. COMPOSITIONS WITH OXIDIZERS

1. Thermites are thermite-incendiarys in which the chief oxidizer is a metal oxide.
2. Compositions in which the oxidizer is an oxygen-containing salt.

II. COMPOSITIONS WITHOUT OXIDIZER

1. Petroleum products.
2. The "electron" alloy.
3. Phosphorus and its compounds.
4. Other incendiary substances and mixtures.

Occasionally, flame-thrower mixtures and substances which self-ignite on contact with water or with atmospheric oxygen (sodium metal, etc.) are separated from group II. All types of incendiary compositions must meet the following requirements:

1. Have the highest possible combustion temperature; the combustion temperature of any incendiary composition should be no lower than 800-1000°C.

2. Burn at a fixed rate best suited for the ignition of the given materials to be kindled. It should be noted that the combustion rate of incendiary compositions changes over a very wide range with the function and structure of the object to be ignited; thus, the combustion of compositions placed in incendiary bullets takes place almost instantaneously; the combustion rate of pressed thermite compositions is measured in millimeters per second; liquid petroleum products burn even less rapidly.

3. Be easily ignited by ordinary ignition compositions, since the use of intermediate compositions is undesirable in many cases.

3. THERMITES

In the 1860's, the eminent Russian scientist N. N. Beketov carried out a reaction between barium oxide and aluminum, and this reaction, and his further studies aimed at the preparation of alkali metals by reacting metallic aluminum with their compounds, laid the foundation for a new branch of metallurgy called aluminothermy. Reactions occurring according to the equation:

$$MO + M_1 = M_1O + M + Q \text{ kcal}$$

where MO is the metal oxide and M_1 is the metal used for the reduction (aluminum), have been called *aluminothermic reactions*, and the reactive mixtures of a metal oxide with another metal were named *thermites*. As an example, we can cite the well-known combustion reaction of iron-aluminum thermite:

$$Fe_2O_3 + 2Al = Al_2O_3 + 2Fe + 205 \text{ kcal}$$

Aluminothermic reactions have found extensive applications in civil industry in the production of large amounts of pure carbon-free metals: chromium, manganese, etc. [201]. Iron-aluminum thermite is widely used for welding ferrous metals (aluminothermic welding of rails) [283, 204].

The characteristic features distinguishing the process of combustion of thermites from the combustion of other pyrotechnic compositions are:

1) an almost complete absence of gaseous reaction products during combustion, causing a flameless combustion;
2) a high temperature of the combustion reaction; for most thermites used, it lies in the range 2000-2800°C;
3) formation of molten slags during the combustion.

We should also point out the difficulty of igniting thermites (the self-ignition temperature of all aluminum thermites is about 800°C, and the self-ignition temperature of iron-aluminum thermite is 1300°C). Thermites have a high density, since they are prepared with oxides of high density (for example, Fe_2O_3, which has a density of 5.1).

Thermites used as incendiary compositions must meet the following requirements:

1) evolve the maximum quantity of heat in combustion;
2) give a high combustion temperature;
3) the slags formed during burning should spread easily, be low-melting and non-volatile;
4) ignite easily;
5) be difficult to quench with ordinary fire-quenching substances;
6) their linear combustion rate should be expressed in units of millimeters per second.

The fuel (metal) used in thermites must:

1) evolve the maximum amount of heat during its combustion;
2) form a low melting and involatile oxide;
3) have a high density.

According to S. F. Zhemchuzhnyy's experiments, the quantity of heat evolved by the combustion of thermites should be no less than 0.55 kcal per gram of composition; otherwise, the combustion reaction takes place with difficulty and does not go to completion. On this basis, and considering the large amount of heat that must be consumed in decomposing the metal oxide, it becomes evident that only high-energy fuels can be used in thermites.

The properties of certain simple substances characterizing their applicability to thermites are listed in Table 17.1.

Table 17.1 COMBUSTIBLES FOR THERMITES

Combustible	Density	Formula of Oxide	Heat of Formation of Oxide 1g-atom Oxygen, kcal	Formula of Thermite, %		Heat of Combustion of 1g of Thermite, kcal
				Fe_2O_3	Combustible	
Al	2.7	Al_2O_3	133	75	25	0.93
Mg	1.7	MgO	144	69	32	1.05
Ca	1.5	CaO	152	57	43	0.93
Ti	4.5	TiO_2	109	69	31	0.57
Si	2.3	SiO_2	104	79	21	0.58
B	2.3	B_2O_3	101	88	12	0.59

The most suitable fuel for thermites in calorific power and substantial density as well as in a comparatively low fusion temperature (2050°C) is aluminum.

It is usually assumed that aluminum is always trivalent in its compounds. However, there also exist compounds of univalent aluminum (Al_2O, Al_2S, AlF, etc.) [196] which are stable in the gaseous state at elevated temperatures and which on cooling decompose into the metal and the corresponding compounds of trivalent aluminum.

Dautzenbery [208] indicates that the vaporization of the combustion products of iron-aluminum thermite is chiefly due to the formation of aluminum sub-oxide Al_2O, which is volatile at 1450°C. It is formed by the reaction of hot aluminum oxide Al_2O_3 with unreacted aluminum:

$$Al_2O_3 + 4Al = 3Al_2O$$

then oxidizes to Al_2O_3 in air. The smoke (particularly noticeable during the burning of large amounts of thermite) formed by the combustion of iron thermite consists chiefly of alumina Al_2O_3.

In addition to economic considerations and its low density, the use of magnesium is also prevented by the high fusion temperature of its oxide (2800°C). Experiments have established that iron-magnesium thermite does not produce spreading liquid slags.

The use of Ca, Ti, Si, and B as individual fuels in thermites is undesirable, but their alloys may be of some interest in this respect.

Goldschmidt [7, p. 206] points out that a thermite in which the alloy CaSi was used as the fuel (in the proportion of 2:1 by weight) produced readily fusible slags.

The solidification temperature of slags is lowered by the formation of calcium silicate $CaSiO_3$, which has a melting point of 1512°C.

An oxide used for the preparation of thermites should meet the following requirements:

1) have a minimum heat of formation;
2) contain a sufficient quantity of oxygen (no less than 25-30%);
3) have the highest density possible;
4) be reduced to a metal having a low melting point and high boiling point.

The properties of certain oxides are listed in Table 17.2. For data on the properties of the reduction of products of oxides see Ch. 2 and 3. Chromium melts at 1083°C and boils at 2360°C. Oxides of elements of low atomic weight are of little use in thermites, since they have appreciable heats of formation and low densities.

The use of oxides of metals with a high atomic weight (for example, Pb_3O_4) is prevented by their low oxygen content; thermites prepared with their participation contain a small amount of fuel and evolve an insufficient quantity of heat during combustion.

The use of oxides of metals having an intermediate atomic weight (approximately from 40 to 80) should be considered the most desirable for use in thermites (see Table 17.2).

Table 17.2 PROPERTIES OF CERTAIN OXIDES

Oxide	Heat of Formation 1g-atom Oxygen, kcal	Oxygen Content of Oxide, %	Density	Formula of Thermite %		Heat of Combustion 1g of Thermite, kcal
				Oxide Content	Al Content	
B_2O_3	101	69	1.8	56	44	0.73
SiO_2	104	53	2.2	63	37	0.56
Cr_2O_3	90	32	5.2	74	26	0.60
MnO_2	62	37	5.0	71	29	1.12
Fe_2O_3	66	30	5.1	75	25	0.93
Fe_3O_4	66	28	5.2	76	24	0.85
CuO	38	20	6.4	81	19	0.94
Pb_3O_4	43	9	9.1	90	10	0.47

In the presence of suitable reductants, copper oxide readily gives up its oxygen; the combustion of copper-aluminum thermite takes place at high rates and resembles an explosion (a similar case is that of Pb_3O_4, and particularly, PbO_2).

Metallic manganese formed during the combustion of manganese-aluminum thermite has a lower boiling temperature (2000°C) than iron; it is vigorously vaporized during the combustion:

$$3 MnO_2 + 4Al = 3Mn + 2 Al_2O_3 + 425 \text{ kcal}$$

Chromium-aluminum thermite burns comparatively slowly, but its combustion evolves a much smaller quantity of heat than other thermites:

$$Cr_2O_3 + 2Al = 2Cr + Al_2O_3 + 130 \text{ kcal}$$

Most acceptable from all points of view should be considered the use of iron-aluminum thermite as an incendiary composition.

The addition of SiO_2 (sand) to iron-aluminum thermite causes a certain lowering of the solidification temperature of its slags as a result of the formation of FeSi alloy, but this is associated with a certain decrease in the calorific power of the thermite. The fusion temperature of FeSi alloy containing 22% Si is 1250°C.

Iron-aluminum thermite is most frequently prepared by using iron scale Fe_3O_4 (ferrosoferric oxide) rather than ferric oxide (Fe_2O_3). The equation of the combustion reaction of thermite in this case is:

$$3Fe_3O_4 + 8Al = 4 Al_2O_3 + 9Fe + 802 \text{ kcal}$$
$$76\% \quad 24\% \quad 45\% \quad 55\%$$

One of the drawbacks of iron-aluminum thermite is the low fluidity and rapid solidification of the slags formed by its combustion.

Powdered iron-aluminum thermite has a gravimetric density of 1.8-2.0, and pressed thermite (containing a few percent of binder), 3-3.4.

On powerful presses developing pressures of the order of 3000-6000 kg/cm², thermite compacts well even without the addition of a binder; the pressed thermite has a high mechanical strength.

Thermite is prepared by taking iron scale and coarsely ground powder (sieve No. 8-10); the use of finely dispersed aluminum is not admissible, since this would accelerate the combustion of thermite. One-kg samples of thermite of ordinary size reduction without pressing burn up in 15-20 sec; the same samples, but pressed at a pressure of 200 kg/cm² burn up in 35-50 sec.

According to Gorlov's report [199], a thermite briquet weighing 1 kg in the shape of a cylinder 15.5 cm high and 5.5 cm in diameter burns up in 40 sec. According to his paper, 50 g of thermite melts sheet iron 2 mm thick in a few seconds.

Pure iron-aluminum thermite containing no admixtures has no explosive properties, is not sensitive to the impact of bullets, and has very little sensitivity to mechanical and thermal effect. Iron-aluminum thermite cannot be ignited with matches, quick match, or ordinary compositions.

Several different mixtures have been proposed for igniting powdered thermite. They all contain magnesium powder or finely divided aluminum dust as the fuel: 1) MnO_2 68.0%, aluminum powder 7.5%, aluminum dust 7.5%, magnesium powder 17%; 2) BaO_2 88%, magnesium 12%; and 3) BaO_2 31%, Fe_3O_4 29%, aluminum fine powder 40%.

In addition, ordinary illumination compositions may be used to ignite powdered thermite.

Pressed thermite ignites with much more difficulty than powdered thermite; its ignition is accomplished with intermediate compositions containing 40-60% thermite (Fig. 17.1).

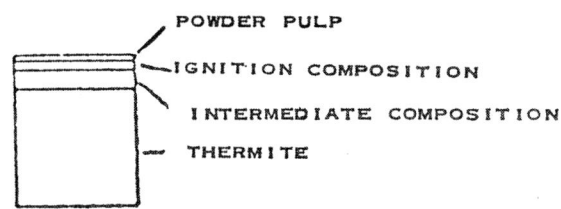

Fig. 17.1. Diagram of ignition of pressed thermite.

4. THERMITE INCENDIARY COMPOSITIONS

In contrast to thermites, thermite incendiary compositions are multicomponent mixtures which in addition to thermite contain other substances (admixtures). The thermite content of such thermite compositions most frequently amounts to 50-80%.

The introduction of various admixtures to thermite is aimed at increasing its heat of combustion, producing a flame during its combustion, facilitate the ignition of thermite, speed up (or slow down) the process of its combustion, and impart strength to the pressed thermite compositions. As an example of a thermite composition containing an additional salt oxidizer, we can cite the following formula used for loading incendiary aerial bombs:

Barium nitrate	26%
Iron scale	50%
Aluminum	24%

The heat of combustion of this composition is approximately 1.1 kcal/g. The combustion of the composition forms 2.5% by weight of gaseous products and 97.5% of solid slags.

Russian 77mm shells are loaded with the following pressed thermite incendiary composition:

Barium nitrate	44%	Aluminum	13%
Potassium nitrate	6%	Magnesium	12%
Iron oxide	21%	Binders	4%

The quantity of gaseous products evolved by the combustion of such a composition is large; the flash point of the composition is in the range 600-700°C. Addition of nitrates to thermite increases its thermal effect, produces a flame during combustion, and lowers its flashpoint, but at the same time, makes the composition more sensitive to mechanical effects.

Among other salt oxidizers, barium or calcium sulfates may also be introduced into thermite compositions.

Sulfur, liquid glass, or organic resins are introduced into thermite compositions as binders.

Ellern [9] cites the following composition formula: thermite 69%, barium nitrate 29%, sulfur 2%.

The combustion of thermites containing sulfur is associated with the formation of a flame due to partial combustion of sulfur to SO_2; the slags obtained are more fusible and fluid, since in addition to aluminum oxide, they contain sulfur compounds (Table 17.3).

Table 17.3 PROPERTIES OF SULFUR COMPOUNDS

Compound	Molecular Weight	Heat of Formation, kcal/g-mole	Remarks
Al_2S_3	150	140	Melts at 1100°C; is decomposed by water to form H_2S
MgS	56	84	Decomposed by water
CaS	72	111	Same
FeS	88	23	Insoluble in water; melts at 1193°C
MnS	87	44	Sparingly soluble in water
SiS_2	92	34	--

As was pointed out by Gorlov [199], in order to achieve a uniform mixing of thermite with molten sulfur and avoid the formation of lumps, the components must be heated to 120-140°C, and the possibility of their rapid cooling during mixing must be removed. This fact makes the compaction of thermites a very time-consuming operation requiring considerable care.

The introduction of water glass (15% aqueous solution of sodium silicate) into thermite compositions as a binder makes it necessary to dry the composition afterward. The sodium silicate solution has a sharply alkaline reaction; during the drying, an intensive corrosion of aluminum powder may be observed which in some cases is associated with a significant heating up of the composition (up to 100°C).

Among organic binders used in thermite incendiary compositions are asphalts, colophony, drying oil, Bakelite, etc.

5. INCENDIARY COMPOSITIONS WITH SALT OXIDIZERS

Most effective are mixtures with a high content of magnesium or aluminum powders:

I		II		III	
Potassium nitrate	65%	Potassium perchlorate	56%	Potassium perchlorate	50%
Aluminum	26%	Aluminum	34%	Magnesium	50%
Wood charcoal	9%				

Usually, thermite incendiary compositions are loaded into large-sized objects (medium-caliber shells, aerial bombs), whereas compositions with salt oxidizers are more suitable for the loading of small-caliber shells or incendiary bullets designed to ignite liquid fuel. Compositions similar to mixture III produce no liquid slags during their combustion, and their incendiary action is based exclusively on the direct effect of the flame. They must meet these requirements:

1) The flame formed by the combustion of the compositions should have the maximum temperature and an appreciable size;
2) The action of the flame on the liquid fuel should take place in a fairly short period of time (for example, a few tenths of a second).

Such compositions can be activated by a mechanical impulse (impact on armor) and by initiation with the aid of explosives placed in the same object. In the latter case, the combustion of the compositions may take place at very high rates expressed in tens, hundreds, or even thousands of meters per second.

Mixtures of explosives with aluminum powder or dust may also be regarded as a kind of incendiary mixtures.

During World War II, it was found that the incendiary effect of an explosive is markedly increased by adding metal powders to the explosive. As an example, we can cite the explosive mixture for German 20mm shells: 76% hexogen, 20% aluminum, 4% deterrent.

6. LIQUID PETROLEUM PRODUCTS AND SOLIDIFIED FUELS. FLAMETHROWER MIXTURES.

LIQUID PETROLEUM PRODUCTS. They found extensive applications in various incendiary products during World War II. For example, incendiary bombs were filled with naphtha, fuel oil, kerosene, gasoline, and other petroleum products. In some cases, in order to extend the combustion time and decrease the spraying of these liquids, they were used to impregnate cotton, cotton wool, or cotton waste.

The chief advantages of liquid petroleum products used as incendiary substances are:

1) a large thermal effect (the combustion of 1 g of kerosene yields 10 kcal, and that of thermite, 0.8 kcal);
2) formation of a large flame during combustion, and hence, generation of a large fire center;
3) low combustion rate;
4) a comparatively low cost and fairly broad raw material base.

Their disadvantages include:

1) a low combustion temperature (700-900°C; thermite develops a temperature of 2400°C during combustion);
2) low density (0.7-0.8; the density of pressed thermite is 3-3.4);
3) absence of solid combustion products, i.e., slags;
4) an extremely high mobility and spreadability;
5) excessive ease of evaporation, resulting in a vigorous combustion process when a sufficiently strong thermal impulse is imparted to the liquid.

The latter two disadvantages may be eliminated to a certain extent by using liquid petroleum products in the so-called solidified form.

SOLIDIFIED FUELS. These are fuel fuels (hydrocarbons) converted to a jelled state by a suitable treatment. A solidified fuel can be used, both individually and in combination with thermite, for filling heavy bombs designed to ignite large wooden structures (Fig. 17.2). Solidified fuels can be produced by dissolving stearic acid in the petroleum products, then treating the solution obtained with an alcohol solution of sodium hydroxide.

The reaction taking place during mixing of these two solutions may be expressed by the equation

$$C_{17}H_{35}COOH + NaOH = C_{17}H_{35}COONa + H_2O.$$
Stearic acid Soap

The hot liquid is poured through tubing into the shells of incendiary bombs, where it is converted into a jellied mass on cooling. This method is used to obtain the "solid naphtha", "solid kerosene", etc.

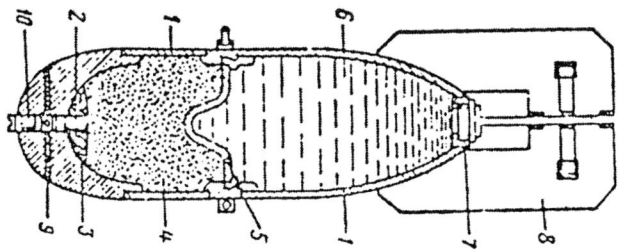

Fig. 17.2. ZAB-50 composite-load incendiary bomb.
1 - shell, 2 - ignition pellet, 3 - intermediate composition, 4 - principle thermite incendiary composition, 5 - diaphragm, 6 - solidified fuel, 7 - bottom plug, 8 - stabilizer, 9 - gas discharge openings, 10 - fuse socket.

In practice, in order to obtain solidified fuels, use may be made, not of stearic acid, but of commercial stearin, which is a mixture of stearic and palmitic acids, the latter having the formula $C_{16}H_{32}O_2$. For the properties of these acids see Ch. 4. Solidified fuels must meet the following special requirements:

1) low volatility;
2) ease of inflammation;
3) maximum heat of combustion;
4) high combustion temperature;
5) constant consistency during transportation and storage;
6) absence of syneresis phenomena in the presence of sharp temperature fluctuations (from -30 to +40°C).

Gasoline and kerosene solidified in the usual manner are easily ignited by a burning match.

The calorific value of a solidified fuel is close to that of the corresponding petroleum product, but nevertheless somewhat lower, particularly when the quantity of solidifying additives exceeds 10-15% of the total weight of the fuel. The melting point of solidified fuels is ~60°C, and their density is close to that of the corresponding petroleum product. Another type of solidified fuel is the "thickened" fuel. In the USA, the thickened fuel has been named napalm [209]. This name is applied in the USA to the thickened fuel and to the powder thickening it. The thickening powder, i.e., napalm, is a mixture of aluminum salts of oleic, naphthenic and other organic acids. The raw materials for its production are:

1) oleic acid, naphthenic acid, and the fatty acid of coconut oil;
2) aqueous solution of NaOH;
3) aqueous solution of alum.

When these substances are mixed (Fig. 17.3), the aluminum salts precipitate out; the suspension obtained is centrifuged, and its water content is thereby reduced to 35%.

The moist napalm is then fed into a rotary drier, where it is subjected to the action of hot air having a temperature of about 160°C. After the drying, the water content of the napalm is 0.4-0.8%.

The dry napalm is ground into a powder in mills. The napalm powder is used for thickening gasoline; the amount of napalm introduced into gasoline ranges from 4 to 11%, depending on the degree of thickening required. The thickened gasoline is used for flame throwing purposes, and also to fill the shells of incendiary bombs. The distinctive properties of thickened napalm are its considerable viscosity and stickiness, making it a suitable agent for kindling wooden structures.

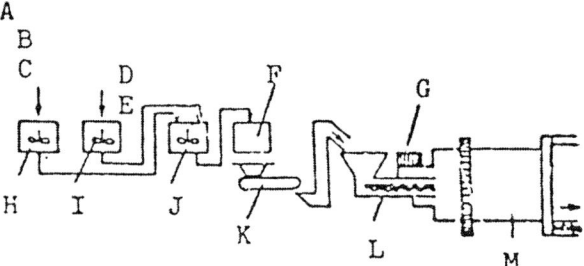

Fig. 17.3. Diagram of napalm production.
A - Fatty acids, B - NaOH, C - Water, D - Alum, E - Water, F - Centrifuge, G - Hot air supply, N - Soap solution, I - Stirrer, J - Mixer, K - Conveyer, L - Screw conveyer, M - Rotary drier.

The production, transportation, storage, and use of "solid gasoline" in the national economy is described in [202]. The hardener agents are casein, urea-formaldehyde resins, or polyvinyl alcohol solutions. The hardener for polyvinyl alcohol is formaldehyde. The gasoline content by volume in "solid gasoline" is 87-90%.

FLAMETHROWER MIXTURES. The flamethrower was first used in combat during World War I in 1915. Flamethrowers projected a jet of flame. The range of the flamethrower jet did not exceed 40 m.

In World War II, the Germans also used flamethrowers in 1940. The effective radius of a flamethrower jet ejected by a light tank was 50 to 75 m according to reports in the foreign press.

Flamethrower liquids must satisfy the following requirements:

1) have the highest possible density (to cover the necessary effective range of the jet);
2) be readily ignited by a suitable ignition device;
3) develop a maximum temperature during their combustion;
4) not burn too fast in air (rapid combustion would cause only a slight amount of the liquid to reach the target).

Mixtures of various liquid hydrocarbons or oils are most frequently used as flamethrower liquids. For example, a flamethrower mixture containing 25% gasoline, 25% kerosene, and 50% naphtha has been used in the USA.

Thickened napalm-type fuel liquids have also been used.

A solution of phosphorus in CS_2 has not found any significant application as a flamethrower liquid. The temperature of the flame produced by the combustion of a flamethrower jet of liquid petroleum products does not exceed 700-900°C.

According to patent disclosures (British Patent 656089, 1951), an increase in the flame temperature of a flamethrower jet may be achieved by introducing an appreciable amount of magnesium powder (up to 50-60%) to the petroleum products, and simultaneously adding 10% of anhydrous sodium sulfate.

The ignition of a flamethrower jet is accomplished with the aid of a special ignition cartridge.

Gorlov [199] indicates the possibility of adding tetranitromethane to flamethrower mixtures. An admixture of $C(NO_2)_4$ lowers the flash point of the mixture, increases its combustion rate, and makes it explosive.

7. THE "ELECTRON" ALLOY

The light "electron" alloy contains approximately 90% magnesium as well as aluminum, zinc, manganese and other metals. An example of an electron composition for preparing shells of incendiary bombs follows:

Magnesium	90-94%
Aluminum	0.5-8%
Zinc and manganese	1.5-5%

The electron alloy has a density of 1.80-1.83 and high mechanical qualities. Its fusion temperature is 630-635°C. It is completely stable toward alkaline solutions, but corrodes easily under the influence of even weak solutions of acids or ammonium salts. On burning, electron develops a high temperature (up to 2000°C) and evolves a considerable amount of heat (~6 kcal/g).

As an incendiary agent, electron is used for the preparation of various incendiary articles and mainly in the manufacture of shells of small aerial bombs (weighing about 1 kg). In these articles, electron burns by combining with atmospheric oxygen. This forms a small, dazzlingly bright, white flame and evolves a certain amount of white smoke of magnesium oxide. Electron is ignited by means of pressed or powdered thermites or thermite incendiary compositions placed inside the shells of electron bombs (Fig. 17.4).

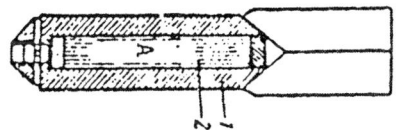

Fig. 17.4. German 1-kg electron bomb.
1 - electron shell, 2 - thermite composition. A - Thermite.

When electron burns, its oxidation may be due not only to atmospheric oxygen (and nitrogen), but also to the oxygen present in the wood that is in contact with it.

8. PHOSPHORUS AND ITS COMPOUNDS

Usually, phosphorus, its solutions and compounds are used to kindle readily ignitable materials. The chief advantage of white phosphorus over other incendiary substances is that in the finely divided state, it self-ignites and burns in air:

$$4P + 5O_2 = 2P_2O_5.$$

The combustion forms a yellowish-white flame and evolves a large quantity of a white smoke, phosphorus pentoxide, that is very stable in air.

As was indicated by Gorlov [199], phosphorus mines or hand grenades have proven very effective in trench warfare. By producing a large amount of smoke, the explosion of such mines has a demoralizing effect on the adversary. The finest splashes of molten burning phosphorus burn through the clothing and injure the body by inflicting serious damage to the skin. Among other advantages of white phosphorus as an incendiary agent is its ability to self-ignite again after being extinguished. White phosphorus is a soft, waxlike, pale yellow substance. Its density is 1.83, melting point, 44°C, and boiling point, 290°C.

The chief disadvantages of white phosphorus as an incendiary substance are its low combustion temperature (not above 1000°C) and the difficulties involved in loading it in suitable articles; to prevent its self-ignition, pouring of white phosphorus into ignition devices should be carried out under water. It should be noted also that white phosphorus is very poisonous (a dose of 0.1 g of white phosphorus is lethal).

Active red phosphorus is very seldom used as an incendiary substance. In some cases, however, incendiary devices are loaded with a mixture of red and white phosphorus.

PHOSPHORUS SOLUTIONS. The best solvent for white phosphorus is carbon disulfide (100 g of saturated phosphorus at 0°C contains 81 g of phosphorus); in addition, white phosphorus is soluble in many organic solvents, for example, in benzene, turpentine, etc.

When a solution of phosphorus in CS_2 (a very volatile solvent) is evaporated, the fine particles of phosphorus remaining on the drenched object readily ignite in air and kindle the surrounding carbon disulfide vapor. In order to slow down the combustion process, liquid petroleum products, tar oil, nitro compounds and other substances are sometimes added to the solution of phosphorus in carbon disulfide.

Red phosphorus is not soluble in carbon disulfide.

PHOSPHORUS COMPOUNDS. Among phosphorus compounds, phosphorus sulfides and mainly the sesquisulfide P_4S_3 are most frequently used as incendiary substances. Some properties of phosphorus sulfides are listed in Table 17.4. The sesquisulfide P_4S_3 is soluble in CS_2 (100 g of CS_2 at 0°C dissolves 27 g of P_4S_3), and more sparingly in benzine. Pure P_4S_3 at room temperature is stable toward water, and decomposes in boiling water, evolving hydrogen sulfide.

The higher sulfides P_4S_7 and P_4S_{10} are not stable toward water (decompose with the evolution of H_2S); they dissolve more sparingly than P_4S_3 in carbon disulfide and other solvents.

When white phosphorus is mixed with sulfur at room temperature, a liquid alloy is formed; the phosphorus content of the eutectic, which solidifies at -7°C, amounts to approximately 75%.

Table 17.4 PROPERTIES OF PHOSPHORUS SULFIDES

Formula of Compound	Phosphorus Content,%	Density	Temperature, Deg. C	
			Melting Point	Boiling Point
P_4S_{10}	27.9	2.06	290	514
P_4S_7	35.6	2.19	310	523
P_4S_3	56.3	2.09	172	408

The melt of P_4S_3 with phosphorus forms a eutectic which solidifies at -40°C. Figure 17.5 shows the sulfur-phosphorus phase diagram [48, vol. 8].

PHOSPHIDES. Calcium phosphide is completely stable in a dry atmosphere, but in moist air, or when moistened by water, it decomposes into calcium hydroxide and phosphine:

$$Ca_3P_2 + 6H_2O = 3Ca(OH)_2 + 2PH_3$$

Because of the presence of other phosphines (P_2H_2) in addition to PH_3, hydrogen phosphide ignited spontaneously in air.

In external appearance, Ca_3P_2 consists of reddish-brown crystals with a density of 2.5. The preparation of calcium phosphide by combustion of the mixture $Ca_3(PO_4)_2 + Al$ and the use of Ca_3P_2 in producing a secondary fire center have been described in a German patent [210].

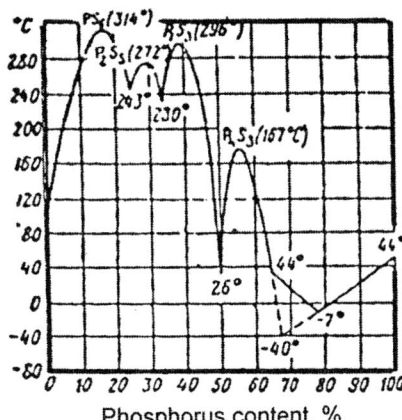

Fig. 17.5. Phase diagram of the compounds and alloys of the phosphorus-sulfur system.

At the present time, calcium phosphide is most frequently used, not as an incendiary agent, but for loading special signaling devices used in the Navy.

Magnesium phosphide is similar in properties to calcium phosphide, but its decomposition by water evolves a slightly smaller amount of heat.

9. HALOGEN COMPOUNDS OF FLUORINE

Free fluorine combines very vigorously with organic substances; this reaction forms a large quantity of heat and involves the inflammation of fuel materials. However, the use of free fluorine as an incendiary substance is practically impossible, since fluorine is a gas difficult to liquefy (boiling point at p=1 atm is –187°C).

Halogen compounds of fluorine (halogen fluorides), which in many cases have a sufficiently high chemical activity, also have a higher boiling point and are either liquids or gases that are easily liquefied. Such compounds may be evaluated on the basis of the following indices:

1. Fluorine content (in weight %).
2. Boiling point at p=1 atm.
3. Density in the liquid state (at t=25°C).
4. Heat of formation from the elements in kcal/g.

Table 17.5 lists the physical properties of halogen fluorides [207], [94].

Of greatest interest to pyrotechnists is chlorine trifluoride ClF_3, since, of all the halogen fluorides, it has the maximum reactivity. Chlorine monofluoride contains the lowest percentage of fluorine and is difficult to liquefy.

Compounds of iodine have an appreciable heat of formation and are therefore less reactive.

The practical importance of halogen fluorides for industry is their use as fluorinating agents.

Table 17.5

Compound	Fluorine Content, %	Boiling Point, °C	Density at 25°C g/cm^3	Heat of Formation, kcal		Remarks
				per mole	per g	
ClF_3	61.6	11.7	1.81	37.9	0.41	Gas
ClF	34.9	-101	--	13.3	0.25	Gas
BrF_5	54.3	40	2.46	124(ra3)	0.71	Liquid
BrF_3	41.6	125	2.80	75	0.55	Liquid
BrF	19.2	20	--	18.4	0.19	Liquid
IF_7	51.2	4	2.75	232	0.89	Gas
IF_5	42.8	100	3.19	204.7	0.92	Gas

Chlorine trifluoride ClF_3 is a colorless, easily liquefiable gas. Its melting point is -76.3°C. The temperature dependence of ClF_3 vapor pressure is expressed by the equation

$$\log p \, mm = 7.42 - \frac{1292}{T} \qquad (17.1)$$

The original temperature of ClF_3 lies in the range 154-174°C, and its calculated critical pressure is 57 atm. The molar heat capacity at +5°C is 28.0 cal/mole deg.

The heat of decomposition

$$ClF_3 = ClF + F_2 \text{ is} -26.6 \text{ kcal}$$

Soft steel is stable toward ClF_3 (up to 250°C); even more stable are copper (up to 400°C) and nickel (up to 750°C); ClF_3 reacts vigorously with water

$$3H_2O + 4\,ClF_3 = 6HF + 3F_2O + Cl_2. \qquad (17.2)$$

As a rule, organic substances react with ClF_3 with inflammation. ClF_3 is obtained by the direct reaction of chlorine and fluorine.

During World War II, a semi-industrial production of ClF_3 was organized in Germany for the purpose of using it as an incendiary substance.

In recent years, ClF_3 has been widely advertised in the USA as a fluorinating agent and as an oxidizer in jet engines. It should be noted that because of the great reactivity of ClF_3, work with this compound involves major difficulties and is dangerous for the experimenter; one must constantly remember the toxicity of ClF_3 and its decomposition products, as well as the possibility of its explosion on contact with many substances.

10. OTHER INCENDIARY SUBSTANCES AND MIXTURES

Among elements, in addition to magnesium and phosphorus, alkali metals, i.e., potassium and particularly sodium, have found application as incendiary substances.

The chief advantage of metallic sodium over other incendiary substances is that it reacts vigorously with water, evolving hydrogen:

$$2Na + 2H_2O = 2NaOH + H_2 + 135 \text{ kcal}$$

Under certain conditions, hydrogen may form with air a detonating gas whose explosion scatters the particles of water and burning alkali metal in all directions and increases the number of fire centers.

Disadvantages of alkali metals as incendiary substances include: 1) a low density (sodium,

0.97; potassium, 0.86) and 2) a considerable inertness toward dry air. Alkali metals ignite with difficulty in dry air, and once ignited, can be easily extinguished. The combustion temperature of alkali metals in air is moderate (not above 1000°C).

Alkali metals in incendiary devices are used exclusively in combination with other incendiary substances or compositions: phosphorus, liquid petroleum products, solidified fuel, and thermite.

The technology of loading incendiary articles with metallic sodium or potassium is somewhat complicated by the fact that these metals have to be stored in kerosene to prevent their oxidation in humid air.

Metallic sodium melts at 98°C and boils at 877°C; metallic potassium melts at 63°C and boils at 762°C. A eutectic alloy of metallic sodium and potassium is a liquid containing 23% sodium in the alloy and solidifying at 12.5°C [67].

The literature describes many different substances which ignite spontaneously in air and which, if desired, may be used in incendiary devices.

In addition to the above-described white phosphorus and phosphines, these substances include silanes, formed by the reaction of metal silicides (for example, Mg_2Si) with hydrochloric acid.

Similarly, boron hydrides (boranes) are obtained by reacting metal borides (for example, Mg_3B_2) with dilute acids. Self-igniting substances also include the most diverse alkylmetal compounds, for example, dimethylzinc, alkyl derivatives of magnesium, aluminum, etc. Even more active are alkyl compounds of alkali metals.

Pyrophoric metals may also be classified among materials self-igniting in air. In addition to their very fine particle size, metals in the pyrophoric state are distinguished by the fact that because of a special method of preparation (reducing atmosphere), their surface does not have an oxide film preventing oxidation. The majority of metals (except noble ones) can be obtained in the pyrophoric state by some method.

11. METHODS OF TESTING INCENDIARY COMPOSITIONS. EXTINCTION OF INCENDIARY COMPOSITIONS.

The transfer of heat to the object being ignited is accomplished during the combustion of an incendiary composition with the aid of solid or liquid incandescent slags as well as by the direct action of the flame. The total amount of heat transferred from the burning ignition composition to the object being ignited will depend on:

1) the average difference between the temperature of the slag and flame of the incendiary composition and the temperature of the object being ignited, ΔT_m;
2) the surface of contact of the slags and flame with the object being ignited, F;
3) the time of their contact t;
4) the coefficient k of heat transfer from the combustion products of the incendiary composition to the material being ignited:

$$Q = k \cdot \Delta T_m \cdot F \cdot t.$$

In some cases, it is useful to distinguish the amount of heat Q' transferred to the ignited object by the solid or liquid slag from the amount of heat Q" transferred to the ignited object by the gaseous reaction products, i.e., the flame:

$$Q = Q' + Q''. \quad (17.3)$$

$$Q = k' \cdot \Delta T_m \cdot F' \cdot t' + k'' \cdot \Delta T'' \cdot F'' \cdot t''. \quad (17.4)$$

Usually, for a composition producing any significant amount of slag during its combustion, Q' is much greater than Q"; this means that a large proportion of the heat is transferred to the ignited object by the slags, not by the flame. This takes place because

$$k' > k'' \text{ and } t' > t''.$$

In view of the fact that the coefficient k in the case of heat transfer from the combustion products of incendiary composition to wood and other ignited materials is unknown, any thermal calculations based on the above formulas is impossible.

In addition to the amount of heat transferred to the ignited material, of great importance is also the "thermal head", i.e., the amount of heat transferred from the composition to the ignited material per unit time, Q/t kcal/sec. It is evident that the "thermal head" during combustion of thermite will be much greater than for example during the combustion of solidified fuel.

Fig. 17.6 Diagram of calorimetric apparatus.

The experimental determination of the effectiveness of incendiary compositions can be reduced to the determination of the amount of heat given up by the combustion of 1 g of composition to a plane surface of any material. To obtain accurate results, it is desirable that the heat conductivity of the material chosen be close to that of the materials being ignited (for example, wood).

The magnitude of the so-called <u>gram heat transfer</u> of a composition will obviously depend to a considerable extent on the testing conditions (material of the plate, weight of the sample of composition and its arrangement on the plate), and therefore only results obtained under the same conditions can be compared. Figure 17.6 schematically shows the calorimetric apparatus for determining the "gram heat transfer of incendiary compositions".

Ya. M. Paushkin determined the gram transfer of a series of pressed incendiary compositions and fuels during their combustion in an open steel cup placed in the calorimeter. These data, which are of considerable interest and enable one to make a quantitative comparison of the effectiveness of incendiary compositions, are listed in Table 17.6.

Simultaneously with these experiments, Paushkin conducted tests on the combustion of the most effective incendiary substances and compositions on a flat wooden object placed in the calorimetric vessel. Despite the fact that, as was indicated by the author himself, these data must be considered tentative because of a certain burning off of the wood, they are of considerable interest, since they give an idea of the amount of heat transferred from incendiary substances under actual conditions (Table 17.7).

It is evident from Tables 17.6 and 17.7 that the most effective incendiary substances in gram heat transfer are magnesium (or electron), followed by petroleum products, then by iron-aluminum thermite.

Table 17.6

Incendiary Substance or Composition	Heat Transfer on Wood, kcal/g	Heat Utilization Factor, %	Iron-aluminum Thermite
KNO_3+Mg	0.17	1.8	10
$Ba(NO_3)_2+Mg$	0.49	1.6	31
$KClO_4+Mg$	0.42	2.4	17
BaO_2+Mg	0.33	0.5	63
Fe_2O_3+Mg	0.62	1.1	60
Fe_2O_3+Al	0.63	0.9	75
Cr_2O_3+Al	0.48	0.6	80
MnO_2+Al	0.47	1.1	42
Magnesium	1.90	6.1	30
Kerosene	1.50	10.0	15

Among other special laboratory tests aimed at determining the effectiveness of incendiary compositions, I. I. Vernidub and V. A. Sukhikh indicate the following:

1) determination of the amount of solid slags (in percent of initial weight of composition) remaining on the surface of the limited material:
2) determination of the spreadability of the slags, i.e., measurement of the area occupied by the slags after the combustion of a given amount of composition on a horizontal and an inclined plane;
3) determination of the solidification temperature of the slags; for low-melting slags (melting point below 1500°C), this determination may be carried out by means of a thermoelectric pyrometer (thermocouple).
4) determination of the dimensions of the flame obtained from the combustion of the composition (photographing of the flame);
5) testing of incendiary compositions for their direct ignition of suitable types of fuel materials (liquid fuel, etc.).

Table 17.7

Incendiary Substance	Heat transfer on wood, kcal/g	Heat Utilization Factor, %
Iron-aluminum thermite	0.15	17
Kerosene	0.40	4
Magnesium	0.50	8

For certain types of incendiary compositions (for example, thermites), tests are sometimes conducted for the melting of sheet iron of a given thickness and for fusing slags onto suitable metal articles.

QUENCHING OF INCENDIARY COMPOSITIONS. The quenching of fires resulting from the combustion of incendiary compositions can in most cases be achieved with ordinary fire-quenching agents. The quenching of the incendiary compositions themselves is seriously complicated by the short combustion time of modern incendiary substances (not more than 5-10 min.), and therefore the success of the operation substantially depends on the timeliness of the start of quenching.

In quenching of thermite incendiary compositions by small amounts of water, the latter may decompose and form explosive mixtures of hydrogen and air. It has been found experimentally [199], however, that water supplied in large quantities in a strong jet under pressure constitutes the best means of quenching thermite incendiary and electron bombs, since it then causes a sharp local temperature drop and interrupts the combustion process.

Thermite incendiary and electron incendiary bombs up to 5 kg in weight may be quenched by immersing them in the largest possible amount of water. To limit the effect of incendiary bombs, sand should be used.

In quenching thermite compositions containing sulfur, it should be considered that the action of water on a burning composition forms large amounts of hydrogen sulfide.

The use of water or aqueous solutions is inadmissible in quenching of metallic sodium or potassium. The most reliable substances for their quenching are sand and dry soda.

The best quenching agents for phosphorus are aqueous solutions of $KMnO_4$ or $CuSO_4$. The action of $KMnO_4$ solution is based on the formation of a of a protective coating of manganese dioxide on the surface of phosphorus; a copper sulfate solution forms a very dense protective layer of metallic copper.

The quenching of solidified fuels is accomplished by the same means as that of liquid hydrocarbons. Most suitable are thick-foam flame quenchers, atomized water, and concentrated aqueous solutions of ammonium salts.

Modern methods of quenching of fuel liquids are described in detail in the monograph [197].

# Chapter 18

COMPOSITIONS OF SMOKE SCREENS

1. GENERAL DATA ON AEROSOLS

In true solutions, the solute particles consist of individual molecules, atoms, or ions; in colloidal systems, however, the number of atoms or molecules making up a particle amounts to many hundreds and thousands. Colloidal systems consist of a dispersion medium and a finely divided substance therein, i.e., a dispersed phase. If the dispersion medium is air, the colloidal system is called an aerosol. If the dispersed phase in air is a liquid, such a system is called a fog, and if the dispersed phase is in the solid state, such an aerosol is called a smoke. Consequently, smoke is the name given to the finest suspension of a solid substance in air. The particle sizes of various smokes and fogs range from 10μ to $1\, m\mu$ (from 1×10^{-3} to 1×10^{-7} cm). The radius of aerosol particles used for smoke screen formation ranges over narrower limits: 8×10^{-5} to 2×10^{-5} cm.

Dust is the name usually given to a suspension in air of coarser particles of a solid measuring from 1×10^{-3} to 1×10^{-2} cm.

In industry, smokes constitute an undesirable phenomenon in most cases, and are studied chiefly for the purpose of finding ways of their elimination; smokes are removed either with the aid of a strong electric field, or with the aid of various filters.

In military engineering, smokes and fogs are used for the production of neutral smoke screens; colored smokes are used for daytime signaling (see Chapter 19). The concentration of smokes and fogs may be expressed in two ways:

1) in terms of weight, in g per 1 m^3 (or in mg per liter), C_p;
2) in terms of the quantity of smoke or fog particles per unit volume (per 1 ml), C_n; the latter quantity is called the particle concentration.

The weight concentration C_p of smokes used for producing a smoke screen is usually expressed in tenths of mg/l, which corresponds to a content of several million smoke particles in one ml of air.

Smoke or fog particles are in continuous movement in air. This movement is determined by the force of gravity, which forces the particles to fall. The rate of fall of particles whose radius is less than 1×10^{-3} cm remains constant and may be expressed by Stokes' formula:

$$v = \frac{2}{9} r^2 g \frac{(\rho - \rho')}{\eta} \text{ cm/sec}$$

where v is the rate of fall of a particle in cm/sec;
r is the particle radius in cm;
g is the coefficient of acceleration due to gravity, equal to 981 cm/sec^2;
ρ is the density of the particle;
ρ' is the density of the medium (in this case air);
η is the viscosity of air expressed in poises, equal to 1.81×10^{-4} g/cm sec.

[Note: The Stokes formula is derived by assuming that the aerosol particles settle without coagulating and are spherical in shape; the latter assumption is valid only for fogs, but smoke particles may have the most varied shapes. For particles with radii of less than 4×10^{-5} cm, the assumptions on the homogeneity of the medium and absence of slip which are used as the basis of the Stokes formula cease to be valid. To calculate the rate of fall of such small particles, use must be made of Cunningham's formula, see [211].

The quantity ρ' for air is so small that it may be safely assumed to be equal to zero. Replacing $2g/9\eta$ by the constant k, we obtain Stokes formula for an aerosol in the simpler form

$$v = k \cdot r^2 \rho.$$

where $k = 1.2 \times 10^6$ cm^2/g sec.

The path S (in cm) traveled by a particle under the influence of the force of gravity in time t is calculated from the formula

$$S = k \cdot r^2 \cdot \rho \cdot t.$$

Smoke (or fog) particles simultaneously undergo a constant random movement called Brownian movement, resulting from impacts received by the aerosol particles from molecules of the gaseous medium. The most vigorous Brownian movement is exhibited by the smallest particles of smoke or fog. Smoke particles move simultaneously with the medium itself, i.e., together with the air currents. The largest smoke clouds, whose particle size does not exceed 1×10^{-4} cm, are usually dispersed by atmospheric air currents without having had the time to settle under the influence of the force of gravity or to disperse under the influence of shocks of molecules of the gaseous medium. Therefore, in studying the physicochemical (optical, electrical, etc.) properties of smoke, to guard from the dispersing influence of the convection currents of air, use is frequently made of a chamber with glass walls which enable one to observe the processes taking place in the system. Occasionally, for purposes of thermal insulation, double chamber walls are made with a layer of air between them.

One of the chief problems of a physicochemical study of a smoke system is the determination of the smoke particle size. For this purpose, the aerosol particles are caught on glass plates, then examined visually under a microscope or photographed. This makes it possible to count the number of particles in the field of view.

The Deryagin-Vlasenko flow ultramicrophotometer is used to determine particle concentrations in air [57]. The number of particles is counted in reflected light in a microscopically small volume through which the aerosol is drawn by suction.

This instrument makes it possible under field conditions to take into account aerosol concentrations containing more than 10 million particles per cm^3. In addition, there exist methods of determining the size and shape of smoke particles (MgO, ZnO, Al$_2$O$_3$, etc.) by means of the electron microscope.

In most cases, smokes and fogs are polydisperse systems, i.e., they contain particles of various sizes (Fig. 18.1).

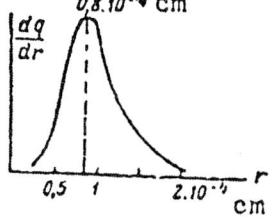

Fig. 18.1. Size distribution of ammonium chloride smoke particles.
dq/dr - percentage by weight of smoke particles having values of radius r plotted along the abscissa.

One must also know the electrical properties of smokes and fogs, since the signs and magnitudes of particle charges determine the duration of the existence of a smoke or fog.

In aerosols, the phenomenon of coagulation is observed, starting with the instant of their formation. The <u>coagulation</u> of an aerosol is the term applied to the enlargement (sticking together) of smoke or fog particles. The formation of very large particles causes their rapid precipitation from a gaseous medium. The coagulation rate of aerosols is directly proportional to the square of their particle concentration. For smokes, the coagulation process is frequently termed flocculation (from the English word flocks). The capacity of smokes to flocculate is decreased in the presence on the particles of a like electric charge or film of adsorbed gas; the absence of a charge, or, in particular, the presence of unlike charges on the particles facilitates the flocculation of smokes.

Smoke particles can get their charge:

a) as a result of friction against the gaseous medium;
b) as a result of dissociation of the particles when the smoke is formed;
c) by trapping gas ions from the dispersion medium.

2. OPTICAL PROPERTIES OF AEROSOLS

The presence of smoke or fog in air decreases its penetrability to light. The incoming light beam passes only partially through it; a considerable portion of the light is absorbed or scattered by the smoke or fog particles.

The phenomenon of scattering of light by colloidal particles is called the Tyndall effect. The intensity of scattered light is determined with an instrument called the Tyndall meter.

In smokes whose particle sizes are greater than the wavelength of light, the intensity of scattered light I_s is expressed by the formula

$$I_s = k \cdot \frac{C}{r}$$

where k is the proportionality coefficient;
C is the weight concentration of the aerosol in mg/l;
r is the particle radius of the aerosol.

However, when the particle size of the aerosol is much smaller than the wavelength of the light passing through it, the intensity of the scattered light increases rapidly with increasing particle size:

$$I_s = k \cdot C \cdot r^3.$$

Maximum light scattering and hence, the best smoke screen capacity, is displayed by aerosols whose particles have sizes close to the wavelengths of visible light, i.e., 1×10^{-4} to 1×10^{-5} cm.

The intensity of light scattering increases appreciably as the wavelength decreases. According to Rayleigh's formula, the intensity of scattered light for any one dispersed system will be expressed by the formula

$$I_s = I_0 \frac{k \cdot C \cdot r^3}{\lambda^4}$$

where I_0 is the intensity of incident light;
λ is the wavelength of incident light.

Long-wavelength, red and particularly infrared rays are scattered least by smokes and fogs.

The quantitative dependence between the absorption of light by an aerosol, the aerosol concentration, and the thickness of the absorbing layer may be expressed in the case of constant aerosol dispersity by the Lambert-Bouguer-Beer formula:

$$I = I_o \cdot e^{-k \cdot C \cdot l}$$

where I_o is the intensity of the light entering the absorbing layer;
I is the intensity of light leaving the absorbing layer;
l is the thickness of this layer;
e is the base of natural logarithms.

For smoke screens, of great importance is the concept of the "hiding power" of a smoke. The hiding power of a smoke is greater the smaller the ratio of the intensity of transmitted light to the intensity of the light entering the smoke, other things being equal. In practice, the hiding power (optical density) D of smoke is frequently determined from the formula

$$D = \frac{1}{l}$$

where l is the thickness of the smoke layer that makes invisible the light of the lamp serving as the standard in the measurement.

For example, in certain tests, a 25W bulb was used, and the test itself was conducted in a chamber having a width and height of 1.8 m and a length of 2.4 m.

To characterize the comparative value of smoke-forming substances, the so-called total darkening $W = V \cdot D$ is sometimes employed, where V is the volume of the smoke or fog obtained from a unit weight of the smoke-forming substance, and D is the hiding power of the smoke. The quantity W represents the area of the curtain in m^2 obtainable from 1 kg of smoke-forming agent and producing a complete darkening.

Values of total darkening for certain smoke forming substances and fuel mixtures as given by Prentiss are listed in Table 18.1

Table 18.1

Smoke-forming Substance or Mixture	Value of Total Darkening W m^2/kg
White phosphorus	1042
$NH_3 + HCl$	567
American NS mixture with a hexachloroethane base	466
$TiCl_4$	430
$SiCl_4$	340
$So_3 + HCl$	317

3. METHODS OF PREPARATION OF AEROSOLS

Smokes or fogs are prepared by the dispersion method and the condensation method. These two methods differ in that during the dispersion, the specific surface of the initial system increases while it decreases during condensation. The preparation of a smoke or fog by dispersion method consists in comminuting the solid (or liquid) substance by crushing (spraying) or pulverizing it by means of an explosion. The energy investment necessary for the preparation of aerosols by the dispersion method amounts to the execution of a certain amount of mechanical work.

The condensation process proceeds spontaneously and requires only a certain amount of energy to produce supersaturated vapor. When the vapor condenses, individual molecules cluster together, forming large aggregates, i.e., colloidal particles. A supersaturated solution may be obtained in two ways:

1) by cooling heated vapor;
2) by obtaining from the gaseous products a solid or liquid substance whose vapor supersaturates the ambient space.

Usually, the dispersion method produces aerosols of larger particles than the condensation method. This and the use in many cases of fairly complex and cumbersome dispersion apparatus are the reasons why in practice, aerosols are most frequently obtained by condensation. In some cases, dispersion and condensation may be used simultaneously for preparing aerosols, this being called the combined method of aerosol preparation.

In practice, smoke screens are prepared in the following ways:

1. Dispersion in air of volatile liquids (for example, $SnCl_4$), which form smoke by reacting in a finely divided state with atmospheric moisture: $SnCl_4 + 4H_2O = Sn(OH)_4 + 4HCl$.

2. By burning in air various substances whose combustion products condense in air on cooling. White phosphorus is most frequently used for this purpose: $4P + 5O_2 = 2P_2O_5$

3. Phosphorus anhydride vapor reacts with atmospheric moisture to form metaphosphoric and orthophosphoric acids, which have very low vapor pressures at room temperature.

4. By evaporation or incomplete combustion of oils; in the first case, the evaporated oil condenses in air in the form of fine droplets forming a fog; in the second case, incomplete combustion of oil results in the evolution of part of the carbon in the free state in the form of soot, forming black clouds that initially prevent objects from being detected. However, the soot rapidly flocculates. On contact, its particles cluster together and form flocs which steadily increase in size and quickly settle to the ground. Black smoke has poor hiding power. An intermediate method consists in an incomplete combustion of oil with simultaneous evaporation of its excess. In this case, the liquid oil particles coat the black carbon particles and thus initially prevent flocculation. Such a fog of grayish-black color is more constant until one of the ingredients, oil, evaporates. Thus, flocculation of a fog is only slightly postponed.

5. By burning pyrotechnic compositions of smoke screens containing or forming various smoke-forming substances in the course of the combustion reaction.

6. Some physicochemical properties of liquid smoke-forming substances of this type are given in Table 18.2.

Table 18.2

Smoke-forming Substance	Density	Melting Point, °C	Boiling Point, °C	Products of Reaction with Atmospheric Moisture
Tin tetrachloride	2.2	-33	+114	$Sn(OH)_4$; Hcl
Titanium tetrachloride	1.8	-23	+136	$Ti(OH)_4$; HCl
Silicon tetrachloride	1.5	-69	+59	$Si(OH)_4$; HCl
Chlorosulfonic acid	1.8	-80	+158	H_2SO_4; HCl
Sulfur Trioxide	1.9	+17	+45	H_2SO_4

4. COMPOSITIONS OF SMOKE SCREENS AND REQUIREMENTS IMPOSED ON THEM

The following special requirements are imposed on the compositions of neutral smoke screens:

1) The smoke obtained as a result of the combustion reaction should have the maximum hiding power (at least 500 m^2/kg) and be sufficiently stable in air;
2) when burning in smoke pots, the compositions must not produce a flame;
3) The slag formed by the combustion of the compositions must be loose (porous) and must not block the smoke passing through it.

The combustion products of compositions designed to produce a smoke curtain within friendly lines must not have any harmful effect on the health of the troops.

There are two markedly different types of smoke screens. One contains smoke-forming substances in ready form, while in the other, the smoke-forming substances are obtained from combustion (metal chloride smoke compositions). Compositions of the first type contain the following components: oxygen, fuel, and smoke-forming substance.

The smoke-forming substances used are readily volatilizable substances such as ammonium chloride (sal ammoniac) and aromatic hydrocarbons such as naphthalene, anthracene, etc. The process of smoke formation consists in the volatilization of the smoke-forming substances contained in such a composition and formation of smoke as they pass to the solid state as a result of cooling in air.

Ammonium chloride is readily soluble in water, and its density is 1.52. An appreciable sublimation of ammonium chloride begins at 250°C, while at 339°C its vapor pressure is already 760 Hg. The melting point of ammonion chloride in a closed space is 520°C. The heat capacity of ammonium chloride is 0.39 cal/g deg. At elevated temperatures, in addition to sublimating, ammonium chloride dissociates into ammonia and hydrogen chloride:

$$NH_4Cl = NH_3 + HCl.$$

The reaction is reversible and shifts to the left on cooling. [Note: At high temperature, ammonium chloride may be oxidized by atmospheric air:]

$$2NH_4Cl_{solid} + 2O_2 = 4H_2O_{vapor} + N_2 + Cl_2 + 80 \text{ kcal}.$$

Anthracene $C_{14}H_{10}$ consists of white crystals; its density is 1.25, melting point, 218°C, and boiling point, 342°C. Anthracene sublimes readily at about 200°C.

Phenanthrene $C_{14}H_{10}$, an isomer of anthracene, has a melting point of 99°C and boiling point of 240°C; it volatilizes readily at about 200°C.

Another smoke-forming agent frequently used in smoke compositions is commercial anthracene, which consists of a greenish-yellow mass containing anthracene (up to 25%), phenanthrene, carbazole ($C_{12}H_9N$), and other hydrocarbons. The flash point of crude anthracene is 150-162°C.

Wood charcoal is sometimes used as the fuel in smoke screen compositions.

In certain compositions, naphthalene and anthracene, in addition to playing their smoke-forming role, also act as fuels, partially volatilizing and partially burning. Potassium chlorate is usually employed as the oxidizer in smoke screen composition. Typical smoke compositions similar to Yershov's mixture contain:

	Potassium chlorate	20-30%
	Ammonium chloride	50%
	Naphthalene (or anthracene)	20%
	Wood charcoal	0-10%

The low chlorate content and the high content of the low-activity substance ammonium chloride ensure a comparatively low sensitivity of such compositions to mechanical effects.

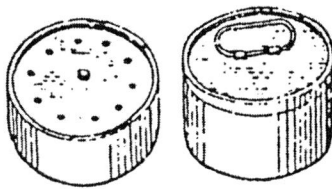

Fig. 18.2. Open and closed smoke pots.

The oxygen balance of smoke mixtures similar to Yershov's mixture is sharply negative, and the gaseous phase formed during combustion contains a significant amount of CO and readily ignitable naphthalene or anthracene vapor. When such mixtures are loaded into objects, they are placed in a shell provided with separate exit and entrance holes, which protects the hot surface of the composition from being penetrated by atmospheric oxygen (Fig. 18.2).

Despite the precautions taken, smoke compositions sometimes flare up during combustion. The uniformity of the combustion is thus disturbed, a flame appears at the pot openings, the amount of smoke evolved decreases markedly, and the smoke acquires a greyish tinge. For this reason, special substances, i.e., flash hiders, are added to excessively active smoke compositions. The best flash hiders are carbonate salts such as soda, chalk, and magnesium carbonate; their decomposition evolves carbon dioxide, which dilutes the gaseous reaction products capable of burning in air. Depending on the properties of a smoke composition, its content of flash hiders may be as high as 10-15% in certain cases.

$$CCl_4 + 2Zn = C + 2ZnCl_2.$$

Compositions of the second type are metal chloride smoke compositions. In such compositions, the smoke-forming substances are obtained as a result of a combustion reaction. The smoke formed in this case is gray, since in addition to the white particles of zinc chloride, it contains black particles of carbon.

Zinc chloride consists of white, very hygroscopic crystals; the density of zinc chloride is 2.91, its melting point 365°C, and its boiling point, 732°C. Zinc chloride partially hydrolyzes in moist air. For the properties of CCl_4, see [7, p. 235].

In addition to the two main components - chlorinated organic compound and metal powder - metal chloride smoke mixtures contain a number of components playing an auxiliary part. All the mixtures with CCl_4 contain an absorbent for this liquid. The absorbent most frequently used is diatomaceous earth (SiO_2), zinc oxide, or magnesium carbonate. For more information on the absorbents, see [211, p. 159].

Occasionally, an additional oxidizer such as $NaClO_3$, $KClO_3$ or $KClO_4$, and an additional smoke-forming agent such as ammonium chloride is also introduced into metal chloride smoke mixtures.

In the presence of an additional oxidizer - chlorate (or perchlorate) in the mixture, the smoke obtained is much whiter, since the carbon formed by the combustion of the mixture burns, forming CO (or CO_2).

As was pointed out by Ellern [9, p. 48], mixtures with CCl_4 are no longer used at the present time. [Note: On the use of smoke screens in the US Army, see Pavlov [212].]

In modern smoke mixtures, carbon tetrachloride has been replaced by solid chlorinated organic compounds such as hexachloroethane, hexachlorobenzene, hexachlorocyclohexane, etc. The elimination of liquid CCl_4 from the mixture makes it unnecessary to introduce absorbents into the composition.

An example of a smoke-forming mixture consisting only of solid ingredients is as follows:

Hexachloroethane	30%
Zinc	38%
Potassium chlorate	29%
Wood charcoal	3%

During World War II, German smoke charges were loaded with a composition consisting of the following ingredients:

Hexachloroethane	53%
Zinc	44%
Magnesium oxide	3%

One of the disadvantages of hexachloroethane is its significant volatility at room temperature, which surpasses even the volatility of naphthalene. Thus, at 60°C, the vapor pressure of naphthalene is equal to 2 mm, and that of hexachloroethane, 6 mm.

Much better properties in this respect are those of hexachlorobenzene.

Gordon and Campbell [213] studied a ternary mixture A consisting of zinc, C_6Cl_6 and $KClO_4$, and binary mixtures of the same components. According to their data, the heat of combustion of mixture A is 0.52 kcal/g, and the volume of gas is v_0 = 77 ml/g (85% CO and 15% CO_2).

The initiation temperature of mixture A is 325°C, and that of binary mixture B, Zn/ $KClO_4$ (65/35), 520°C. The heat of combustion of mixture B is 0.82 kcal/g. On the basis of their studies, the authors concluded that the preinitiation, initiation, and combustion of mixture A are due to the occurrence of three reactions:

1. $3Zn + C_6Cl_6 \rightarrow 3ZnCl_2 + 6C$.
2. $4C + KClO_4 \rightarrow KCl + 4CO(+CO_2)$.
3. $4Zn + KClO_4 \rightarrow KCl + 4ZnO$.

In his book, Ellern [9] mentions that metal chloride mixtures are completely stable when stored in dry form, but in some of them in the presence of moisture, spontaneous heating and even self-ignition may occur. For this reason, the mixture (C_2Cl_6 + Zn) is no longer used at the present time.

Some properties of solid chlorinated organic compounds are listed in Table 18.4.

Table 18.4

Compound & Formula	Percent of Chlorine in Compound	Molecular Weight	Density	Melting Point, °C	Boiling Point, °C
Hexachloroethane	90	237	2.1	185	Volatilizes rapidly at 185°
Hexachlorobenzene	75	285	1.6	227	323
Pentachlorobenzene	71	250	1.8	86	277
Octachloronaphthlene	70	404	--	203	--

In addition to zinc dust, powders of aluminum, iron, and calcium silicide may be used in metal chloride smoke compositions. The combustion of such mixtures takes place with the formation of chlorides of the corresponding metals:

$$C_2Cl_6 + 2Al = 2C + 2AlCl_3.$$

$$C_2Cl_6 + 2Fe = 2C + 2FeCl_3.$$

Given below are formulas of binary mixtures of metal powders with hexachloroethane, obtained by calculation assuming a complete oxidation of the fuel:

Hexachloroethane	81.4%	Hexachloroethane	67.9%
Aluminum	18.6%	Iron	32.1%

The mixture of hexachloroethane and aluminum burns vigorously; in order to slow down its combustion, a significant amount of substances are added which simultaneously function as additional smoke-forming agents (ammonium chloride, zinc oxide, etc.). For example, the HC white smoke composition has the formula [9]:

C_2Cl_6	45-47%
ZnO	47-48%
Al	5-8%
$ZnCO_3$	0-1%
Zinc borate	0-1%

The combustion of a mixture containing a small amount of aluminum involves the reaction [9]:

$$2Al + 9ZnO + 3C_2Cl_6 = Al_2O_3 + 9ZnCl_2 + 6CO.$$

However, the smoke formed by the combustion of such a mixture contains a small amount of soot, and its color is more grey than white.

New American formulas contain C_2Cl_6, ZnO, and NH_4ClO_4 and are wetted with polyvinyl acetate solution. Such mixtures can be loaded without pressure, and after the evaporation of the solvent, become strong and elastic at the same time.

Aluminum chloride, formed by the combustion of mixtures containing C_2Cl_6 and a substantial percentage of aluminum, is a white, readily subliming and extremely hygroscopic substance.

A mixture of hexachloroethane with iron powder burns slowly and rather inactively, evolving a brown smoke of ferric chloride. Anhydrous ferric chloride $FeCl_3$ sublimes at 250°C and boils with partial decomposition at 316°C. In air, it eagerly absorbs moisture, forming the crystal hexahydrate $FeCl_3.6H_2O$.

In the presence of an explosive initial impulse (detonating cap No. 8) and a strong shell, an explosion may take place in binary mixtures of CCl_4 or C_2Cl_6 with magnesium or aluminum, but, judging from the data obtained (Table 18.5) and considering the amount of mixture tested, the explosive force is rather weak.

Table 18.5 RESULTS OF TESTING OF METAL CHLORIDE MIXTURES IN A TRAUZL BOMB*

Composition of Mixture %	Expansion in Trauzl bomb CM^3
CCl_4 76; Mg 24	215
CCl_4 81; Al 19	156
C_2Cl_6 81.5; Al 18.5	64

Weighed portion of mixture, 20g; instead of the usual immersion in sand, a lead rod was placed in the bomb channel.

The reaction of hexachloroethane with calcium silicide forms volatile silicon tetrachloride $SiCl_4$, which on reacting with moist air forms silicic acid smoke.

As an example of a smoke mixture with calcium silicide, we can cite the following:

        Hexachloroethane   45%
        Zinc oxide   45%
        Calcium silicide   10%

Chapter 19

COMPOSITIONS OF COLORED SMOKES

Compositions of colored smokes are used in warfare for signaling in the daytime; they are also used in adjustment fire and for target designation in bombing and artillery fire. In transportation, they are used for distress (emergency) signals on land and particularly at sea.

1. SIGNALING BY MEANS OF COLORED CLOUDS AND METHODS OF THEIR PREPARATION

In daytime signaling, use is made chiefly of smoke: red, yellow, green, and blue.

Indications are that it is also possible to use black smoke for daytime signaling, the black being given either as a fifth signal or to replace one of the aforementioned colors (blue or green). Blue smoke and green smoke are the most difficult to tell apart, particularly at great distances.

The chief disadvantage of daytime signaling by means of colored smoke is that the accuracy of observation and the reliability of determination of color depend on the weather conditions.

The factors exerting the greatest influence on the results of observation of smoke signals are:

1) size and shape of the colored smoke signal;
2) brightness and color of the background on which the smoke signal is projected;
3) height of the sun above the horizon and position of the observer with respect to the sun and smoke cloud;
4) wind speed;
5) atmospheric precipitation (snow, rain, fog).

The heat visibility and color discrimination of smoke clouds exist in clear calm weather at a wind speed of not more than 2-3 m/sec.

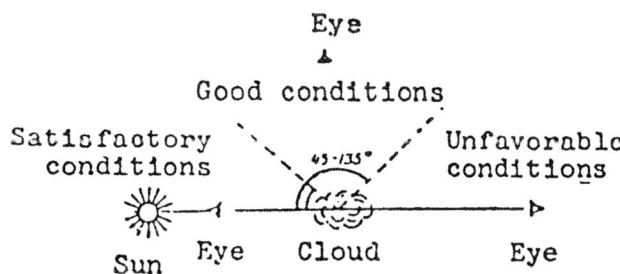

Fig. 19.1. Zones of different conditions of observation of colored signal smokes.

At a wind speed of over 6 m/sec, a smoke cloud dissipates rapidly. The color of a smoke cloud is perceived worst when the latter is located on a straight line between the sun and the observer's eye (Fig. 19.1); when the observation is made under such conditions, many colored clouds appear almost white. Conversely, when the sun is behind the observer's back, a colored cloud acquires a dark color. The best conditions for observing the cloud color exist when the an gle formed by the sun, the smoke cloud and the observer's eye are within a 45-135° range. The stability of a signal cloud in air is greatly affected by its size and shape.

Parachute signal devices usually produce a colored smoke band. This band is usually scattered by air currents, but fresh portions of smoke continually pour out of the smoke pot into the atmosphere.

In devices with no parachute, smoke formation must take place almost instantaneously: a compact color cloud is formed in the air, and an accurate observation of this cloud, depending on the amount of smoke composition used and the weather conditions, is possible in a time period ranging from several tens of seconds to several minutes.

Colored signal smokes used to be produced by dispersion by atomizing finely divided mineral paints with explosives or powder. The mineral dyes used were minimum, cinnabar, and ultramarine blue.

However, practice has shown that in order to obtain a colored cloud having a satisfactory quality and observation time, it is necessary to consume an excessive quantity of dyes (1-2 kg). It has also been observed that colored smoke obtained by mechanical atomization of mineral dyes is dispersed very quickly, since its particles are too large (of the order of 1×10^{-2} to 1×10^{-3} cm) and settle rapidly.

Production of colored smokes by condensation through chemical reactions taking place between inorganic substances has not had any particular success either.

Relatively good results have been obtained only with black smoke compositions. It has been suggested that black smoke be obtained by burning metal chloride mixtures containing carbon-rich compounds of the aromatic series as the third component.

As an example, a black signal smoke composition of the following formula can be cited:

Hexachloroethane	60%
Magnesium	19%
Naphthalene (or anthracene)	21%

According to the literature data, this mixture burns at the rate of 4 mm/sec to form a thick black smoke and is sensitive to impact. Its drawbacks include a rapid volatilization of hexachloroethane and naphthalene, and a low melting point of the hexachloro-ethane-naphthalene mixture. A mixture of 8 parts of naphthalene to 20 parts of hexachloroethane melts at 53°C. It has therefore been found desirable to replace naphthalene with anthracene.

The combustion of such mixtures should take place with free access of air, otherwise naphthalene or anthracene will not burn, but sublime, and the smoke obtained will not be black, but gray.

Black smoke compositions containing no chlorinated organic compounds have the following formulas:

Potassium chlorate	45%
Naphthalene	40%
Wood charcoal	15%

or

Potassium chlorate	55%
Anthracene	45%

A characteristic feature of such mixtures of black smoke in comparison with compositions of masking white smoke is the absence of ammonium chloride and high content of potassium chlorate. In this case, the combustion temperature of the mixture rises substantially, and its burning involves not the sublimation of naphthalene, but its incomplete combustion, associated with the evolution of a large amount of soot.

Of the colored smokes obtained by chemical reaction, only the yellow smoke composition appears to deserve some attention [9]: Mg 16%, C_2Cl_6 48%, Fe_2O_3 36%. Ferric chloride $FeCl_3$ formed by the combustion imparts color to the flame.

Highest-quality signal smokes of all colors (red, yellow, green, blue) have been obtained by subliming organic dyes.

In compositions containing an oxidizer, a fuel, and an organic dye, the latter is converted to the vapor state by the heat evolved by he combustion, and is driven out by the gaseous reaction products into the atmosphere, where its vapors condense to form a colored smoke.

An example of mixtures in which the formation of smoke is based on sublimation of organic dyes is the composition of blue smoke:

Potassium chlorate	35%
Lactose	25%
Synthetic indigo	40%

2. DYES USED IN SIGNAL SMOKE COMPOSITIONS

Organic dyes used in signal smoke compositions must meet the following special requirements:

1) They must sublime rapidly starting at 400-500°C;
2) Their sublimation must be associated with minimum decomposition of the dye;
3) The colored smoke formed by condensation of dye vapor in air should have a distinct and specific color (red, yellow, blue) and be sufficiently stable in air.

The sublimation rate of the dye depends chiefly on its vapor pressure at high temperatures. The vapor pressure of certain dyes at 260°C and their latent heats of sublimation are given in [7, p. 243].

Obviously, the sublimation rate will also depend on the heat capacity and latent heat of sublimation of the dyes; the dye will sublime faster, the less heat is required for converting it to the vapor state.

The average specific heat of the majority of dyes employed in the 20-200°C range amounts to 0.3-0.4 cal/deg.

A rapid sublimation of organic dyes is necessary because the latter decompose when exposed to high temperatures for a long time.

In qualitative testing of a dye for its thermal stability and capacity to form colored vapors, a pinch of dye is dropped into a porcelain crucible heated to 400-600°C on an ordinary gas burner. If the dye forms a colored vapor, this indicates that it is relatively stable to heat and it should be tested further in pyrotechnic compositions. If however, the dye decomposes during the test, evolving colorless or dirty gray vapors, it is unsuitable for use in signal smoke compositions.

It has been established experimentally that the dyes most stable to heat have a simple chemical structure. In addition, it has been shown that certain chemical groups in an organic dye may be responsible for the fact that no colored vapors are formed when the dye is heated. Groups whose presence in the dye prevents its conversion to the vapor state include:

1. The sulfo group in the form of $R-SO_3H$ or $R-SO_3K$, or $R-SO_3Na$.
2. The R-O-Me group (Me being a metal).
3. The benzidine group when present in azo dyes:

$$\equiv N-\langle\ \rangle-\langle\ \rangle-N\equiv$$

Water-insoluble sulfur dyes do not produce colored vapors either, and hence are unsuitable for signal smoke formation.

Colored smokes are produced most frequently by using mononitroamino compounds (intermediate products), azo dyes, and diphenylmethane, triphenylmethane, thiazine and anthraquinone dyes (for a classification of dyes, see [214].) The following intermediate products and dyes are most frequently used in colored smoke compositions:

1. Paranitroaniline - yellow smoke.
2. Auramine O - yellow smoke (used with an admixture of the brown dye chrysoidine).
3. Sudan I (phenylazobetanaphthol) - orange smoke.
4. Rhodamine B - raspberry red smoke.
5. Sudan red (2-anisidine-azo-beta-naphthol) - red smoke.
6. Methylene blue - blue smoke.
7. Indigo - blue smoke.

Anthraquinone dyes may also be used for the preparation of colored smokes. An example of this class of dyes is 1,4-dimethylamincanthraquinone

which on subliming gives a good smoke of blue color.

Anthraquinone dyes also include "Sudan blue G", which is 1-methylamino-4-paratoluene-anthraquinone

$$C_{16}H_{16}N_2O_2$$

which was used in German blue smoke compositions.

1-Methylaminoanthraquinone is used for the preparation of red smoke.

3. COLORED SMOKE COMPOSITIONS

The sublimation of organic dyes for the purpose of producing colored signal smokes takes place as a result of the heat evolved by the combustion of the so-called "thermic mixture" made up of the oxidizer and fuel.

The thermic mixture must evolve the amount of heat necessary to convert the dye to the vapor state, but must not develop a high temperature during combustion so as not to decompose the dye. In addition, the combustion of the thermic mixture must evolve a significant quantity of gaseous reaction products that will promote a rapid expulsion of the vapors of the subliming dye from the zone of the combustion reaction. For these reasons, metal powders can not be used as fuels in colored smoke compositions.

Among fuels, the most suitable in this case are organic substances. In practice, carbohydrates are most frequently used as fuels in signal smoke compositions, since their combustion forms a large quantity of gaseous reaction products and does not evolve too much heat (Table 19.1). It is evident from Table 19.1 that the best indices are those of mixture 2, where the combustion of lactose proceeds only up to carbon monoxide. Experience shows that when such a thermic mixture, designed for incomplete combustion of the fuel, is used in colored smoke compositions, the best quality of smoke can be obtained.

Among carbohydrates, lactose or beet sugar, starch, and sawdust can be used in colored smoke compositions. [Note: These substances are products of the food industry; they can be replaced by fuels used in insecticide smoke compositions, for example, dicyandiamide (see Ch. 3).]

In colored smoke compositions, potassium chlorate is used almost exclusively as the oxidizer; however, the foreign literature occasionally gives formulas of compositions with potassium perchlorate as the oxidizer. Becher [8] noted that a thermal mixture of potassium chlorate with carbohydrates may be replaced by a nitrocellulose mixture if desired.

Table 19.1 HEAT OF COMBUSTION REACTION & SPECIFIC VOLUME OF GASEOUS REACTION PRODUCTS OF BINARY MIXTURES

No.	Combustion Reaction of Mixture	Content of Combustible in Mixture, %	Heat of Combustion, kcal/g	Specific Volume Gaseous Reaction Products, cm^3/g
1	$8KClO_3 + C_{12}H_{22}O_{11} \cdot H_2O$ $= 8KCl + 12CO_2 + 12H_2O$	26.9	1.06	401
2	$4KClO_3 + C_{12}H_{22}O_{11} \cdot H_2O$ $= 4KCl + 12CO + 12H_2O$	42.3	0.63	632

The amount of dye in colored signal smoke compositions varies from 40 to 60%, depending on the chemical nature of the dye itself and on the required combustion rate of the composition.

In some cases, a certain amount of substance preventing the inflammation of the dye vapor is introduced. Such substances have been named flash hiders. One such substance is $NaHCO_3$. The role of a flash hider is to lower the temperature in the composition and to form an unreactive gas (CO_2) during the decomposition which reduces the contact between the dye vapor and air.

As a colored smoke composition with a "plastic binder", Ellern [9] cites the following:

Organic dye	51%
Sugar (cane)	18
$KClO_3$	23
$NaHCO_3$	8

Above 100% of the mixture, 2.2 parts of polyvinyl acetate dissolved in 50 parts of dichloromethane (DCM is driven off as the composition dries) is introduced into the mixture.

The content of the ingredients of chlorate mixtures of colored signal smokes containing no flash hider is approximately as follows:

Potassium chlorate	30	±10%
Carbohydrate	20	±5%
Dye	50	±5%
Resin binder	0 - 5%	

The introduction of a binder is necessary only when granulation of smoke compositions is carried out. The granulation operation is performed in order to increase the combustion surface of the composition and thus reduce the total time of smoke formation. In addition, granulation of the composition improves the smoke color, since in a granulated composition, the dye vapor can be more rapidly expelled from the reaction zone and must not pass through the slag, which has been heated to a high temperature and has a very dense consistency in some cases. The diameter of chlorate composition granules being prepared ranges from 2 to 5 mm. Granulation of the composition prevents its stratification during transportation, loading, and application.

Signal smoke compositions must burn without access of atmospheric oxygen. For this purpose, the composition is placed in a porous shell, i.e., burlap or cardboard, or a sheet iron cup with a row of holes to allow the smoke to escape. When atmospheric oxygen has access to the mixture, its combustion temperature rises appreciably as a result of the additional combustion of carbon monoxide $2CO+O_2 = 2CO_2+136$ kcal, which is associated with the formation of a flame and caused the colored dye vapors to burn up almost completely.

However, even when the components of the composition interact without forming a flame, partial decomposition of the dyes still takes place. The amount of smoke formed by the combustion of chlorate smoke compositions usually amounts to 30 to 70% of the initial weight of the dye, and in addition, the smoke itself contains only 60-80% of the dye (the remaining 20-40% corresponds to its decomposition products, i.e., tars). For formulas of chlorate signal smoke compositions, see [7, p. 252]. Among compositions containing nitrates, the following can be indicated:

Sodium nitrate	25%
Sawdust	35%
Rhodamine	40%

An important disadvantage of this composition is its great hygroscopicity.

In artillery ranging under combat conditions (particularly in the case of simultaneous firing of many weapons), it frequently becomes necessary to note the impact of high explosive shells or their air bursts. In the daytime, this is accomplished by the formation of a cloud of vividly colored smoke as the shell explodes. Such shells can also be used in antiaircraft protection.

It is well known that mixtures of organic dyes with various explosives are used in the formation of colored smoke explosions in artillery fire. When the charge explodes, the organic dye sublimes and a colored smoke is formed.

In 1941-45, German target-designating shells were loaded with mixtures of high explosives (PETN, hexogen) with azo dyes. The dye content of such systems was 55-65%.

An American patent [218] gives the following description of the design and loading of 3- and 5-inch target-designating shells. The colored substances (organic dyes) are arranged in the shell case so that the explosion converts them to the vapor state; on coming in contact with cold air, the sublimed dyes condense as a thin, bright-colored smoke. Dyes making up to about 10% by weight of the explosive charge should not be mixed with explosives, but distributed between the walls of the shell and the central explosive charge. For low-melting dyes (melting point below 150°C), loading by centrifugal filling is possible; to lower the melting point, the dye should be mixed with paraffin or a thermosetting resin.

Particularly suitable as explosives, since they do not produce a dark cloud of soot upon exploding, are: 1) nitroguanidine, 2) a mixture of ammonium picrate and ammonium nitrate (NH_4NO_3 from 40 to 75%).

Dyes which have shown good results are:

1. Red or orange smoke: Sudan I and diethyl-m-aminophenol-pathalein hydrochloride;
2. Yellow smoke: aminoazotoluene and benzeneazodimethyl-aniline;
3. Blue smoke: quinizarin blue, anthraquinone violet.

Explosions of orange-red smoke are observed in air best.

Ellern [9] indicates that EC smokeless powder can be used for the preparation of colored smokes, and gives the following composition formula:

Organic dye	50%
EC powder	50%

EC powder is a mixture of partly gelatinized nitrocellulose (NC) with an admixture of nitrates, namely:

NC	80.4%
KNO_3	8.0%
$Ba(NO_3)_2$	8.0%
Starch	3.0%
Diphenylamine	0.6%

For formulas of signal smoke compositions used during World War II, see also [216].

4. METHODS OF TESTING COLORED SMOKE COMPOSITIONS

Special studies of colored smoke compositions consist in investigating the properties of the colored smoke formed by their combustion. The following tests are most frequently carried out:

1) determination of the total amount of smoke obtained from the combustion of 1 g of smoke composition;
2) determination of the stability of smoke in air, and determination of the sizes of smoke particles;
3) determination of smoke chromaticity.

These tests are conducted in the same type of smoke chamber as that for studying the properties of masking powder.

The determination of the total amount of smoke formed by the combustion of a smoke composition in the chamber is made by weighing glass plates (9 x 12 cm) placed on the bottom and side walls of the chamber before and after the experiment.

The stability of colored smoke in air may be determined by any of the methods used for studying the stability of aerosols.

The chromaticity of smokes may be determined by the methods used for measuring the color of stained tissues and generally of all colored materials. Although in this case it is necessary to consider only the determination of smoke color in reflected light, this determination is of the greatest value.

A determination of the color characteristics of signal smokes in reflected light on the Pulfrich spectrophotometer gave the r results listed in Table 19.2.

Table 19.2 COLOR CHARACTERISTICS OF SIGNAL SMOKES

Composition of Signal Smoke, %	Color Characteristics of Smoke		
	Hue	Color Purity P, %	Brightness of Smoke in Relative Units
Potassium chlorate 40 Aramine 26 Crysoidine 14 Lactose 20	584	75	12.6
Potassium chlorate 40 Sudan I 20 Rhodamine 20 Lactose 20	600	54	8.9
Potassium chlorate 40 Ammonium oxalate 7 Methylene blue 40 Lactose 13	487	11	14.1

Studies dealing with the determination of smoke chromaticity have also been carried out abroad [217].

The determination of smoke color in transmitted light is of no particular interest to pyrotechnists and does not permit an evaluation of the quality of the smoke obtained.

Chapter 20

SOLID PYROTECHNIC FUELS

1. CLASSIFICATION AND ENERGY CHARACTERISTICS

[Note: The term "solid pyrotechnic fuel" is not generally accepted. It was introduced by the author in order to distinguish from the numerous types of jet engine solid fuels those which are similar in composition and properties to certain pyrotechnic compositions.]

Various solid pyrotechnic fuels have now been developed. According to their function, they can be divided into four characteristic types:

1) fuels for ramjet engines (RJE) or rocket-ramjet engines (RRE);
2) fuels for hydro jet engines (HJE);
3) fuels for composite engines (CE);
4) composite fuels for rocket engines (RE) are also similar to solid pyrotechnic fuels.

The use of a fuel determines its composition and properties. In all jet engines, the energy source is the energy evolved by the combustion of the fuel. When the reaction products flow from the jet engine nozzle, the energy of the fuel is converted into the kinetic energy of the jet.

To determine the energy characteristics of solid fuels (as well as the characteristics of their combustion products), a high degree of accuracy of the calculations is necessary, and all the physicochemical processes involved in the combustion of the fuel and in the flow of the combustion products out of the nozzle must be taken into account to the maximum possible extent. It is usually assumed that the process of discharge of the combustion products from the nozzle takes place isentropically, i.e., $\Delta S = 0$.

For this reason, the change in the kinetic energy of a jet being discharged from the nozzle takes place as a result of a change in the heat content of the combustion products of the fuel:

$$\Delta I = \frac{G_f}{2g} (u_n^2 - u_{cc}^2),$$

where ΔI is the change of heat content of the combustion products in kcal/kg;
G_f is the flow rate of fuel in kg/sec;
u_n is the gas discharge velocity at the exit from the nozzle in m/sec;
u_{cc} is the gas discharge velocity at the end of the combustion chamber in m/sec;
g is the acceleration due to gravity.

Thus, the problem amounts to determining the heat content of the combustion products of the fuel in the combustion chamber and at the nozzle exit. Special thermodynamic calculations are carried out for this purpose, based on the equality of the total heat content of the fuel to the total heat content of the reaction products at a given pressure, and on the principle of complete thermodynamic equilibrium of the combustion products at a given temperature and pressure [224, 221, 232, 229].

At the same time, one takes into consideration the heat of formation of the fuel components, the change in the thermodynamic characteristics of the combustion products with changing temperature [233], and the dissociation of the combustion products in the combustion chamber at high temperature.

Such calculations are used to determine the temperature of the combustion products of the fuel at various pressures, the composition and characteristics of the combustion products, and the specific thrust (specific impulse) at a given pressure in the chamber. Results of these calculations form the basis of subsequent calculations of the engine and of various processes occurring in the engine during combustion of the fuel, namely, calculation of the nozzle, geometry of the solid fuel charge, conditions of flow of the combustion products in the engine chamber or along the channel of the charge, and calculations of the heat insulation of the engine and heat losses.

Energy transformation occur differently in different jet engines.

1. A solid-fuel ramjet engine develops a thrust in flight as a result of a change in the momentum of the air jet flowing through the engine. A diagram of this engine is shown in Fig. 20.1.

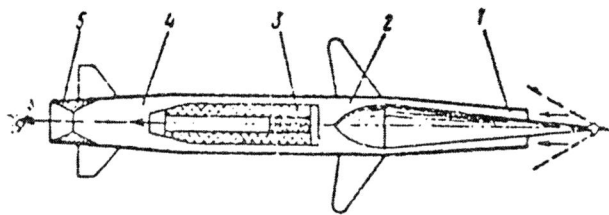

Fig. 20.1. Diagram of solid-fuel ramjet engine.
1 - supersonic part of diffuser, 2 - subsonic part of diffuser, 3 - solid fuel charge, 4 - combustion chamber, 5 - exhaust nozzle.

A change in the momentum of the air jet in the engine takes place as a result of the influx of heat liberated by the combustion of fuel in the combustion chamber. Thus, the air entering through a special air intake (diffuser) in this case acts as the oxidizer and the mass carrier.

The specific impulse and thrust coefficient are usually employed as the basic characteristics of a ramjet engine. The specific thrust I characterizes the economy of the engine. The magnitude of the thrust coefficient C_R determines the maximum thrust that can be obtained from a given fuel in a given engine. It is known from the theory of RJE that the magnitude of the specific thrust and the thrust coefficient depend on the flight velocity and altitude [222, 219, 230]:

$$I = \frac{R}{G_f} = \frac{\alpha \cdot L_0 \cdot V_H}{g} \left(\beta\sqrt{\sigma} - 1 \right),$$

$$C_R = \frac{R}{S_p \dfrac{\gamma_H V_H^2}{2g}} = 2\left(1 - \frac{1}{\beta\sqrt{\sigma}}\right)$$

where R is the engine thrust in kg;
G_f is the flow rate of fuel in kg/sec;
α is the excess air coefficient;
L_0 is the stoichiometric constant;
V_H is the flight velocity in m/sec;
β is the mass increment coefficient, equal to the ratio of the flow rate of combustion products to the flow rate of air;

PRINCIPLES OF PYROTECHNICS

σ is the relative heating of air;
S_p is the cross sectional area of the rocket in m²;
γ_H is the density of air at the flight altitude in kg/m³.

The stoichiometric constant L_0 of fuel is the amount of air in kg theoretically necessary for the combustion of 1 kg of fuel. The relative heating σ in solid fuel operation of the engine depends on the calorific power of the fuel H_u and the ratio of the flow rate of air to the flow rate of fuel αL_0

$$\sigma = \frac{T_b}{T_a} = 1 + \frac{H_u}{(1+\alpha L_0)C_p T_a}$$

where T_b is the stagnation temperature of the combustion products in °K;
T_a is the stagnation temperature of the incident flow of air in °K;
C_p is the heat capacity of the combustion products in kcal/kg.

Therefore, the highest economy of RJE or the maximum specific thrust may be obtained by using fuels having the maximum calorific power (in kcal/kg) when they burn in air.

Example 1. Calculate the specific thrust I of RJE at V_H = 500 m/sec and an altitude of 10 km; the flow rate of air through the engine is 20 kg/sec, G_f = 2 kg/sec. The fuel is magnesium. For the given V_H and a flight altitude of 10 km, T_a = 355°K. The heat capacity of the combustion products may be assumed equal to C_p = 0.3 kcal/kg deg. The calorific power of magnesium is 5880 kcal/kg.

$$\text{Complex } \alpha L_0 = \frac{G_a}{G_f} = 20/2 = 10$$

where G_a is the flow rate of air.

The relative heating

$$\sigma = 1 + \frac{5880}{(1+10)0.3 \cdot 355} = 6$$

$$\beta = \frac{20+2}{20} = 1.1$$

The specific thrust

$$I = \frac{10 \cdot 500}{9.81}(1.1\sqrt{6-1}) = 865 \frac{\text{kg sec}}{\text{kg}}$$

Example 2. Calculate the specific thrust I under the same conditions as in Example 1. The fuel is aluminum. Calorific power of aluminum, 7350n kcal/kg.

$$\sigma = 1 + \frac{7350}{(1+10)0.3 \cdot 355} = 7.27$$

The specific thrust

$$I = \frac{10 \cdot 500}{9.81}(1.1\sqrt{7.27-1}) = 1000 \frac{\text{kg sec}}{\text{kg}}$$

To obtain a large thrust coefficient C_R, which is required in a case of flight at a supersonic speed, a propellant with a high heating power is needed:

$$V_H = \frac{H_u}{1+L_0}.$$

The calorific efficiency is the amount of heat corresponding to 1 kg of combustion products for a stoichiometric composition of the mixture of fuel and air. For this reason, solid fuels for RJE should, in addition to a high calorific power, have the lowest possible value of the stoichiometric constant L_0. In particular, liquid hydrocarbon fuels of high calorific power, 9000-10,000 kcal/kg, do not meet the latter requirement, since they also have a high stoichiometric constant L_0 equal to 13-15 [222]. Usually, solid fuels for RJE consist of a metallic fuel, an organic binder fuel and special additives [258]. Certain fuels for RJE also contain an oxidizer. Light metals, i.e., Mg, Al, and their alloys, are used as metal fuels [258, 268]. These metals have a relatively high calorific power and efficiency. The best theoretical characteristics are those of Be, but the objections usually raised concern the high cost and incompleteness of combustion of this metal [260].

The use of metals in fuels for RJE is particularly necessary at high flight velocities (when a very high temperature is developed in the combustion chamber), since the combustion products of the metals have a high thermal stability.

It is known from literature data that there exists a fast burning fuel for RJE which consists of a magnesium-aluminum alloy, a small amount of binder, and a small amount of oxidizer [258]. Charges made of this fuel are prepared by pressing at a high specific pressure.

In addition to the metal and oxidizer, slow burning fuels for RJE contain up to 50% of epoxy resin. Charges of such fuel are prepared by pouring into a shell or chamber where the fuel solidifies. Typical molds of solid fuel charges for RJE are charges of either radial or end combustion [253, 265].

In some cases, it is desirable to use a rocket ramjet engine. This engine constitutes an organic union of a rocket engine (RE) and a ramjet engine into a single structure [258]. In addition to a high calorific power, solid fuels for such engines must have a sufficiently high intrinsic specific thrust. It is also desirable to introduce readily ignitable dispersed metal particles into the composition of a solid fuel for RRE [230]. At the same time, fuels for RRE must have a high combustion temperature and yield a sufficient quantity of gaseous reaction products on burning. The characteristics of ramjet engines substantially depend on the degree of combustion of the fuel in the engine [222, 219]; it is characterized by the combustion efficiency coefficient ϕ_b. <u>The combustion efficiency coefficient is the ratio of the heat content increment of the combustion products to the chemical energy of the fuel introduced into the chamber</u>. The achievement of a high combustion efficiency requires the solution of such important problems as mixing of the combustion products of fuel and air, and afterburning of gaseous and solid particles under the combustion chamber conditions. Tests of fuels based on an aluminum-magnesium alloy and epoxy resin gave a combustion efficiency of 0.4 to 0.8 depending on the testing conditions [258].

2. Hydro jet engines (HJE) use overboard seawater as the oxidizer. A diagram of such an engine is shown in Fig. 20.2. Hydro jet engines are used in torpedoes [256]; during the motion of a torpedo, seawater enters the engine through special water inlets and is used as the oxidizer and mass carrier. In the combustion chamber, the heat energy of the fuel is consumed in vaporizing the water. The mixture of water vapor and fuel combustion products is expelled from the nozzle and thus provides the necessary thrust. For underwater engines, the volume specific thrust, i.e., the thrust developed by the engine as a result of combustion of 1 liter of fuel per second, is the determining thrust. Therefore, the most important characteristic of a hydro jet fuel is the heat liberated by the combustion of 1 liter of fuel reacting with water, and the amount of water L_0 (in kg) necessary for the combustion of 1 kg of fuel.

The high specific heat (in kcal/l) of the reaction with water is one of the chief requirements placed on a hydro jet fuel. In addition, the substances used as fuels should vigorously react with water [264].

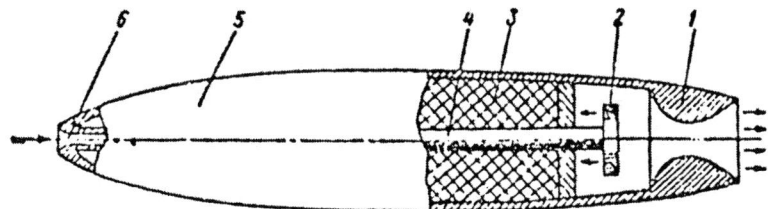

Fig. 20.2. Diagram of solid-fuel hydro jet engine.
1 - exhaust nozzle, 2 - collector and head with nozzles for atomizing water,
3 - solid-fuel charge, 4 - channel for introduction of water, 5 - engine body, 6 - water inlet.

Some metals such as Al, Zr, Mg, as well as Li and its alloys possess very high energy characteristics both when overboard water is used as the oxidizer and when an oxidizer entering into the fuel composition is used [256, 261].

The necessity of ensuring a high combustion efficiency of the fuel in the engine requires special ways of solving the problem. One of the methods of preparing the charge for a solid hydro jet fuel consists in pressing a mixture of metal and oxidizer powders [256]. Such a fuel should have a negative oxygen balance; perchlorates may be taken as the oxidizers. Occasionally, in order to increase the rate of combustion of the fuel, admixtures of finely divided powders of metals such as cobalt, copper, ferrovanadium, lead, or tin, are introduced into the perchlorates. To prepare the charges, a thoroughly stirred mixture of the fuel components is pressed at a high specific pressure [248]. However, the highest specific thrust in a ramjet HJE may be obtained by using pure metals as fuel. The magnitude of the specific thrust of HJE depends on the ratio of the flow rate of overboard oxidizer G_a to the flow rate of fuel G_f

$$\frac{G_a}{G_f} = \alpha L_o$$

where α is the excess water coefficient.

The energy possibilities of hydro jet fuels are shown in Fig. 20.3 [256].

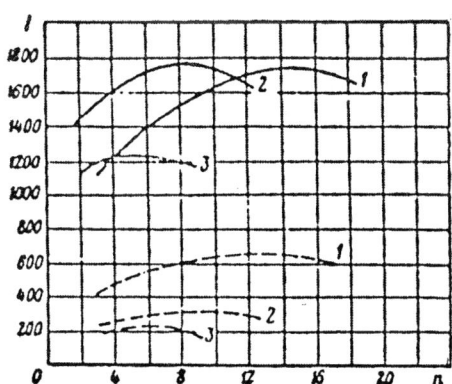

Fig. 20.3. Specific impulse (I) of HJE versus excess water in moles (n).
1) $2Al + n\ H_2O$; 2) $Zr + n\ H_2O$; 3) $ZrH_2 + n\ H_2O$.
— 1 volume specific impulse in kg sec/l; --- 1 weight specific impulse in kg sec/kg.
Pressure in the engine chamber, 23.2 kg/cm²; pressure at nozzle exit, 1.14 kg/cm².

3. Fuels for composite engines, CE, are a comparatively new type of solid fuels. CE operate on solid fuel and liquid oxidizer and are a type of rocket engine. At a given pressure in the chamber of a rocket engine, the specific thrust is independent of the flight conditions and surrounding

medium, and is determined only by the properties of the solid fuel and liquid oxidizer employed. Rocket engines operating on mixed fuel (solid-liquid), i.e., CE, have considerable advantages over RE, since their specific thrust is higher, and they permit regulation of the engine thrust and its repeated starting up by changing the flow rate and cutting off the liquid component [252]. One of the possible layouts of a composite engine is shown in Fig. 20.4. The liquid oxidizers used are nitric acid HNO_3, nitrogen oxides (for example, N_2O_4, perchloryl fluoride $FClO_4$, hydrogen peroxide H_2O_2, or chlorine trifluoride ClF_3.

From the standpoint of operation of the engine, the best characteristics are those of fuels capable of self-igniting in contact with a liquid oxidizer. At the present time, multicomponent solid fuels are known which in the presence of a liquid oxidizer generate a sufficiently high specific impulse [252]. Such a fuel consists of a metal powder, thermoplastics, a hypergolic component, and its plasticizer. Among metals, Al, Be, Mg, or Li is used, and the metal particles should be no larger than 10μ.

The thermoplastic used is polychlorovinyl plasticized with butyl phthalate, rubber, or polyurethane [252]. The hypergolic components used are solid amines, for example, paratoluidine, paraphenylenediamines, 1,5-diaminonaphthalene, etc. [252].

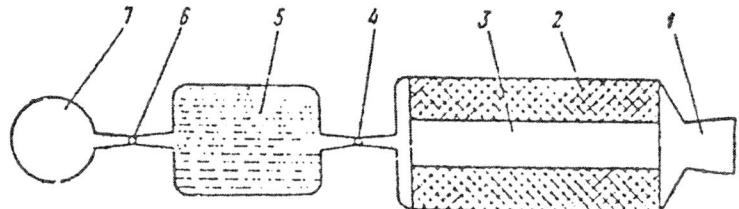

Fig. 20.4. Diagram of composite rocket engine.
1 - exhaust nozzle, 2 - solid fuel charge, 3 - channel, 4 - shutoff valve and head with nozzles for atomizing the liquid oxidizer, 5 - liquid oxidizer tank, 6 - nozzle, 7 - compressed gas tank.

The introduction of up to 20% ammonium perchlorate into the solid phase leads to a reduction of the ignition time lag and to the formation of a cleaner free combustion surface, thus improving the combustion characteristics [252].

4. Solid fuel rocket engines - SFRE [244, 251, 266], have lately undergone a considerable development owing to the advent of high-energy composite fuels (propellants). A typical diagram of SFRE is shown in Fig. 20.5. The chief advantages of SFRE are: a comparative simplicity of design, instant readiness, convenience and relative safety of operation, and high degree of operational dependability. Combustion of the fuel in SFRE liberates a large quantity of high-temperature gases. A pressure of the order of several tens of atmospheres is thus developed in the chamber. When they are discharged from the nozzle, the gaseous reaction products expand to the external pressure. The expansion process is associated with an acceleration of the motion of the gas in the nozzle, i.e., thermal energy is converted into kinetic energy of motion of the gases. The engine thrust is determined from the equation of conservation of momentum

$$R = \frac{G_f}{g} \cdot u_n + (p_n - p_H)S_n,$$

where   R is the engine thrust in kg;
   G_f is the flow rate of fuel in kg/sec;
   u_n is the gas discharge velocity at the exit from the nozzle in cm/sec;
   p_n is the pressure at the nozzle exit in km/cm^2;
   p_H is the atmospheric pressure in kg/cm^2;
   S_n is the area of the exhaust cross section of the nozzle in cm^2.

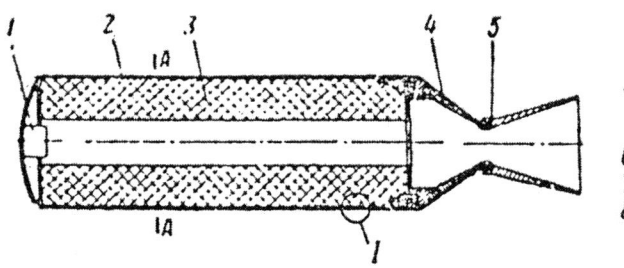 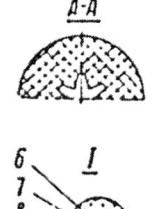

Fig. 20.5. Diagram of solid-fuel rocket engine.
1 - igniter, 2 - engine body,
3 - solid fuel charge,
4 - nozzle, 5 - graphite lining,
6 - fuel, 7 - armor protection,
8 - clearance,
9 - thermal insulation and engine wall.

The main characteristic of a fuel is the specific thrust, i.e., the thrust developed by the engine when fuel in the amount of 1 kg/sec is consumed:

$$I = \frac{R}{G_f} = \sqrt{\frac{2k}{k-1} \cdot \frac{R_{gas}}{M \cdot g} T \left[1 - \left(\frac{P_c}{P}\right)^{\frac{k-1}{k}}\right]}$$

where k is the specific heat ratio of the gaseous combustion products;
M is the average molecular weight of the gaseous combustion product;
R_{gas} is the universal gas constant in kg m/deg mole;
T and p are the temperature and pressure in the chamber.

Example. Calculate the specific thrust of a composite fuel at p = 40 kg/ cm²; combustion temperature of fuel T = 2790°K, M = 25.5, k = 1.22, R_{gas} = 848.

$$I = \sqrt{\frac{2 \cdot 1.22}{1.22 - 1} \cdot \frac{848}{25.5 \cdot 9.81} 2790 \left[1 - \left(\frac{1}{40}\right)^{\frac{1.22-1}{1.22}}\right]} = 227 \text{ sec.}$$

The methods of calculation and measurement of specific thrust have been adequately described in [224, 221, 232, 229]. The highest specific thrust will be that of fuels whose combustion products have the lowest molecular weight and a high combustion temperature. Of all the types of solid fuels for rocket engines, composite rocket fuels meet these requirements to the fullest extent. These fuels usually consist of thoroughly mixed organic binder fuels and solid crystalline oxidizers [230]. To increase the specific impulse of such fuels, 10-20% of light metals or their alloys - Al, Mg, Li, Be, are usually introduced into their composition [221, 247, 265]. Possessing a high heat of reaction and forming thermally stable products on burning, these metals make it possible to raise the temperature of the fuel combustion products and hence, the specific thrust of the fuel. However, the products of the reaction of these metals with oxygen and chlorine, being high-melting and high-boiling substances (see Ch. 3), are in condensed form to a significant degree, even at the combustion temperature. For this reason, the introduction of a considerable amount of metal into the composition of rocket fuels is inadvisable.

The greatest increase in specific thrust is obtained by introducing binary alloys or mixtures of light metals into the rocket fuel. The amount of metal (for example, Al) combining with oxygen and the amount of metal (for example, Li) combining with the halogen should correspond to the content of oxygen and halogen in the oxidizer used [241]. The optimum value of the specific thrust is obtained in fuels having a negative oxygen balance as a result of a reduced value of the average molecular weight of the combustion products and lower dissipation energy less than in stoichiometric fuels.

The fuels used in practice usually contain 80-85% of oxidizer; at the same time, satisfactory processing properties of the fuel can be obtained, and in some cases, the detonating tendency of perchlorate fuel can be eliminated [221]. Composite fuels may have the most diverse compositions and have different mechanical properties. Brief data on the composition and properties of composite fuels are given by Kurov and Dolzhanskiy [226].

A more detailed analysis of these fuels, indicating the mixtures as well as the properties of the individual components, is given in the book of Paushkin [228], Zaehringer's handbook [262], the books [220], [225], and the papers of Maxwell and Young [227], Tavernier [259], and many other papers published in the foreign press [236, 245, 241, 243, 247, 254] and in abstracts [239].

Table 20.1 lists certain typical composite fuels for SFRE.

Table 20.1

Components of Fuel	Specific Thrust, sec.	Physico-mechanical Properties	Proposed Use in	Reference
NH_4ClO_4; Butadiene-acrylic acid copolymer; Al	230-250 ($p = 70 kg/cm^2$)	good	Minuteman 1st stage; Scout 2nd stage; Nike-Zues;Saybar	(266)(265)
NH_4ClO_4; 80%; Al up to 20%; Butadiene-acrylic acid copolymer; Plasticizer hardener	245	good	Rocketdyne	(267)
NH_4ClO_4; Polyvinyl chloride; Plasticizer oil; Al up to 20%	250 ($p = 70 kg/cm^2$)	good; preserved at low temp.	Sergeant; Lacrosse; Meteor; Bomarc; Falcon	(265) (263)

The characteristics of RE composite fuels containing no metal fuel additives are listed in Table 20.2; they are taken from [221].

Table 20.2

Composition of Fuel	Molecular Weight of Products of Combustion	Adiabatic Exponent	Isobaric Combustion Temp, °K	Specific Thrust, $p = 70 kg/cm^2$	Index γ, $p = 70 kg/cm^2$	Rate of Combustion, cm/sec	Density g/cm³	Temperature Coefficient
NH_4ClO_4 75%; fuels & additives 25%	24	1.24	2420	224	0.4	0.5-1.5	1.66	$1.2-2.4 \cdot 10^{-3}$
NH_4ClO_4 80%; fuels & additives 20%	25.5	1.22	2790	236	0.4	0.8-2.0	1.72	$1.2-2.4 \cdot 10^{-3}$
NH_4ClO_4 80%; fuels & additives 20%	22	1.26	1755	195	0.4	0.2-0.3	1.55	$2.5 \cdot 10^{-3}$

Thus, the above-discussed four types of solid fuels for different jet engines fundamentally represent heterogeneous systems which are mixtures of an inorganic oxidizer, organic fuel, and a metallic fuel. The inorganic oxidizer and the metallic fuel are added to the fuel composition in the form of finely divided powders. The composition of the fuel, the technology of preparation, and the characteristics are determined chiefly by the type of jet engine. In Table 20.3 we give the compositions of various types of solid pyrotechnic fuels, and their characteristics.

PRINCIPLES OF PYROTECHNICS

Table 20.3

Function of Fuel	Special Requirements	Fuel Composition			Manufacturing Process	Specific Thrust	Remarks
		Organic Fuel	Oxidizer	Metal Fuel			
For RJE	High calorific value; high efficiency of combustion in air	Epoxy resin type; up to 50%	Crystalline; with low content in the fuel	Mg; Al; Li; and their alloys. Different contents in the fuel	Casting or pressing	500-800 sec.	I given for the range L_o=8.5 and numbers M-1.5 to 3.2 (Ref.258)
For RRE	Same; significant intrinsic impulse	Same	Same; increased content in the fuel	Same	No data	No data	--
For ramjet HJE	High specific heat of reaction with water; significant rate of reaction w/water	--	Inorganic salts (perchlorates)	Mg; Al; or their alloys 30%	Pressed at high specific pressure	1200-1600 g sec/cm^3	I given for excess coefficient of water = 1.33-5.7 (Ref. 256)
For CE	High specific thrust w/liq. oxidizer; self-ignition on contact w/liquid oxid.	Polychlorovinyl; rubbers; polyurethanes; amines	NH_4ClO_4; up to 20%	Al; Mg; Li or their alloys up to 20%	Casting	250-290 sec at p = 70kg/cm^2	% of metal indicated w/respect to rate of solid fuel & liq. oxidizer
Composite solid fuels for SPRE	High specific thrust; low molecular wt. of combustion products	Rubbers; resins; vinyl polymers; polyurethanes; etc. 15-20%	Nitrates; perchlorates up to 80%	Al; Mg; Li and their alloys 10-20%	Vacuum or pressure die casting	230-250 sec at p = 70kg/cm^2	--

2. SERVICE REQUIREMENTS

In addition to the special requirements discussed above, all types of rocket fuels must meet a whole series of other general requirements.

1. High density of solid fuel. Increasing the fuel density leads in all cases to a decrease in the volume of the combustion chamber and structural weight of the engine. The density of modern fuels does not exceed 1.8 g/cm^3.

2. A sufficient mechanical strength and high elastic modulus of the molded charge. Crumbling of a solid rocket fuel charge in storage or in the course of combustion is inadmissible, since it leads to an increase in the surface of combustion, to an increase of the pressure in the chamber, and to explosion of the chamber.

3. Composition homogeneity for all parts of the charge. Otherwise, a uniform combustion of the charge and hence, the ballistic qualities of the rocket can not be achieved.

4. Stability of the fuel characteristics during storage and changes of the ambient temperature.

5. A stable combustion in the engine chamber at a reduced pressure of the order of 15-35 at [236, 226]. A high pressure in the chamber makes it necessary to increase the thickness of the chamber wall, leading to an increase in structural weight.

6. Minimum dependence of the combustion rate on the initial temperature of the charge. This dependence is characterized by the temperature coefficient of the combustion rate. Solid pyrotechnic fuels have the lowest temperature coefficient (for real values of the temperature coefficient of pyrotechnic compositions, see Ch. 10).

7. The slightest possible dependence of the fuel combustion rate on pressure. Fuels having a slight dependence of the combustion rate on pressure have a low value of the exponent Such fuels provide the most stable intraballistic characteristics, since random pressure fluctuations do not cause any appreciable change in fuel combustion rate in this case.

8. It should be possible to prepare from solid pyrotechnic fuel large-sized charges possessing the necessary physical properties (see para. 2). The process of manufacture of the charge should be as simple as possible.

In most cases, during the combustion of solid fuel charges in an engine, a flow of combustion products exists along the burning surface. If the velocity of this flow is higher than a certain threshold value, there is observed a so-called erosion combustion, associated with an increase in the linear rate of fuel combustion. The regularity of increase in fuel combustion rate u in the course of erosion combustion is approximately determined by the expression:

$$\frac{u_{er}}{u_0} = 1 + K_{er}(v - v_{thr})$$

where u_0 is the normal fuel combustion rate;
K_{er} is a constant for a given fuel;
v is the rate of flow of the combustion products along the burning surface;
v_{thr} is the threshold value of the flow rate of the gases.

For example, if v_{thr}=200 m/sec, K_{er}=0.002 sec/m, and v=450 m/sec, the increase in fuel combustion rate during erosion combustion will be

$$\frac{u_{er}}{u_0} = 1 + 0.002(450-200) = 1.50$$

The erosion effect is usually manifested at a high charging density, and should be considered when designing full-scale solid fuel charges.

3. OXIDIZERS

The choice of oxidizer largely determines the properties of fuels. Potassium, sodium, lithium, and ammonium nitrates and perchlorates find practical application as oxidizers for solid pyrotechnic fuels. Some properties of these oxidizers are listed in Table 20.4 on next page.

A high content of free oxygen and high density of the oxidizer are particularly important for casting fuels, since they make it possible to introduce a large quantity of organic binder into the fuel and impart desirable processing characteristics to it. Ammonium nitrate is cheapest and most convenient to produce on a mass scale. When ammonium nitrate decomposes, depending on the conditions of this process, various reaction products may be obtained.

Table 20.4 PHYSICOCHEMICAL PROPERTIES OF OXIDIZERS FOR SOLID PYROTECHNIC FUELS

Oxidizer	Molecular Wt.	Density g/cm	Melting Point, °C	Heat of Formation, kcal/mole	% Oxygen Total	% Oxygen Active	Hygroscopic Point, 20°C	Solubility in 100g H$_2$O, 25°C
NH$_4$ClO$_4$	117	1.95	--	69	55	27	96	24.9
LiClO$_4$	106	2.4	247	92	61	61	--	59.7
KClO$_4$	139	2.5	610	103.6	46	46	99	2.5
KNO$_3$	101	2.1	336	119	48	40	92.5	38.1
NaNO$_3$	85	2.2	308	111	56	47	77	91
NH$_4$NO$_3$	80	1.7	169	88	60	20	67	180

Given below are two of the many possible equations of decomposition of ammonium nitrate:

$$NH_4NO_3 \nearrow 0.5NO_2 + 2H_2O_{vapor} + 0.75N_2 + 23 \text{ kcal}$$
$$\searrow N_2 + 2H_2O_{vapor} + 0.5O_2 + 27 \text{ kcal}$$

The heat of reaction corresponds to the liberating of 0.29-0.34 kcal/g. It has been found experimentally that NH$_4$NO$_3$ can burn at high pressure [246, 237], and, in the presence of a catalyst, at atmospheric pressure [237].

Thermal decomposition of ammonium nitrate with various catalytic additives has been studied in the last few years by Heiner, Guiochon, and Jacques, by Rozman [166], and also by the author of this book [237]; [119, p. 401].

It has been established that the thermal decomposition of ammonium nitrate is strongly affected by the catalytic action of compounds of chromium (VI): potassium chromate or ammonium bichromate [223, p. 344], and chromium trioxide, and to a lesser extent by cupric chloride. The combustion of ammonium nitrate with catalytic additives was studied by Taylor [260] and by the author of this book [237]. When added in the amount of 5-10% to ammonium nitrate, chromium compounds, which strongly increase the rate of thermal decomposition of NH$_4$NO$_3$ at 200°C, make ammonium nitrate fuel under ordinary temperature and pressure conditions.

At room temperature, ammonium nitrate is not sensitive to friction in a porcelain mortar, and has little sensitivity to impact. It explodes with difficulty, the explosion temperature being around 1000°C; the detonation velocity ranges from 1000 to 1500 m/sec.

None of the solid fuels checked produces a sufficiently high specific thrust with ammonium nitrate. For this reason, ammonium perchlorate, which is more expensive and more dangerous to handle, is usually preferred. The properties of ammonium perchlorate are described in the books of Kast, Escales, and also in Blinov's monograph [19].

As a pure substance, ammonium perchlorate has explosive properties: the explosion temperature is about 1200°C, and the detonation velocity ranges from 2500 to 3500 m/sec. The preparation of perchlorates and their properties and uses are described in Schumacher's monograph [49].

The thermal decomposition of NH$_4$ClO$_4$ with various catalytic additives has been studied in the last few years by Jacobs [249], Arden, Pawling and Smith [240], Shol'moshi and Reves [238], and also by the author of this book in collaboration with Shmagin [237].

The thermal decomposition of ammonium perchlorate may be approximately expressed by the equation [223]:

$$10\ NH_4ClO_4 = 6O_2 + 4N_2O + 2NOCl + 2Cl_2 + HClO_4 + 3HCl + 18H_2O_{vapor}$$

heat being evolved in the amount of 256 kcal/kg. The thermal decomposition of NH_4ClO_4 is markedly accelerated by copper compounds (oxides and chlorides) and according to the work of Birkumshaw [235, p. 329], by manganese dioxide.

The combustion rate of NH_4ClO_4 with catalytic additives has been studied by Friedman, Nugent, et al. [246].

The combustion rate of ammonium perchlorate-based fuels is 0.4-2 cm/sec (at p = 70 kg/cm²). These fuels have a fairly small exponent γ in the equation $u = A + Bp^\gamma$ and are characterized by a low temperature coefficient of the combustion rate.

The decomposition of such oxidizers as lithium and potassium perchlorates constitutes a very weak exothermic process. The over-all reactions of their decomposition may be expressed by the equations

$$LiClO_4 = LiCl = 2O_2 + 5.9\ kcal$$

$$KClO_4 = KCl + 2O_2 + 0.6\ kcal$$

The rate of the latter reaction increases in the presence of potassium halides in the order KCl < KBr < KI [223]. Decomposition of oxidizers $LiClO_4$ and $KClO_4$ forms chlorides which are comparatively high-boiling compounds.

Fuels based on $KClO_4$ have a fairly low specific impulse (180-220 sec). Although these fuels have a high density (1.7-2 g/cm³) and a high combustion rate, from 3 to 6 mm/sec (at p = 70 kg/cm²), their combustion at pressures below 70 kg/cm² is unstable.

Also suggested as oxidizers for solid rocket fuels have been nitronium perchorate NO_2ClO_4 and nitrosyl perchlorate $NOClO_4$, but these compounds are too hygroscopic, and in the presence of moisture in the air hydrolyze to form the free acids [228].

4. ORGANIC AND METALLIC FUELS

From the standpoint of fuel energetics, binder fuels should contain the maximum quantity of hydrogen and have a low heat of formation and high density. For composite fuels for RE, of definite interest are fuels containing oxygen (for example, in C-O-NO_2 groups), since they permit the introduction of a close-to-optimum amount of oxidizer to the fuel composition.

The choice of the binder fuel is most frequently determined by processing considerations and requirements placed on the physicochemical characteristics of the fuel.

For large engines, the most suitable method involves preparation of charges by pouring the fuel directly into the engine body. For this reason, before loading, the binder should be a liquid. This liquid should be sufficiently viscous, so that the oxidizer powder added to it will not settle to the bottom of the mixer. In addition, the mass formed should retain its fluidity in the presence of 75-85% of the solid component.

In order to obtain the necessary properties of the fuel, the freezing point of the binder should be low. A high elastic modulus of the charge is achieved when the binder is a high-molecular compound (polymer).

Polymerization of the binder (with a catalyst admixture) is carried out after the preparation of the fuel (mixing of the binder with the oxidizer) and pouring of the fuel into the engine body; the polymerization process may be carried out both at room temperature and with heating.

The binders used are synthetic rubbers, various (phenolic, polyester, epoxy) resins, and various polymers: polysulfides, polyurethanes, polyacrylate, polyamides, polyethylene, polybutadiene, polyisobutylene, etc. [221].

Polysulfide rubber is obtained by polycondensation of dihaloide alkyls with alkali metal polysulfides, for example, dichloroethane with sodium tetrasulfide:

$$x Na_2S_4 + x C_2H_4Cl_2 = [-CH_2-CH_2-\overset{S}{\underset{\|}{S}}\overset{S}{\underset{\|}{S}}-]_x + 2x\, NaCl$$

Polysulfide rubbers ("thiokols"), obtained by using dichloroethyl ether, have the empirical formula $C_4H_8OS_4$ and a calorific value of 5600 kcal/kg.

The starting materials for the preparation of polyurethanes is diisocyanate (for example, hexamethylene diisocyanate $O=N=C(CH_2)_6\text{-}C\text{-}N=O$) and diatomic alcohols (for example, butanediol $H\text{-}O\text{-}(CH_2)_4\text{-}O\text{-}H$). The polymerization reaction between these compounds begins at room temperature and ends at a higher temperature (up to 200°C). The molecular weight of the material obtained increases with rising temperature and reaches 10,000-15,000. The elastic properties of polyurethanes are retained at low temperatures.

Polyurethanes contain the radicals $-(CH_2)_n$ -and the groups NH and O=C<.

According to Zaehringer [262], the structure of the polymer chain of polyurethanes may be represented as

$$(-R-NH-\overset{O}{\underset{\|}{C}}-O-R'-O-\overset{O}{\underset{\|}{C}}-NH-)_x$$

Assuming the composition of radicals R and R' to be $(CH_2)_6$ and $(CH_2)_4$, we obtain the empirical formula $C_{12}H_{22}O_4N_2$ for polyurethane. The calorific value of polyurethanes is 7250 kcal/kg.

The introduction into the composition of rocket fuels of up to 20% metallic fuel in the form of powders of light metals or their alloys increases the specific thrust of solid rocket fuel. In such fuels as those of RJE or HJE, the metallic fuel is the main component. In all cases where such fuels are used, a particularly serious problem is that of combustion efficiency of the metallic fuel. Success in this area is directly related to the results of study of the combustion mechanism of solid pyrotechnic fuels. Problems of combustion of fuels and metals are adequately treated in [223, 225, 221, 229, 234]. The fundamental rules of combustion of fuels are very close to those of various pyrotechnic compositions.

Chapter 21

IGNITION COMPOSITIONS.
OTHER TYPES OF PYROTECHNIC COMPOSITIONS.

1. IGNITION COMPOSITIONS AND REQUIREMENTS PLACED ON THEM

Ignition compositions are designed to ignite the main pyrotechnic compositions (illumination, smoke, solid rocket fuel and other compositions). [Note: Ignition compositions containing initiating explosives and designed for loading ignition caps are not discussed here. Concerning this question, see [270].] the action of an ignition composition consists in heating up a portion of the main composition up to the ignition point.

It follows that the higher the self-ignition (flash) point of the main composition, the "stronger" the ignition composition required to excite the combustion reaction therein.

The ignition of compositions whose flash point is not above 500-600°C does not involve any special difficulties.

Compositions whose flash points are above 1000°C ignite with great difficulty. Special ignition and intermediate compositions must be selected to ignite such compositions (for example, thermites) particularly if they are in the pressed state.

The following special requirements are placed on ignition compositions:

1) they must be ignited easily by a comparatively minor thermal impulse; the flash point of any ignition composition should not be above 500°C;
2)
3) the combustion temperature of ignition compositions should be several hundred degrees above the flash point of the main compositions being ignited by it.

In addition, in some cases it is required that the ignition composition emit little light during combustion so as to avoid unmasking the troops.

Occasionally, ignition compositions must also meet the requirement of liberating the minimum amount of gases during combustion; such ignition compositions are somewhat inaccurately termed "gasless" ignition compositions.

The igniting action of ignition compositions is chiefly determined by the amount of heat transferred to the main composition from the slags formed by the combustion. Thus, the igniting action of an ignition composition will be stronger the higher its combustion temperature and the more slag remains after its combustion on the surface of the main composition being ignited.

In liquid slags, the surface of contact with the ignited composition will be greater than in the formation of nonmelting solid slags, and hence, the amount of heat transferred by the slags to the main composition per unit time will also be greater.

It has been established experimentally that the best igniting action is that of slowly burning ignition compositions, which permit a sufficient time for the transfer of heat to the main composition being ignited. For this reason, in pyrotechnic devices, the ignition compositions are almost always used in pressed form.

In designing ignition compositions, use should be made of oxidizer which give up their oxygen readily and at as low a temperature as possible, and which readily oxidize the fuel.

There are many oxidizers which readily give up their oxygen. They include alkali metal permanganates, chlorates, nitrates, etc. However, on the basis of the requirement of low sensitivity of the composition to mechanical effects, potassium nitrate is employed as the oxidizer in ignition compositions much more often than other compounds.

As readily oxidizable fuels, wood charcoal, iditol, followed by antimony, etc., are frequently used.

For easily igniting main compositions (chlorate compositions of signal smokes or chlorate compositions of signal flares), the use of ignition compositions close to smoke powders in their formulas is entirely possible.

For example, the following formula may be cited:

Potassium nitrate	75%
Wood charcoal	15%
Iditol	10%

To increase the igniting effect of ignition compositions, an additional fuel, magnesium powder, which after burning leaves a solid magnesium oxide slag, is introduced into these compositions.

Bystrov [2] points out that the following composition may be used for the ignition of many illumination compositions:

Potassium nitrate	75%
Magnesium	15%
Iditol	10%

To ignite compositions containing aluminum as the main fuel and having a high flash point, it is necessary to use ignition compositions containing no less than 15-20% magnesium powder.

In the preparation of ignition compositions for tracers, barium peroxide BaO_2 is used as the oxidizer more frequently than other compounds.

Barium peroxide gives up its oxygen at a higher temperature than potassium nitrate, but the process of its decomposition requires the consumption of a very small amount of heat:

$$BaO_2 = BaO + 0.5O_2 - 17 \text{ kcal}$$

In the decomposition of BaO_2, the weight of the solid residue amounts to 91% of the weight of the oxidizer.

The following formulas may be cited as examples of ignition compositions for tracers:

I		II	
Barium peroxide	80%	Barium nitrate	48%
Magnesium	18%	Barium peroxide	30%
Binder	2%	Magnesium	13%
		Iditol	9%

Many ignition compositions used for tracers during World War II contained the following substances in addition to barium peroxide, magnesium, and binders: barium nitrate, red lead, strontium picrate, tetranitrocarbazole, calcium silicide, sodium oxalate, graphite, calcium and zinc stearates, etc.

A strong igniting effect is displayed by thermite-type compositions:

I		II	
Ferric oxide	69%	Silicon dioxide	55%
Magnesium	31%	Magnesium	45%

However, the ignition of these compositions (particularly, composition II) does not take place easily, since they have high ignition temperatures.

Compositions containing zirconium powder ignite easily and display a satisfactory igniting action. Examples of such compositions used for igniting a trail may be [12]:

1. Black powder 75%; KNO_3 12%; Zr 13%.

2. KNO_3 48%; Zr 52%.

Ignition compositions for a trail must not contain a large amount of easily fusible substances; otherwise, when the shell moves along the bore, the liquid will be expelled from the burning layer during combustion.

When the main pyrotechnic composition can not be ignited even by intensified ignition compositions, so-called transitional or intermediate compositions are employed.

Intermediate compositions are obtained by mixing the ignition and main compositions in certain proportions. One of the cases of application of an intermediate composition has already been discussed in the example of ignition of pressed iron-aluminum thermite (see Fig. 17.1).

To kindle certain main compositions that are particularly difficult to ignite, it is sometimes necessary to use several intermediate compositions simultaneously, of which the intermediate composition in direct contact with the main composition contains the smallest amount of ignition composition.

A particular form of ignition compositions are friction compositions, whose ignition is brought about by rubbing some object ("grater") against them, the grater also being coated with a special pyrotechnic composition.

The formulas of friction compositions are close to those of compositions used in match production, for example, one of the friction compositions contains 60% potassium chlorate, 30% antimony sulfide, and 10% resin.

The striker is coated with a phosphorus paste consisting of 56% red phosphorus, 24% glass powder, and 20% iditol.

Friction compositions are very sensitive to friction and very dangerous in handling.

2. IGNITION COMPOSITIONS FOR ROCKET ENGINES

It is well known that the reliability of operation of a rocket engine substantially depends on the presence of an effective ignition system. Earlier igniters based on the use of black powder did not meet the requirements placed on the ignition of composite solid fuel and ballistic powders. In this connection, American companies have worked out a series of designs of pyrotechnic igniters which are now being used for solid and liquid fuel rocket engines.

The pyrotechnic ignition system is the most common one and is used in rocket engines of most diverse types, in particular, in the engines of the Titan, Polaris, and Minuteman rockets [266]. Such a system is reliable and safe and provides for a fast startup of the engine at high altitudes, a low-power energy source being required to turn it on.

The flame tongue formed as a result of the operation of a pyrotechnic igniter has a high temperature and sufficiently large dimensions (large ignition area).

The literature describes a large number of designs of pyrotechnic igniters: roller, basket, "rocket within a rocket", etc. [274].

In the first stage of development, binary mixtures, for example, a mixture of aluminum powder with $KClO_4$, were used in pyrotechnic igniters [266]. Known as "Alclo", this is a mixture of high calorific power (see Ch. 6), which burns intensively, and its ignition temperature is about 360°C.

Later, an ethylcellulose binder began to be introduced into the Alclo mixture.

The following pyrotechnic mixture is used for igniting composite fuels with a high NH_4ClO_4 content. $KClO_4$, 26-50%; $Ba(NO_3)_2$, 15-17%; Zr/Ni (50/50) alloy, 32-54%; ethylcellulose, 3% [273].

Ignition compositions containing boron, for example KNO_3; 71% B (amorphous), 24%; rubber, 5%, are also finding applications.

The development of pyrotechnic igniters involves the necessity of conducting numerous tests to determine their processing characteristics under different conditions.

The main points in the development of pyrotechnic igniters are [266]:

1) a specified size;
2) the method of installation in the rocket engine;
3) the minimum required temperature of the flame tongue;
4) characteristic of the combustion products (maximum permissible amount of condensed oxides, degree of dispersity of solid particles, etc.);
5) conditions of operation (vibration, overloads, pressure of ambient medium, maximum and minimum operating temperatures, etc.).

3. OTHER TYPES OF PYROTECHNIC COMPOSITIONS: GASLESS, SIMULATION, AND HISSING COMPOSITIONS

In addition to the types of pyrotechnic compositions discussed above, there are also pyrotechnic compositions used for various special purposes, sometimes similar in formulas to the ignition, illumination, or other compositions already described; in other cases, their special use makes it necessary to create a new special formula.

The following types of compositions are most frequently used for special purposes:

1. "Gasless" or low-gas compositions used in time fuses and detonators, in delayed-action electric primers and in certain special heating articles.

2. Simulation and hissing compositions whose uses are obvious from their names.

In addition, there are special formulas for pyrotechnic compositions used in the preparation of fireworks and in filmmaking [284].

Firework pyrotechnic compositions are not discussed in this book; a description of their preparation and properties is given in specialized literature [6, 15].

"GASLESS" AND LOW-GAS COMPOSITIONS. A distinctive property characterizing the majority of this type of compositions is the slight dependence of the combustion rate on the outside pressure [113]. This property is very valuable when these compositions are used in time fuses and detonators.

In the latter case, the temperature coefficient (see Ch. 10, Sec. 2) of the combustion rate should also be low; however, far from all "gasless" compositions meet this requirement.

In addition, compositions for time fuses must satisfy all the requirements placed on ordinary tubular powders; particular mention should be made of the ready ignitability of compositions and dependability of their ignition of the next link of the fire chain (for example, ignition of the percussion cap).

The term "gasless" compositions are put in quotes because at the instant of the reaction taking place at high temperature these compositions form a certain quantity of gaseous products which on cooling revert to solids. However, the amount of gaseous products is so slight that they may be referred to as low-gas compositions, this time without quotes.

The substances most frequently used as oxidizers in low-gas compositions are compounds of lead, chromium, and manganese, for example, Pb_3O_4, PbO, $PbCrO_4$, $BaCrO_4$, $K_2Cr_2O_7$, $KMnO_4$ and also BaO_2, which on decomposing form metallic lead (boiling point 1744°C) and the involatile oxides Cr_2O_3, BaO, MnO, etc.

The fuels used in low-gas compositions are Zr, Sb, Si, Mn, FeSi, etc.

The patent literature indicates many formulas of low-gas compositions. Six formulas are given below as examples (for the patent numbers, see [7, p. 260].

1. CuO 60%, Zr 40%.
2. Pb_3O_4 35 to 80%, Zr 65 to 20%.
3. $PbCrO_4$ 40%, PbO 44%, $FeSi_2$ 16%.
4. Pb_3O_4 76 to 84%, Si 18 to 22%, nitrocellulose 3 to 5%.
5. $PbCrO_4$ 63%, Pb_3O_4 25%, Si 12%, nitrocellulose 0.3% (above 100%).
6. Pb_3O_4 70%, Si 30%.

In the preparation of inhibitors for delayed-action electric primers, according to the literature data [260], use is made of a gasless composition containing 56% $KMnO_4$ and 44% Sb, whose combustion reaction may be represented as follows:

$$3KMnO_4 + 3Sb = K_3SbO_4 + 3MnO + Sb_2O_5 + 269 \text{ kcal}$$

Use may be made for the same purpose of a composition containing 5% potassium chlorate, 47% selenium, and 48% metallic bismuth (U.S. patent 2607672, 1952).

"Gasless" compositions also find applications in the heating of food (preserves, soup, etc.) under combat conditions. During World War II, the British used the following composition for this purpose: magnetic iron oxide 81% and calcium silicide 19%, whose combustion may be expressed by the equation

$$5Fe_3O_4 + 4CaSi_2 = 4CaSiO_3 + 4SiO_2 + 15Fe.$$

It should be noted in conclusion that many "gasless" or low gas compositions are similar in properties to thermites or thermite incendiary compositions, but differ from them in a great case of inflammation, and in some cases, in a high combustion rate.

Simulation compositions are used for loading various scientific pyrotechnic articles whose action is supposed to create an external effect similar to that of present military articles. Simulation articles do not have a damaging effect: this is achieved by making their envelope out of cardboard, plastic, or similar materials.

The purpose of certain simulation articles is to imitate the effect produced by the explosion of high explosive shells, rifle grenades, etc. Such simulation articles must produce a loud noise similar to that of the explosion of a high-explosive shell or other object. In addition, it is desirable that the action of simulation articles be associated with the formation of a certain amount of smoke similar in color to the smoke formed by the explosion of explosives.

To obtain a strong sound effect, simulation articles are loaded with grain powder or with a mixture of potassium perchlorate and aluminum dust. In particular, use may be made for this purpose of a composition of 70% potassium perchlorate and 30% aluminum, similar in properties to photomixture compositions. Such a composition, taken in the powdered state, when acted upon by an ordinary thermal impulse, burns up almost instantaneously with a strong sound effect and a flash of light, while evolving a certain amount of white smoke.

Simulation compositions whose combustion evolves gray or black smoke should contain a certain amount of carbon-rich organic compounds (naphthalene, anthracene, etc.).

There are also simulation articles whose action evolves a colored smoke; such articles are used during maneuvers for conventional designation of the point of impact of chemical or incendiary shells or bombs, explosion of mine fields, etc. Loading of these articles involves the use of ordinary colored smoke compositions. Their combustion is slowed down by occasionally adding a certain amount of sawdust.

HISSING PYROTECHNIC COMPOSITIONS. A sound effect similar to a hiss may be obtained by burning a composition containing 75% potassium chlorate and 25% gallic acid $C_6H_2(OH)_3COOH$. The sound effect is explained by the high combustion rate of the composition.

Becher [8] notes that a hiss during combustion is produced by mixtures of chlorates (K or Ba) with phenol derivatives, and that, in addition to gallic acid, the use of resorcinol, phloroglucinol as well as picric acid is possible for the same purposes.

However, mixtures containing picric acid and, in particular, potassium picrate (for example, a mixture of 60% potassium picrate $C_6H_3N_3O_7K$ and 40% KNO_3) are too dangerous to handle and are not being used at the present time.

Experiments performed by Maxwell [271] showed that the frequency of acoustic vibration obtainable from the combustion of these compositions is lower the longer the tube (cardboard) into which the composition is pressed. At the same time, the vibration frequency is higher the faster the combustion rate of the composition.

Chapter 22

USE OF PYROTECHNIC COMPOSITIONS IN THE NATIONAL ECONOMY. MATCH COMPOSITIONS.

The use of pyrotechnic agents and compositions in the national economy, art, and research is becoming more diversified each year and increasingly more significant.

Thermite compositions find very extensive and diverse applications. For many years they have been used in the production of rare and valuable metals [291, 208]. Aluminothermic processes can be used to obtain the following metals and alloys: Cr, Mn, Ti, W, Mo, Ni, Co, Zr, V, etc.

Ellern [9] points out that when the production of a metal by means of an ordinary aluminothermic reaction is difficult, several variants may be recommended to carry out these processes:

1. Addition of a more active oxidizer to a lower oxide, for example, chromates to chromium oxide.

2. Addition of strong oxidizers which increase the thermal effect of the reaction and facilitate the process, for example, BaO_2, PbO_2, $Na_2S_2O_8$, $KClO_3$, etc.

3. Addition to aluminum of other strong reductants, for example, Ca, Mg, Si, or CaC_2.

4.
5. Carrying out the process with preheating (in a furnace). This variant is recommended for the preparation of titanium and zirconium when obtained by reduction of the oxides with magnesium.

The variant with preheating is discussed in detail for various mixtures by Venturini [109]. The preparation of metals by means of metallothermic reactions under laboratory conditions is described in [20].

Niobium and tantalum can be obtained by reduction of their sulfides with aluminum powder. Separation of the reaction products is achieved by vaporizing Al_2S_3 above 1550°C.

The use of thermite compositions for welding railroad tracks is generally known [283]. New thermite compositions used for welding telephone and telegraph wires have been developed comparatively recently.

In describing methods of welding of steel wires by means of powder charges, A. Kazhdan [7, p. 277] indicates that "a squib (see Fig. 22.1a) is a cylindrical cartridge with a longitudinal opening, pressed out of magnesium thermite and an ignition composition and provided with an ignition cap". Squibs are small in size, and their weight is only 7 to 16 g, depending on the diameter of the wires being welded. The combustion of magnesium thermite develops a high temperature, completely sufficient for welding the wires, and the slag remaining after its combustion retains the shape of the squib (Fig. 22.1, b and c).

As is evident from Fig. 22.2, during the welding, the wire ends are brought together by means of special pliers. The thermal muffle charge, slipped on the point of contact, is heated. When it burns, the ends of the wires are pressed together with the pliers. After the wires cool, a strong joint weld is formed.

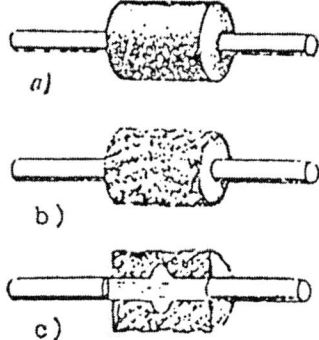

Fig. 22.1. Welding squib before and after welding.

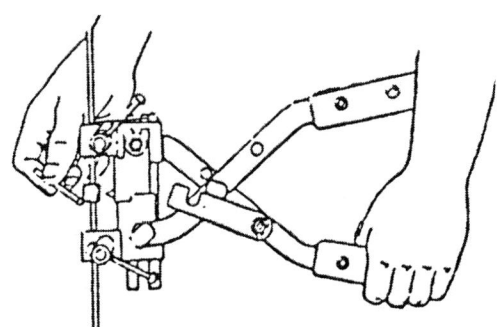

Fig. 22.2. Device for welding wires by means of a squib.

A formula of a copper thermite worked out at the All-Union Scientific Research Institute of Railroad Transportation, consisting of 64% CuO, 16% ferromanganese (80% Mn), and 20% CuAl alloy (46% Al), has found wide application in the welding of rail butt joints. "The reduced metal" obtained by combustion "constituted a pore-free manganese bronze which welds the end of the butt joint to the rail head". [Note: A. N. Kukin. New Types of Thermite Welding. Transzheldorizdat, 1955.] In welding grounding conductors to metal structures, a thermite consisting of 72.5% Fe_3O_4, 18% Al, 4.5% Mg, and 5% FeMn alloy (~1:1) is employed.

It has also been suggested that thermite be used in soil engineering to thaw frozen ground in winter [286]. However, calculation shows [288] that for a volume ground moisture of 25%, not less than 40 kg of thermite is required to that 1 m³ of frozen ground. In Kiselev's view [282], a practical consumption of thermite for partial thawing of the ground amounts to about 8 kg per 1 m³ of ground.

Members of the Moscow Mining Institute carried out experiments on secondary crushing of ore with briquetted thermite [280]. During the war of 1941-45, food preserves were heated by using thermite compositions placed in a metal cartridge inside the tin can. According to [208], thermite can be used to heat water, food, and preheat soldering irons and rivet bolts. In some cases, it is also desirable to use thermite against ice jams.

In 1958, thermite was used to produce a sodium cloud in the upper layers of the atmosphere. Astronomer I. S. Shklovskiy proposed the use of photometric observations of the diffusion rate of a sodium cloud as a method of determining atmospheric density. The corresponding experiment was carried out in the USSR on 19 September 1958. Iron-aluminum thermite was used to vaporize sodium metal. The sodium was crushed into small fragments, mixed with thermite, and the entire mixture was placed in a steel cylinder. During the combustion of thermite, sodium was vaporized, and its vapor was expelled through an exhaust opening in the cylinder [290].

Smoke compositions containing insecticides (or forming them in the course of combustion) are widely used to control insect pests [276, pp. 118-145; 293]. Smoke pots are loaded with these smoke compositions. The composition contains a toxic agent and a thermic mixture (cf. Ch. 19). The toxic agent volatilizes during the combustion of the thermic mixture.

The oxidizer most frequently used is $KClO_3$, and the fuel is anthracene, dicyandiamide (DCDA), urotropin, etc. The toxic agent used is hexachlorane (hexachlorocyclohexane, abbreviated HCCH), hexachloroethane C_2Cl_6, DDT, etc. [287]. An example is the composition HCCH 50-52%, $KClO_3$ 23-26%, NH_4Cl 9-12%, anthracene 9-12%, DCDA 4-6%.

Also known are fungicide compositions. These are mixtures containing substances used against fungal, bacterial and viral plant diseases. An example of such a smoke mixture is the following composition (in percent): dichloronaphthoquinone 58, $KClO_3$ 22, NH_4Cl 10, DCDA 5, and anthracene 5.

Special acaricide smoke compositions have also been developed (acaricides are chemical agents used against ticks). For example, the following composition is used (in %): commercial tedion [25, Vol. 2, p. 82] 50, $KClO_3$ 20, DCDA 30. The smoke formation is flameless, and the temperature developed by the combustion is about 250°C. Greenhouses and fruit and vegetable storage areas are decontaminated with sulfur dioxide SO_2, obtained by burning shell-less pyrotechnic pellets (weighing 500 and 100 g) consisting of 75% sulfur, 17% KNO_3, and 8% diatomaceous earth (loosening agent and combustion catalyst), developed by Sidorov [279]. This has a sharply negative oxygen balance and therefore requires a free access of air during its combustion.

Smoke compositions are also used to protect orchards (particularly citrus) from frost [276, pp. 104-126] and for studying air currents in the atmosphere and in various installations.

Compositions of colored lights and compositions of white, black, and colored smokes are widely used in cinematography. Today, color cinematography would be unthinkable without the use of pyrotechnic agents [284]. Various pyrotechnic signaling agents have found application in marine and inland fleets and in aerial and railroad transportation.

To decrease shrinkage and prevent the formation of cracks during steel teeming, a pyrocomposition containing 70% FeSi (1:3), 10% aluminum powder and 20% $NaNO_3$ is poured into the upper part of the mold above the liquid metal. Recently, pyrotechnic anti-hail rockets and cartridges have been developed.

Studies conducted in the last few years in the Soviet Union and abroad have shown that microphysical processes can be affected by causing artificial crystallization of supercooled water droplets in clouds. This creates the fundamental possibility of controlling natural processes for the purpose of preventing hail, increasing the precipitation and dispersing supercooled clouds and fog. The most active substances causing crystallization of supercooled water droplets are silver iodide (AgI) and lead iodide (PbI+). Theoretical interpretations of artificial modification of supercooled clouds and fogs are given in [285, 278].

The ordinary non-pyrotechnic method of using lead and silver iodides consists in introducing colloidal solutions into the cloud or fog by spraying them from an airplane.

Two variants are possible when pyrotechnic compositions are used: 1) AgI or PbI_2 is present in ready form in the composition; 2) these substances are formed by a chemical reaction during the combustion of the composition. The formula for the composition of the first variant is AgI or PbI_2 40-60%, NH_4ClO_4 24-45%, Iditol 10-15%, and graphite (or industrial oil) 1.5-2%.

Compositions used in the second variant include: lead powder and iodine-containing compounds NH_4I, CHI_3 (iodoform), or $C_6I_4O_2$ (iodanyl). To these substances is added a thermic mixture of fuel and oxidizer: iditol + NH_4ClO_4. An example is a composition of the following formula: lead powder 20-25%, NH_4I 25-34%, NH_4ClO_4 20-30%, resin (iditol) 10-20%.

Gaseous oxygen of sufficient purity used for breathing, for example, in submarines or aircraft, may be obtained by burning pyrotechnic compositions [294]. Devices used for this purpose are called chlorate candles or chlorate briquets; the main component in compositions used for chlorate candles is sodium chlorate (or potassium chlorate). Since the decomposition of sodium chlorate evolves little heat,

$$2NaClO_3 = 2NaCl + 3O_2 + 24 \text{ kcal}$$

in order to increase the thermal effect (which is necessary for stability of combustion), a small amount of fuel is added. In selecting a fuel, it is necessary to keep in mind that its combustion must not form any toxic gaseous products (a small amount of CO_2 is obviously permissible). Frequently, iron powder is used as the fuel; no gases are formed by its combustion, and ferric oxide Fe_2O_3 formed by the combustion is a good catalyst for the decomposition of chlorates.

The literature [294] cites the following formulas: $NaClO_3$ 74-80%, Fe 10%, BaO_2 4%, glass fiber 12-6%. BaO_2 is added in order to prevent the liberation of free chlorine during combustion.

A composition containing 80% $NaClO_3$ in pressed form has a density of 2.45 ± 0.05 g/cm^3; the weight of the gaseous oxygen obtained amounts to 34% of the weight of the candle. The heat of combustion of such mixtures is approximately 200 cal/g of composition, and the combustion temperature (measured with a thermocouple) is 500-700°C. Such mixtures can be loaded by pouring (melting point of $NaClO_3$ 255°C). The ignition composition contains the same ingredients, but in different proportions, providing for a combustion temperature of 900-1000°C; $NaClO_3$ 60%, Fe 20%, BaO_2 10%, and glass fiber 10%. There are literature indications to the effect that the pyrotechnic mixture ($CeClO_4$ + Al) can be used to prepare cesium plasma [292].

In conclusion, in evaluating further prospects for the use of pyrotechnic compositions in the national economy, we should note that the development may follow two fundamentally different lines. First, utilization of energy (heat, light, and other forms) evolved by the combustion of pyrotechnic compositions. Second, utilization of the substances (elements or compounds) which are the combustion products of pyrotechnic compositions.

Thus far, this second variant is being used on a comparatively limited scale: aluminothermy (production of metals and artificial corundum Al_2O_3), production of oxygen (oxygen briquets), production of SO_2 in insecticide compositions, and preparations of AgI and PbI_2 in anti-hail compositions. However, a number of gases such as H_2 [9], Cl_2 (possibly, also F_2), N_2O, etc., as well as many inorganic salts and metal oxides can be prepared by using the pyrotechnic method. This little-studied area is still awaiting investigations by researchers and experimenters.

MATCH COMPOSITIONS. At the present time (1964), so-called safety matches are being produced all over the world, i.e., matches which are ignited only by friction against the coating on the matchbox. Safety matches were first made in Sweden in 1855, and were therefore named Swedish matches.

The world production of matches is expressed by a figure of the order of 10 billion boxes per year (since the population of the globe is approximately 3 billion, the consumption of matches per person per year is about 10 boxes). About 1 g of match paste is consumed per match box. Hence, the consumption of match paste on the world scale amounts to tens of thousands of tons per year.

It is evident that on such a scale (tens of billions of boxes), the production of matches must be automated. Indeed, highly advanced automatons are now in operation in match factories. The chemistry of ignition of modern matches is as follows: friction of the match head against the coating on the box gives rise to an exothermic reaction between red phosphorus (in the coating) and potassium chlorate (in the match head). The heat of this reaction ignites the match head.

As in all other pyrotechnic compositions, the basis of the match head composition is an oxidizer-fuel system. The oxidizer so far has been exclusively $KClO_3$, a substance that readily gives up its oxygen. There are usually two fuels in a match composition: sulfur and animal (bone) glue. The choice of fuels is justified both chemically and technologically: the $KClO_3$ + S mixture has a comparatively low flash point (220°C) and provides for an easy inflammation of the composition; the animal glue is introduced into the composition in the form of an aqueous solution (the water is removed by drying), imparts the necessary consistency to the paste, and protects the match heads from moisture in storage. However, incendiary match compositions contain six to eight ingredients, since they also include additional oxidizers ($K_2Cr_2O_7$), catalysts accelerating the decomposition of potassium chlorate (MnO_2), fillers (iron ocher Fe_3O_4, zinc white ZnO, etc.), substances increasing the sensitivity of the match head to friction (ground glass), and a small amount (fraction of a percent) of organic dyes (rhodamine, acid green, etc.). As an example, the following composition can be cited, in percent:

$KClO_3$	51
Ground glass	15
Animal glue	11
ZnO	7
$Fe^{+3+}O_4$	6
S	5
MnO_2	4
$K_2Cr_2O_7$	1

Phosphorus paste (coating on the box) has the following composition (in %):

Red Phosphorus	37.2
Sb_2S_3	33.5
Animal glue	9.3
Iron ochre	7.0
Dextrin	7.0
MnO_2	3.4
$CaCO_3$	2.0
Ground glass	0.6

A modern American formula of an incendiary match paste is [9] (in %):

$KClO_3$	45-55
Silicate fillers	15-32
Animal glue	9-11
Diatomaceous earth	5-6
S	3-5
ZnO or $CaCO_3$	3
Starch or Dextrin	2-3

An American phosphorus paste has the formula [9] (in %):

Red phosphorus	50
Ground glass	25
Animal glue (or casein)	16
ZnO or $CaCO_3$	5
Soot	4

The quality of matches is usually evaluated:

1) from the result of their verification ignition (which simultaneously checks the quality of the match and box coating);
2) by determining the sensitivity (on a special instrument);
3) by determining the moisture resistance of matches kept for 24 hours at room temperature and at 99.4% relative humidity; then the ignition is checked;
4) by determining the flash point of match heads;
5) by determining the adhesion between the match head and the match stick.

To prevent afterglow prior to the preparation of matches, the match stick is impregnated with a 1.5% solution of H_3PO_4, then coated with paraffin (by dipping in molten paraffin). In addition to ordinary (household) matches, special matches are produced:

1. Storm matches, which burn in the wind, in dampness, and in the rain.
2. Thermic matches, developing a higher temperature on burning the combustion of their head yields a greater quantity of heat.
3. Signaling matches, whose heads produce a colored flame on burning.
4. Photographic matches producing an instantaneous bright flash used in photography, etc.

A detailed knowledge of the chemistry and technology of match production can be obtained from the literature [277, 289]. In the USA, matches are also produced that ignite when struck against any rough surface (so-called SAW, or strike-anywhere matches). A characteristic feature of such compositions is the fact that they contain the sesquisulfide P_4S_3. For example [9]:

$KClO_3$	32
Ground glass and other fillers	33
P_4S_3	10
Animal glue	11
ZnO	6
Colophony	4
Extender	4

Chapter 23

PRINCIPLES OF TECHNOLOGY OF PYROTECHNIC PRODUCTION

Because of the stringent requirements imposed on the production quality of the compositions and their fire and explosion hazard, the production of pyrotechnics constitutes a fairly complex technological process. It should be carried out while observing the following conditions:

1. High quality of the ingredients (usually, grade A).
2. Cleanliness of production facilities and equipment.
3. Low relative humidity (not above 65%) at the work stations.
4. Thorough grinding and screening of the ingredients.
5. Accurate batching of the ingredients.
6. High quality of mixing of the ingredients.
7. Observation of precautions and safety measures and regulations.

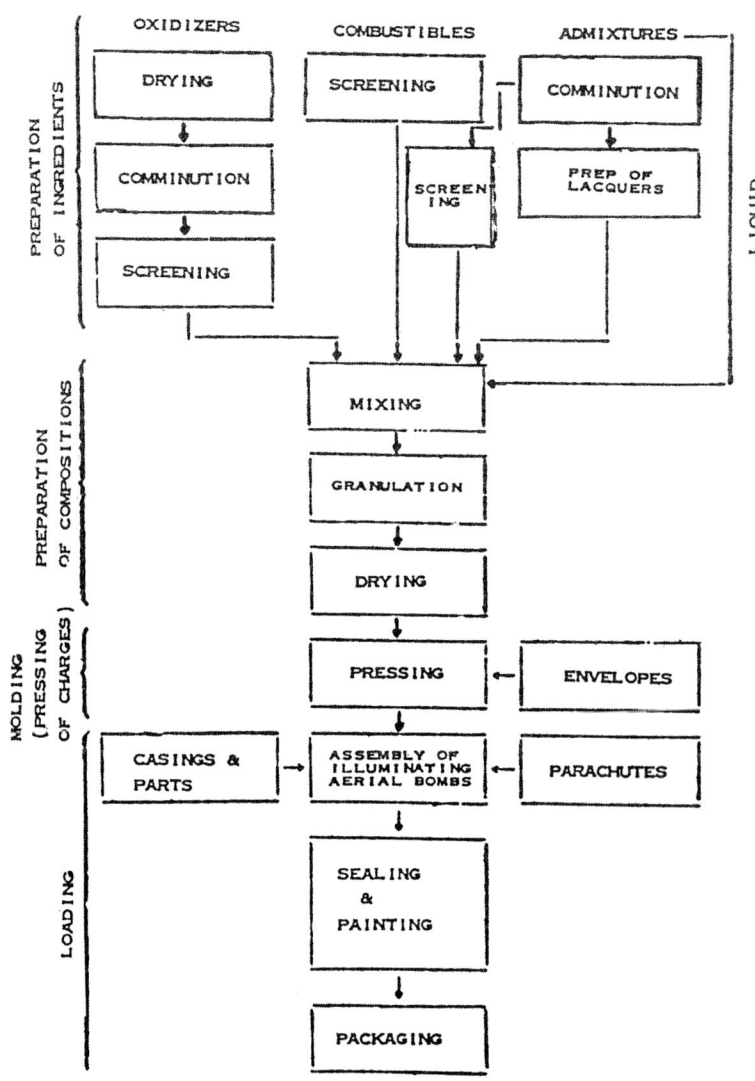

The technological process consists of the following phases:

1. Preparation of ingredients: drying, grinding and screening.
2. Preparation of compositions: batching, mixing of ingredients, granulation and drying of compositions.
3. Compaction and molding of compositions: batching and compaction by pressing, conveyor loading, packing (or pouring).
4. Loading of articles: preparation of casings and parts, assembly of articles, packaging.

As a typical technological process, the preceding page shows a diagram of the production of aerial illumination bombs.

1. PREPARATION OF INGREDIENTS

The ingredients of the compositions and other materials are accepted into production if the data of their chemical analysis meet the GOST (All-Union State Standard) requirements or the TU (Technical Specifications). Ingredients which do not satisfy the moisture requirements are dried. Usually, the moisture content of the ingredients is standardized at 0.1-0.5%.

DRYING OF INGREDIENTS

Oxidizers are dried most frequently. Fuels and binders usually contain little moisture and require no drying. The drying of oxidizers is usually carried out at 70-80°C, water of a heating system or low-pressure steam being used as the heat carrier. [Note: The drying temperature of oxidizers and all other materials is determined by taking the physicochemical properties of the materials into account.]

Small amounts of ingredients (not more than a few tens of kg) are dried in drying ovens, which are metal structures with shelves for trays, where sections of heating pipes are located under the trays or along the lateral walls. Tubular or vacuum dryers are used for drying large quantities of oxidizers.

Tubular dryers (Fig. 23.1) are fairly economical and efficient, operationally reliable, and easily cleaned by washing. The material loaded into the hopper of the dryer enters the upper part of the hopper through a feeder. Rotating discs with scrapers continuously mix the material, tossing it onto the steam-heated pipes. A certain inclination of the drum ensures the movement of the material toward the dumping hatch.

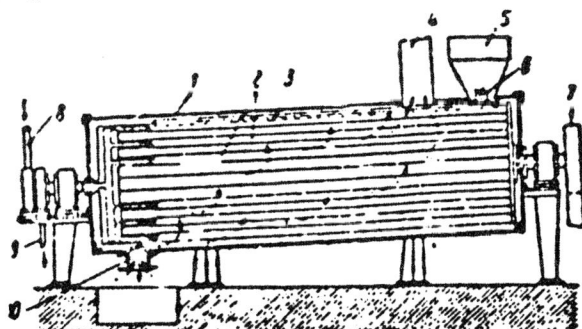

Fig. 23.1. Tubular dryer.
1 - dryer housing, 2 - heating pipes, 3 - scrapers, 4 - outlet pipe, 5 - hopper,
6 - feeder, 7 - drive, 8 - team conduit, 9 - condensate drain, 10 - dumping hatch.

The output of tubular dryers is 200-500 kg/hr. The steam flow rate at maximum load is approximately 50 kg/hr. At 7-8 rpm of the drum and an angle of inclination of 3°, the power consumption does not exceed 2.5 kW.

The evaporated moisture is removed by a natural air draft, which enters through the dumping hatch and exits through the upper outlet. Disadvantages of a tubular dryer include the difficulty of sealing the rotating pipes and the separation of dust from the dumping hatch.

Vacuum dryers (Fig. 23.2). These are horizontally arranged rectangular or cylindrical chambers provided with a row of horizontal hollow plates, through the cavity of which passes steam or hot water.

Trays with the material to be dried are placed on the plates. The chambers are closed with end plates - doors with sealing rubber gaskets and special clamping devices. A reduced pressure is produced by means of RMK No. 2 or No. 3 vacuum pumps. Simultaneous loading of a 3 m³ dryer does not exceed 120-150 kg. Drying at 40-50°C and a pressure of 200-250 Hg lasts about 1 hour for materials of average moisture content.

In comparison with other types of dryers, a vacuum dryer has substantial advantages: a decreased possibility of inflammation of the material being dried (or of the vapor of the fuel liquids driven out of it), no dust being evolved, and acceleration of the drying process. Disadvantages include the periodicity of operation, low output, and considerable investment of manual labor (loading and unloading of trays, pouring the material onto and off them).

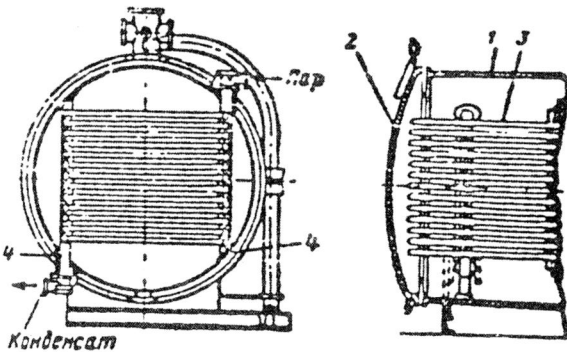

Fig. 23.2. Vacuum dryer.
1 - housing, 2 - lid, 3 - hollow heating plates, 4 - steam conduit for removing evaporated moisture.

COMMINUTION OF INGREDIENTS

The particle size of the ingredients of pyrotechnic compositions must not exceed a few tenths of a millimeter, and sometimes hundredths of a millimeter (i.e., tens of microns). The main ingredients (oxidizers and fuels) are supplied in ground up form, and therefore machines for coarse and medium grinding are not used in pyrotechnic production. As a rule, the oxidizers cluster into lumps during drying, and in addition, the crystal dimensions themselves substantially exceed the dimensions of the particles indicated by the technical specifications.

Metallic fuels are comminuted and screened by the manufacturers. According to the size reduction, the metallic fuels are subdivided into several grades. When the ingredients are prepared, the metallic fuels are subjected only to control screening, and are additionally comminuted only in order to obtain a very fine size reduction (a few microns).

The ingredients are comminuted by using disintegrators and crushers, rotating ball mills and vibrating mills, and less frequently, crusher rolls and roller mills.

A disintegrator (Fig. 23.3) consists of two rotating steel discs enclosed in a cylindrical housing. Steel pins are disposed along the circumference of each disc. Their arrangement is such that each row of pins of each disc enters into the space between the rows of pins of the other disc. The discs are mounted on the ends of shafts rotating in opposite directions. From the hopper, the material to be crushed enters the center of the mill by means of a feeder, then falls into the gaps between the pins and undergoes thorough grinding.

Disintegrators used for preparing the ingredients have a disc diameter of 600-800 mm and a pin length of 100-160 mm. The output of these machines is very large; when the discs rotate at 1000-1200 rpm, the output reaches 1000 kg/hr or more.

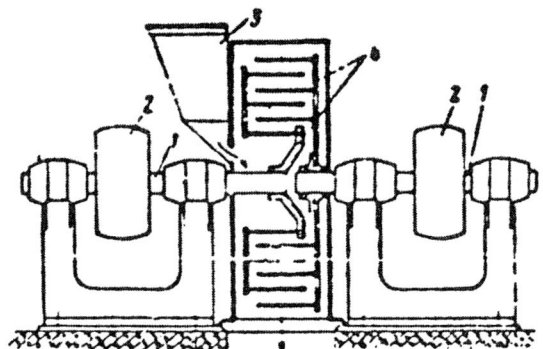

Fig. 23.3. Disintegrator. 1 - shaft, 2 - pulley, 3 - hopper, 4 - discs with rings.

In contrast to a disintegrator, a crusher (Fig. 23.4) has only one rotating disc on which there are scrapers which prevent the crushed material from sticking to the housing.

In addition, a drum with a screen for preliminary screening and holding of the underground material is inserted in the crusher. The presence of friction elements in crushers makes it possible to grind viscous and fibrous materials (iditol, colophony, asbestos, etc.) The output of a crusher at a disc rotation rate of 2500-3000 rpm is 500-1000 kg/hr depending on the properties of the material being ground.

Disadvantages common to a disintegrator and a crusher are:

1. The necessity of closely adjusting and balancing the moving parts, since at high speeds of rotation, even a slight contact between them leads to serious breakdowns.
2. Suction of a large quantity of air, with heavy dust formation requiring special ventilation.
3. Complexity of the disassembly and cleaning, particularly in the case of the disintegrator.

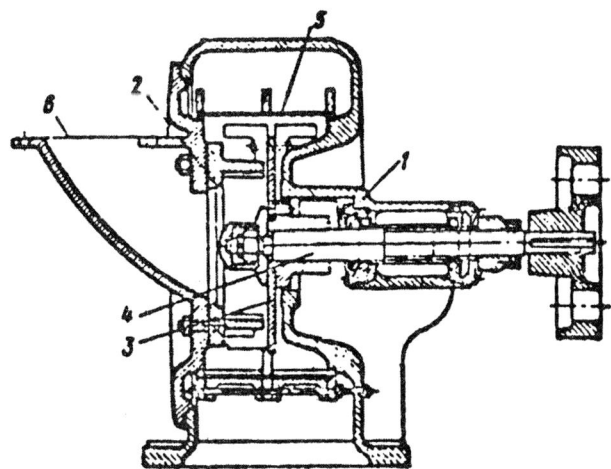

Fig. 23.4. Crusher.
1 - housing, 2 - fixed cast iron disc, 3 - rotating steel disc, 4 - horizontal shaft, 5 - screen, 6 - feed hopper.

Ball mills are rotating drums. Grinding is carried out by means of balls, which as a result of friction against the walls of the drum and against one another periodically rise to a certain height, then roll down again. In the case of grinding of nonexplosive (inert) ingredients, the lining inside the drum is made of a metal, and the balls are made of steel or bronze. When the explosive or fuel materials are ground, the lining is made of leather, and the balls of hard wood (or ceramic).

The grinding time depends on the properties and quantity of the material, speed of rotation of the drum, and size reduction required; the time is usually established experimentally.

Vibrating mills (Fig. 23.5) are not used very much in pyrotechnic production, but they have the following advantages: a large output for the finest size reduction, possibility of grinding metallic fuels in an inert medium, and small size.

The cross section of a vibrating mill is circular or trough-shaped. The balls (milling agents) and the material being ground are loaded into the mill. When the unbalanced shaft rotates, the vibrating mill executes fast vibrations (up to 3000 per minute) perpendicular to the axis of rotation of the shaft. As a result, the loaded material is set into a circular motion in the direction opposite to that of the rotation of the shaft, and the balls rotate around their axis. This causes a vigorous grinding of the material. The chief disadvantages characterizing both vibrating and rotating ball mills are a substantial energy consumption and the loud noise they make. In addition, because of the wear of the balls, a certain contamination of the ground material takes place, this also being undesirable.

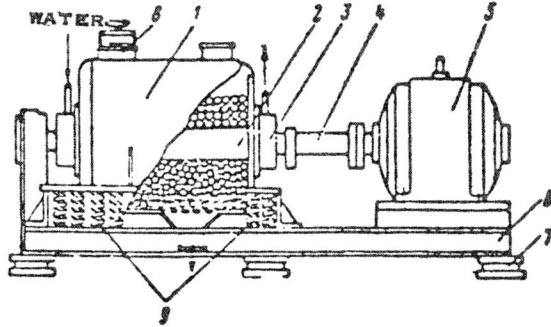

Fig. 23.5. Vibrating mill.
1 - housing, 2 - unbalanced shaft, 3 - cooling system, 4 - flexible coupling, 5 - electric motor, 6 - frame, 7 - shock absorbers, 8 - loading hatch, 9 - springs.

SCREENING OF INGREDIENTS

The ground material is screened in order to obtain a powder with particles relatively similar in size and also to separate random mechanical impurities.

The sieves used are silk or metal screens. For silk sieves, the screen mesh indicates the number of openings per linear centimeter (average in two directions, along the warp and along the weft) and for metal screens, the mesh defines the inside dimension of the openings (in mm).

Standard screens, i.e., shakers and vibratory screens, are used in pyrotechnic production.

A shaker screen (Fig. 23.6) is a wooden box whose lower part is in the shape of a truncated pyramid. A screen applied on a frame is installed in the box at an angle of 3-5°. The box together with the screen is mounted on flat metal springs. The shaker screen is actuated by a motor or transmission through an eccentric, which imparts a reciprocating motion to the screen at the rate of 200-300 vibrations per minute.

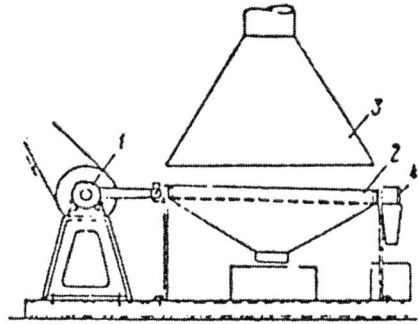

Fig. 23.6. Shaker screen.
1 - crank gear, 2 - screen, 3 - exhaust hood, 4 - collection of siftings, 5 - cone.

A vibratory screen (Fig. 23.7) consists of a metallic table whose top supports an electric motor, an inertial vibrator and a circular hopper with a screen, mounted on rubber plugs inserted in the amount of 2-5 kg, and after closing the lid, the motor is turned on. Through a transmission belt, an electric motor sets in motion the inertial vibrator, which, being rigidly connected to the hopper screen, causes it to vibrate at 1500 to 2000 per minute. The material passed through the screen is poured into a collector box through a funnel.

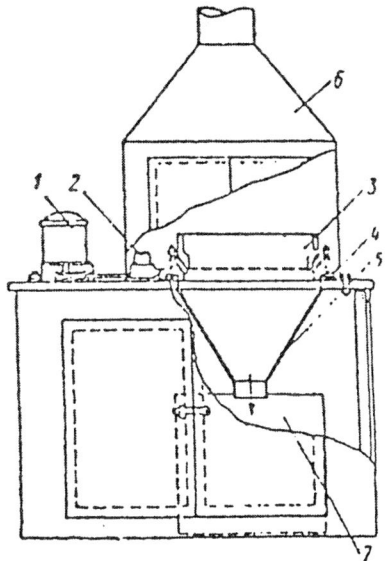

Fig. 23.7. Vibratory screen.
1 - electric motor,
2 - vibrator,
3 - hopper,
4 - rubber plugs,
5 - cone,
6 - exhaust hood,

The chief disadvantages of a shaker screen and vibratory screen are the formation of dust and a considerable investment of manual labor. In carrying out any operation of preparation of the ingredients, it is always necessary to remember that the dust of the oxidizers, by mixing with vapor or dust of organic substances or dust of metallic fuels, may form explosive mixtures. For this reason, all the operations must be carried out in different rooms, and ventilators must always be present.

The preparation of the ingredients frequently involves the use of individual machines for drying, crushing, and sifting, located in different rooms. The machines are loaded and unloaded manually, and the material also is transported either in hand trucks or by hand carrying. All this creates difficult working conditions and a low productivity. It is much more desirable to conduct all the operations in a single machine, i.e., a drying-crushing unit; this ensures a high productivity and considerably improves the working conditions in the plant.

2. PREPARATION OF COMPOSITIONS

A pyrotechnic composition must be homogeneous. A sample take from any part of the composition must correspond to the formula of the composition in its content of the ingredients. To obtain a completely homogeneous mixture is difficult, even after mixing the dry loose materials for a long time, and therefore organic admixtures are frequently introduced in the form of solutions (lacquers); this improves the mixing and at the same time increases the adhesion of the particles to each other. In order to avoid sticking lumps and obtain a homogeneous composition, the latter is subjected to granulation after mixing. To remove the organic solvents, the compositions are dried following granulation.

The preparation of the compositions is an operation involving a fire and explosion hazard. If the safety rules are broken, human casualties are also possible.

The principle safety measures in the course of preparation of the compositions are as follows:

1) The operation of mixing of the ingredients is carried out in embanked isolated buildings separated by a safe distance from the other buildings;
2) The mixers are installed in strong, armed concrete booths, one in each;
3) Control of the mixer (startup, stopping and unloading) is carried out in a safe place and at a suitable distance from the mixing booth;
4) Cleanliness of the facilities and equipment must be observed. Clumps of dust, sticking of the composition to parts of the mixer, and spilling of the composition in the room are inadmissible;
5) Instructions on safety techniques must be strictly observed for each work station;
6) Batching of the oxidizers and fuels must be carried out in different rooms.

Of the large number of various types of mixers used in different industrial sectors, the rotary-blade mixer comes closest to satisfying the conditions of pyrotechnic production.

The rotary-blade mixer (Fig. 23.8) has a trough-shaped housing through which passes a rotating shaft with blades of special design. The housing of the mixer is mounted on a frame in such a way that it can be tilted to pour out the composition. The blades of the mixer are rotated by a belt drive connected to an electric motor via a reduction gear. The motor is explosion-proof.

In order to keep the housing of the mixer clean by removing the composition depositing thereon, hard-rubber tips 5-10 mm thick are attached to the ends of the blades.

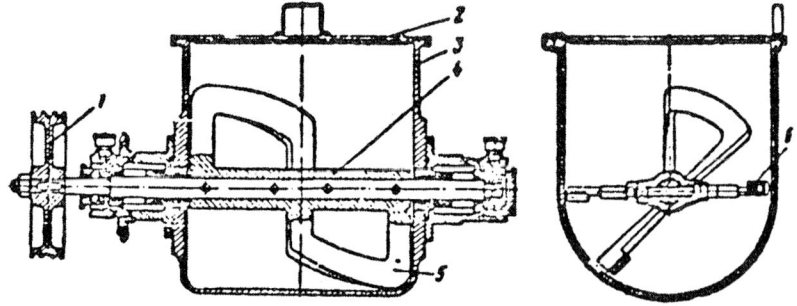

Fig. 23.8. Rotary-blade mixer. 1 - drive pulley, 2 - lid, 3 - housing, 4 - shaft, 5 - blades, 6 - rubber tips.

Far from all compositions can be mixed in a paddle mixer. Some compositions are mixed manually, although in the last few years, many efforts have been expended in trying to design better mixers that would replace the unproductive and dangerous manual labor.

GRANULATION.

This operation consists in rubbing the composition through a screen with a large mesh (10-12 per linear cm) after holding it (drying in the open air) at room temperature for 0.5-1 hour.

For a long time, this operation used to be carried out manually by rubbing the composition through the screen from behind an armor shield by using a rubber stopper mounted at the end of a long stick. This operation, which appears simple at first glance, is very difficult to mechanize, chiefly because of the high viscosity of the composition and its drying during the granulation process; this causes clogging of the screen and sealing of the rubbing elements, and in addition, an increase in friction, which may lead to ignition of the composition. Some positive results of mechanization have been obtained by vibrating the screen or the rubbing elements.

DRYING OF COMPOSITIONS.

This concept is very relative, since the dried ingredients in the course of mixing and granulation absorb practically no moisture; during exposure of the composition in so-called dryers, mainly the organic solvent evaporates until its content is 0.5-0.6%.

In addition, drying chambers are used to store a temporary stock of compositions. The chambers are in individual buildings separated from other buildings by a safe distance and always surrounded by embankments. Considerable amounts of compositions are usually subjected to drying. They are arranged in racks; in the chambers the temperature in the compositions is kept at 30-40°C; under such conditions, the compositions are kept from several hours to several days.

3. COMPACTION (MOLDING) OF COMPOSITIONS

Depending on the design of the article and the required effect, the compositions are given the shape of pellets, etc.; the majority made of compacted compositions are cylindrical in shape. In some cases, the compositions are used to fill the shell of the article directly. In the molding operation, the composition acquires the necessary density and mechanical strength, which play a very important part in producing a suitable pyrotechnic effect. The molding operation is carried out by pressing, screw-conveyor filling, and packing (in the case of rocket compositions, frequently by pouring in the composition) in separate buildings, using special equipment.

PRESSING

In molding of the composition by pressing, a pressing tool is used which most frequently consists of three main parts: a die, a bottom plate, and a plunger (Fig. 23.9).

The pressing tool is manufactured according to high precision standards from high-quality tool steel. Compositions for small articles are dispensed by the volume method directly into the die, into which the bottom plate is inserted; for larger articles, the dispensing is done by weight, and weighed amounts of compositions are placed in cardboard boxes, from which they are poured into dies as required.

The pressing pressure is varied as a function of the physical properties of the composition, and the combustion rate and mechanics strength required.

Tracers for artillery shells and bullets are pressed at a pressure up to 7000 kg/cm^2, illumination and signaling "pellets" at 2000-3000 kg/cm^2, and larger articles, at pressures not in excess of 1000 kg/cm^2.

To obtain a more uniform density and avoid warping the shell, the compositions in many cases are pressed in several ways. To achieve a better cohesion between the individual portions of the composition, grooves of required depth are made on the plungers. In the case of small articles, when the die can not be taken apart, the bottom plate is removed after pressing, and the finished article ("pellet") is pushed out through the bottom opening in the die.

A hydraulic press (Fig. 23.10) consists of a cylinder mounted in an upper block, in which moves a piston with a movable crossbar. Presses with overhead pressure have a return mechanism which returns the movable system to the initial position by a cylinder and a plunger.

The liquid (water or oil) injected into the cylinder at high pressure (150-200 at) lowers the piston with the movable block, pushing the press tool with the composition against the base of the press. By letting the water out of the cylinder and supplying pressure to the upper cylinder, one raises the movable system and releases the press mold.

Presses with underneath pressure also find applications in industry. Although they have a simpler design, they are slow-running and inconvenient to operate. In these presses, the working liquid moves the piston upward, pushing the pressing tool with the composition against the upper fixed frame. When the liquid is let out of the cylinder, the piston with the plate descends under its own weight.

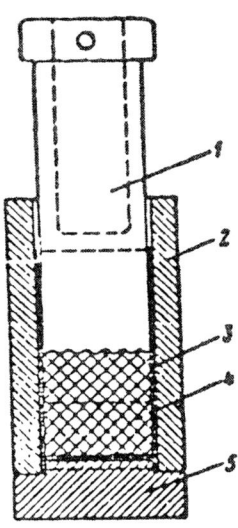

Fig. 23.9. Pressing tool.
1) Die; 2) Matrix;
3) Casing; 4) Pressed composition
5) Bottom plate

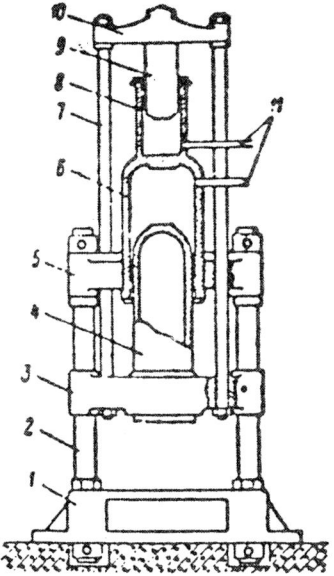

Fig. 23.10. Hydraulic press.
1) Lower block; 2) Columns; 3) Moving crosspiece;
4) Piston; 5) Upper bearing; 6) Cylinder;
7) Pull rod; 8) Upper cylinder; 9) Upper piston;
10) Traverse; 11) Steam lines.

The operation of hydraulic presses requires a unit to deliver the liquid at high pressure. This unit consists of a hydraulic pump and a hydraulic accumulator. The latter equalizes the pressure and flow rate of the liquid.

The power of presses is expressed in tons, as the product of the working pressure in the press cylinder and cross sectional area of the piston. The power of presses used in pyrotechnic production varies widely, from tens to several thousand tons. Pressing of pyrotechnics has the following characteristics:

a) When the liquid is supplied to the working cylinder, the pressure is strictly controlled at the manometer;
b) Holding pressure, i.e., the time during which the composition is pressed, is observed;
c) The operation of the press is followed by means of a system of mirrors or television sets.

Mechanical presses (Fig. 23.11) are used in pyrotechnic production for pressing small-sized articles of fixed height; the group of pressing into so-called "assembly units" is most frequently employed in this case. The mechanical press consists of a housing with an upper fixed block, a lower movable block moving up and down and creating the pressure, a drive, and shock-absorbing springs. The block is set in motion by a spring elbow connected to a flywheel by a link with a crankshaft. The course of the block is regulated by changing the distance between the centers of rotation of the elbow by means of an adjustment screw.

The "assembly units" consist of several dies (from 4 to 80) mounted on a single plate, and as many punches, also mounted into a common plate with a common bottom plate. An "assembly unit" with the composition poured in is placed on the moving block; when the mechanism is turned on, one turn of the flywheel performs the pressing.

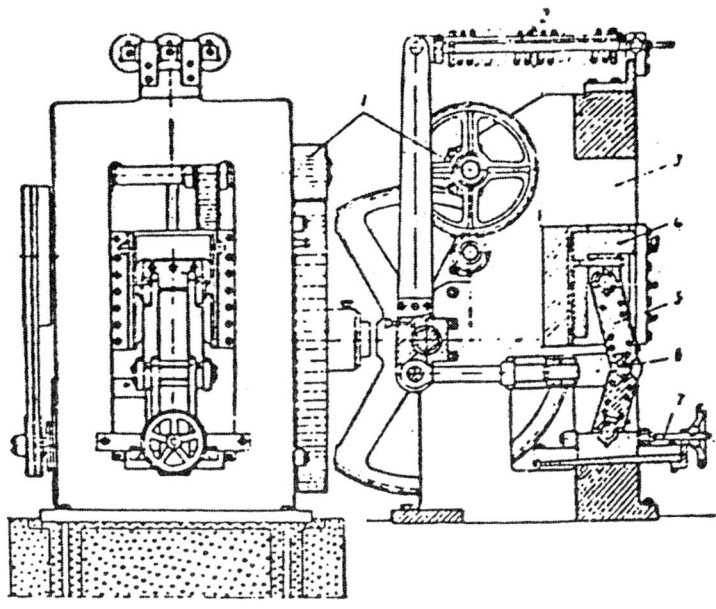

Fig. 23.11. Mechanical press.
1 - drive, 2 - springs, 3 - housing, 4 - movable block, 5 - spring elbow, 6 - link, 7 - adjustment screw.

Mechanical presses have considerable advantages over hydraulic ones in pressing rate and simplicity of design, but the limitation on their power and operation and the impossibility of holding the compositions under pressure during pressing restrict their applications.

4. LOADING OF ARTICLES

In the loading phase, all the parts of an article are prepared, then they are combined into the parts of the structure and the assembly of the article itself is carried out. The loading of each type of article has its characteristic features, which depend on the dimensions and arrangement of the articles. Loading of pyrotechnic articles is usually carried out in the same room on a conveyor or tables, but the assembly hall is always adjacent to a series of booths in which the preparatory operations, which frequently involve a fire or explosion hazard, are carried out.

The preparatory operations common to most articles are:

a) preparation of housings of the articles - unsealing, degreasing, sorting of the parts;
b) preparation of the ejection charge - weighed amount of powder, pouring into bags or celluloid boxes, and sealing them;
c) rigging of parachutes - sorting of canopies and shroud lines, checking of bracing cable, and folding;

d) preparation of pyrotechnic charges - trimming of cardboard sleeves, riveting of obturators, connecting the parachute, etc.
e) selection of cardboard gaskets and bracing parts.

The assembly itself consists in placing all the parts in the prepared housing while observing the sequence specified by the technological process and the blueprint of the article. The most crucial operation of the assembly of many types of articles is the placement of the prepared charge with the parachute in the housing of the article. The presence of a gas check on the pyrotechnic charge and the limitation on the volume occupied by the parachute make it necessary to use a considerable force in ramming them into the housing. Friction or screw presses are used.

A friction press (Fig. 23.12) consists of two columns connected by a crossbar, the drive mechanism being mounted on these columns. Passing through the crossbar is a tightening screw, at the upper end of which is mounted a horizontal friction disc. The tightening screw is moved up or down with the aid of a lever by changing the pressure of the vertical friction discs. Automatic switches are used to restrict the course of the screw.

The operation of ramming the filler into the housing is carried out as follows:

a) The prepared housing with the ejection charge inserted in it is placed in the cradle under the press and secured with a clamp to a bar;
b) The pellet with the gas checks slipped on it is lowered into the housing by hand without any force;
c) The connecting cable of the parachute is rigged;
d) The parachute, previously tightened with auxiliary lines, is placed on the top;

e) The parachute is rammed into the housing, released from the auxiliary shroud lines, and the parachute with the charge are pushed down with the press until they occupy the necessary space.

After this operation, the rear or bottom part is fastened, and the assembly of the article is finished.

For protection against moisture and to obtain a dependable article during extended storage, the joints between the composition and the parts are lacquered, and the joints in the housing are coated with a sealing grease and an oil paint or paraffin.

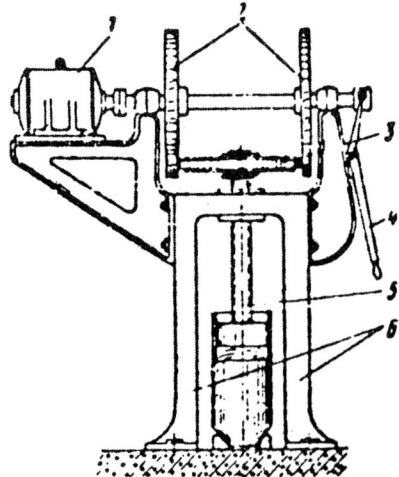

Fig. 23.12. Friction press.
1 - motor, 2 - vertical friction discs, 3 - horizontal friction disc, 4 - lever, 5 - tightening screw, 6 - press columns.

BRIEF SKETCH OF THE HISTORY OF DEVELOPMENT OF PYROTECHNICS IN RUSSIA

The art of creation of fire and its control have their origins in the distant past.

About a century ago, the term "pyrotechnics" was applied to the body of knowledge and methods necessary for the manufacture of powder, explosive mixtures, and various pyrotechnic compositions. However, these areas of military engineering have expanded so much that at the present time, each of them constitutes an independent branch of knowledge.

It is well known that the first pyrotechnic composition used, black powder, was made in Russia in the 15th century in large amounts and that its quality was good.

Under Ivan the Terrible, cannon and powder makers were mostly native Russians. In 1563, the artillery of Ivan the Terrible's army consisted of 200 cannon. At that time, up to 20,000 poods (one pood = 36 English pounds) of black powder was produced each year.

In 1607-1621, gunmaster Onisim Mikhaylov wrote a manual, a copy of which has survived to this day, entitled "Manual of Combat, Cannon, and Other Types of Operations Pertaining to Military Science" A practical handbook for artillerymen, it also gives information on rockets.

In 1674, fireworks were set up and several rockets and firecrackers were set off in the town of Ustyuga.

Under Peter the First, fireworks became an inherent part of festivities associated with any solemn occasions. Peter the First personally prepared the rockets, wheels, and flame patterns; he founded a special "Rocket Institute". On his orders, illumination rockets were used in the army as signaling devices.

At that time, the Okhtenskiy powder plant was founded in Petersburg, and the production of powder considerably increased.

During the years of his stay at the Academy of Sciences, M. V. Lomonosov (1711-65), a scientist of genius, set off many outstanding fireworks. An engraving has been preserved representing a firework made "according to the invention of chief adviser and Professor Lomonosov by head firework specialist Matvey Martynov and burned in Moscow on New Year's Day 1754".

The greatest brightness and beauty were achieved by Russian fireworks in the second half of the 18th century. At that time, the facades of firework fronts were made in huge dimensions: up to 50 sazhens (1 sazhen = 7 feet) in length and up to 25 sazhens in height.

Literature also helped the development of pyrotechnics. In 1779 in Moscow, M. V. Danilov (1722-90) published a handbook "which will enable anyone to prepare and set off fireworks and various illuminations".

In 1824, a book was published by the pyrotechnist F. S. Cheleyev "with an appendix for preparing military firing and incendiary devices" in five parts.

Starting in the 1830's, compositions of colored lights with potassium chlorate began to be produced.

In 1832, a Pyrotechnic Artillery School with a five-year curriculum was founded as part of the Okhtenskiy powder plants. The school taught powder making, military laboratory science and the preparation of fireworks. Rockets with parachutes began to be used, and military rockets were used to launch fireworks.

Beginning in 1847, K. I. Konstantinov conducted systematic studies aimed at designing and producing military rockets.

From 1850 to 1859, Konstantinov headed the Petersburg "Rocket Institute"; during that period, he performed a series of serious scientific experiments and created several new systems of rockets that were successfully used in the Crimean War.

Konstantinov wrote more than 50 scientific papers, including books entitled "On Military Rockets", "Military Rockets in Russia in 1867", and a large paper "The Improvement of Fireworks".

The Mikhaylov Artillery Academy, where Konstantinov lectured, formed in 1855 with the participation of officer classes of the military academy, in which K. I. Konstantinov also lectured, became the scientific center of military technical thought. Special pyrotechnical research was also performed there.

In the second half of the 19th century, they began to introduce many new materials into pyrotechnic compositions.

In 1869 an article by A. I. Plestsov entitled "On the Use of Magnesium in Pyrotechnics" appeared in the Artilleriyskiy zhurnal (Artillery Journal).

The same journal contained articles by Captain A. Ordynskiy, F. F. Matyukevich, and P. S. Tsytovich in 1887-1889.

In the 1890's a number of manuals on the preparation of fireworks were published, among the best of which are the exhaustive works of P. S. Tsytovich and F. V. Stepanov.

At the beginning of the 20th century, thermite compositions, discovered and studied as far back as 1865 by Academician N. N. Beketov, began to be used in pyrotechnics. S. P. Vukolov, a close assistant to D. I. Mendeleyev, worked (since 1907) on the improvement of signaling devices and on the creation of the first samples of tracer shells.

As a result of a significant development of military engineering, by the time of the first world war of 1914-1918, the Russian army was equipped with a number of signaling, illuminating, incendiary and smoke-forming devices.

Russian pyrotechnists Yershov, Sannikov, Pogrebnyakov, Artem'yev, and others, created smoke charges, illuminating and thermite incendiary shells, and other pyrotechnic ammunition.

A bibliography of the works of Russian pyrotechnists of the 19th and beginning of the 20th century is given in the second edition of our book [7, p. 275].

The works published in recent years by Russian pyrotechnists cannot be completely listed in such a brief outline.

Let us note only that the efforts of Soviet scientists and engineers have led to the creation of a large number of new pyrotechnic compositions and pyrotechnic devices which have found uses during the war of 1941-45 and have shown their high qualities; the technology of their manufacture was developed, and new methods of testing of pyrotechnic compositions were created.

APPENDICE

Appendix 1

Heat of Formation of Oxides, Fluorides, Chlorides and Sulfides of Certain Elements

Element	Compound	Molecular Weight of Compound	Heat of Formation, kcal			
			Per g-mole of Compound Q	Per 1 g of Element Q_1	Per 1 g of Compound Q_2	Per g-atom of Compound Q_3
Li	Li_2O	30	143	10.4	4.8	48
	LiF	26	146	20.8	5.5	73
	LiCl	42	96	13.7	2.3	48
	Li_2S	46	112	8.0	2.4	37
Be	BeO	25	142	15.8	5.7	71
	BeF_2	47	240	26.7	5.1	80
	$BeCl_2$	80	122	13.6	1.5	41
	BeS	41	56	6.2	1.4	28
Mg	MgO	40	144	5.9	3.6	72
	MgF_2	62	264	10.9	4.2	88
	$MgCl_2$	95	153	6.4	1.6	90
	MgS	56	84	3.5	1.5	42
Al	Al_2O_3	102	400	7.4	3.9	80
	AlF_3	84	329	12.1	3.9	82
	$AlCl_3$	136	167	6.2	1.2	40
	Al_2S_3	150	140	2.6	0.9	28
Ca	CaO	56	152	3.8	2.7	76
	CaF_2	78	290	7.2	3.7	97
	$CaCl_2$	222	188	4.7	1.7	63
	CaS	73	115	2.9	1.6	57
Ti	TiO_2	80	224	4.7	2,8	75
	TiF_4	124	392	8.2	3.2	78
	$TiCl_4$	190	196	4.1	1.0	39
	TiS_2	112	(80)	(1.7)	(0.7)	(27)
Zr	ZrO_2	123	260	2.9	2.1	87
	ZrF_4	167	445	4.9	2.7	99
	$ZrCl_4$	233	232	2.3	1.0	46
	ZrS_2	155	(148)	(1.6)	(1.0)	(49)
H	H_2O	18	68.4	34.2	3.8	23
	HF	20	64	63.5	3.2	32
	HCl	36	22	21.8	0.6	11
	H_2S	34	5	2.5	0.1	2
C	CO_2	44	94	7.8	2.1	31
	CF_4	88	165	13.7	1.9	18
	CCl_4	154	25	2.1	0.2	5
	CS_2	70	-21	-1.8	-0.3	-7
B	B_2O_3	70	302	14.0	4.3	60
	BF_3	68	270	25.0	4.0	78
	BCl_3	117	103	9.5	0.9	26
	B_2S_3	118	57	5.3	0.5	11

Note: Data obtained by perliminary calculation [94] are given in parentheses.
Table continued on next page

Element	Compound	Molecular Weight of Compound	Heat of Formation, kcal			
			Per g-mole of Compound Q	Per 1 g of Element Q_1	Per 1 g of Compound Q_2	Per g-atom of Compound Q_3
Si	SiO_2	60	208	7.4	3.5	69
	SiF_4	104	360	12.8	3.5	72
	$SiCl_4$	170	150	5.3	0.9	30
	SiS_2	92	34	1.2	0.4	11
P	P_2O_5	142	367	5.9	2.6	52
	PF_3	88	--	--	--	--
	PCl_3	137	79	2.5	0.6	20
	P_2S_5	222	(60)	(2.0)	(0.3)	(9)
Na	Na_2O	62	99	2.2	1.6	33
	NaF	42	136	6.0	3.2	68
	$NaCl$	58	98	4.3	1.7	49
	Na_2S	78	87	1.7	1.1	29
Zn	ZnO	81	83	1.3	1.0	41
	ZnF_2	103	179	2.8	1.7	60
	$ZnCl_2$	136	100	1.5	0.7	33
	ZnS	97	48	0.7	0.5	24
Cu	CuO	80	38	0.6	0.5	19
	CuF_2	102	130	2.0	1.3	43
	$CuCl_2$	134	52	0.8	0.4	17
	CuS	96	12	0.2	0.1	6
Pb	PbO	223	52	0.3	0.2	26
	PbF_2	245	160	0.8	0.6	53
	$PbCl_2$	278	86	0.4	0.3	29
	PbS	239	22	0.1	0.1	11
Hg	HgO	217	21	0.1	0.1	11
	HgF_2	239	(100)	(0.5)	(0.4)	(33)
	$HgCl_2$	271	54	0.3	0.2	18
	HgS	233	16	0.7	0.07	8

Note: Data obtained by perliminary calculation [94] are given in parentheses.

Appendix 2

Formulas of Delay Compositions after Ellern [9]:

1. Zr/Ni alloy   54%
   BaCrO$_4$ 31
   KClO$_4$ 15

When alloys containing 70, 50, and 30% Zr are used, the combustion time of the compositions stands in the ratio 1:2:3. The ration of the oxidizers may also vary.

Formulas with amorphous boron:

2. B/BaCrO$_4$ 5/95
3. B/BaCrO$_4$ 10/90

Combustion time of composition 2, 1 sec/cm; composition 3, 0.2 sec/cm.

Formulas with metallic manganese:

	4	5	6
Manganese	44	37	33
BaCrO$_4$	3	20	31
PbCrO$_4$	53	43	36

The combustion time of these compositions is equal to 1.5, 3, and 5 sec respectively.

"Exotic" Delay Compositions:

	7	8	9	10	11
Niobium	15	18	50	--	--
Tantalum	--	--	--	29	50
BaCrO$_4$	85	82	50	71	50

Experimental compositions are being developed in the USA chiefly for the purpose of creating "gasless" mixtures.

Appendix 3

Average Values of Pressure and Temperature of Air
at Different Heights at Moderate Latitudes

Height Above Sea Level, km	Pressure, mm Hg	Temperature, °C	Height Above Sea Level, km	Pressure, mm Hg	Temperature, °C
0	760	--	30	9.5	-41
2	608	+5	40	2.4	-11
4	486	-6	50	$7.6 \cdot 10^{-1}$	-2
6	372	-20	60	$2.2 \cdot 10^{-1}$	-20
8	281	-35	70	$5.5 \cdot 10^{-1}$	-55
10	205	-48	80	$1.1 \cdot 10^{-1}$	-68
12	152	-50	90	$2.0 \cdot 10^{-1}$	-54
14	114	-50	100	$6.0 \cdot 10^{-1}$	-54
20	42	-60			

Appendix 4

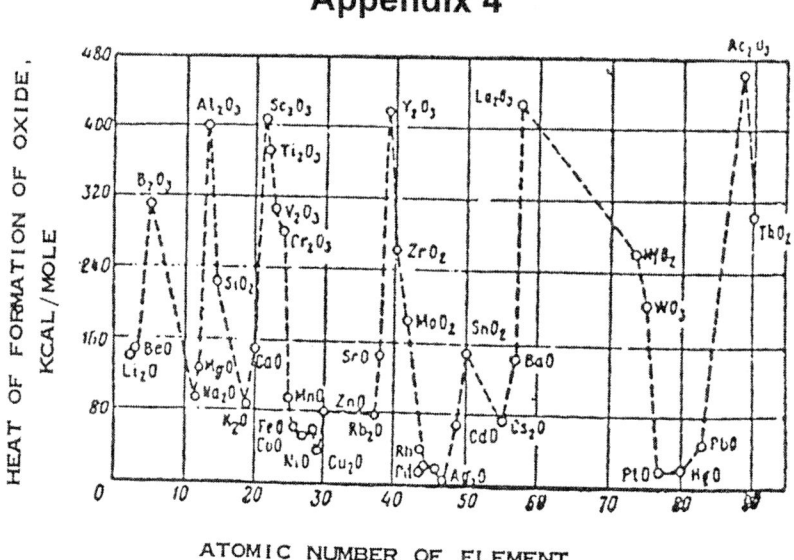

Fig. 1 Heat of formation of oxides as a function of the atomic number of the element

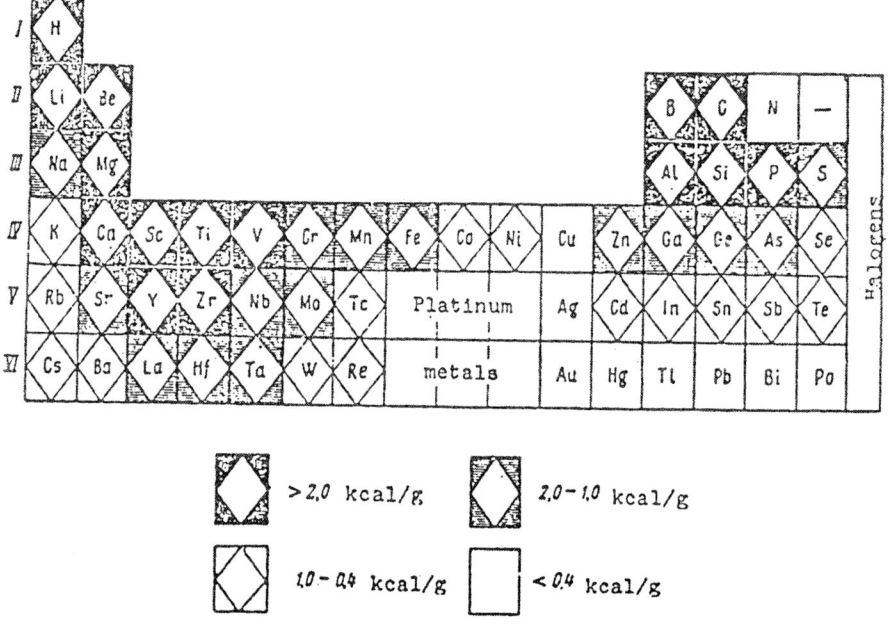

Fig. 2 Heat of formation of oxide (Q_2, kcal/g) versus position of element in the periodic system.

BIBLIOGRAPHY

Chapter 1

1. Budnikov, M.A., Levkovich, I.A., Bystrov, I.V., Sirotinskiy, V.F., and Shekhter, B.I., Vzryvchatyye veshchestva i-porrkha (Explosives and Powder), Oborongiz (State Publishing House for the Defense Ind.), 1955, pp. 216-264.

2. Bystrov, I.V., Kratkiy kurs pirotekhniki (A Short Course in Pyrotechnics), Part I, Oborongiz, 1940.

3. Gorst, A. G., Porokha i vzryvchatyye veshchestva (Powders and Explosives), Oborongiz, 1957, pp. 160-180.

4. Demidov, P.G., Goreniye i svoystva goryuchikh veshchestv (Combustion and Properties of Fuel Substances), Minkomkhoz (Ministry of Municipal Services) RSFSR, 1961.

5. Zhiroz, N.F., Svecheniye pirotekhnicheskikh plamen (Luminosity of Pyrotechnic Flames), Oborongiz, 1939.

6. Solodovnikov, V.M., Piroteknika (proizvodstvo i szhiganiye feyerverkov) (Pyrotechnics (Production and Ignition of Fireworks)), Oborongiz, 1938.

   6A. Terent'yev, A.P., and Yanovskaya, L.A., Knimicheskaya literatura i pol'zovaniye yeyu (Chemical Literature and Its Use), 1964, Moscow, Izd-vo "Khimiya" ("Chemistry" Publishing House).

7. Shidlovskiy, A.A., Osnovy pirotekhniki (Foundations of Pyrotechnics), Oborongiz, 1954, second edition.

8. Becher, W., "Pyrotechnics", article in [51], Band 14, 490-505, 1963.

9. Ellern, H., Modern Pyrotechnics.Foundamentals of Applied Physical Pyrochemistry, N.Y., 1961.

10. Hart, D., "Military Pyrotechnics", article in [46], Vol. 11, pp. 328-332.

11. Hart, D., "Periodicity of Chemical and Thermodynamic Functions", J. Phys. Chem., 1952, 56, 202-214.

12. Izzo, A., Pirotechnica e fuochi artificali, Milano, 1950.

13. Langhans, A., Feuer-Fundamental-Klassification, Band I-II 1930, Stuttgart.

14. Sanderson, R.T., Chemical Periodicity, N.Y.-London, 1960.

15. Weingart, G., Pyrotechnics, N.Y., 1947.

Chapter 2

16. Alekseyenko, L.A., "Effect of Intra-Molecular Polarization Phenomenon on the Thermal Stability of Nitrates and Chlorates", Trudy Tomskogo Universiteta (Transactions of Tomsk University), 1954, 126, 201-226.

17. Belyayev, A.I., and other, Fizicheskaya khimiya rasplavlinnykh soley (Physical Chemistry of Fused Salts), Metallurgizdat (Metallurgical Publishing House), 1957.

18. Billiter, Zh., Promyshlennyy elektroliz vodnykh rastvorov (Industrial Electrolysis of Aqueous Solutions), Goskhimizdat, 1959.

19. Blinov, I.F., Khloratnyye i perkhloratnyye vzryvchatyye veshchestva (Chlorate and Perchlorate Explosives), Oborongiz, 1941.

20. Brauer, G., A Manual on Preparative Inorganic Chemistry (Russian Translation), IL (Foreign Literature Publishing House), 1956.

21. Vol'fkovich, S.I., Yegorov, A.P., and Epshteyn, D.A., Obshchaya khimicheskaya teknologiya (General Chemical Technology), Vols. 1 and 2, Goskhimizdat, 1952-1959.

22. Vorob'yev, A.F., Privalova, N.M., Monayenkova, A.S., and Skuratov, S.M. "Standard Enthalpies of the Formation of Perchloric Acid and Certain Perchlorates", DAN SSSR (Reports of the Academy of Sciences USSR), 1960, 135, No. 6, 1388.

23. Voskresenskaya, N.K., Berul', S.M., "Thermal Stability of a Low-Melting Mixture of the Nitrates and Nitrites of Sodium and Potassium",ZhNKh (Journal of Inorganic Chemistry),1956, 1,No.8.

24. Karyakin, Yu.V., and Angelov, I.I., Chistyye khimicheskiye reaktivy (Pure Chemical Reagents), Khimizdat (Chemistry Publishing House), 1955.

25. Kratkaya khimicheskaya entsiklopediya (Brief Chemical Encyclopedia), (KKnE), 1961-1964, Vols. 1, 2, and 3 (continuing publication)

26. Miniovich, M.A., Soli azotnoy kisloty (nitraty) (Salts of Nitric Acid (Nitrates)), Goskhimizdat, 1946.

27. Nekrasov, B.V., Kurs obshchey khimii (Course in General Chemistry), Goskhimizdat, 1961.

28. Orlova, Ye.Yu., Khimiya i teknologiya brizantnykh vzryvchatykh veshchestv (Chemistry and Technology of High Explosives), Oborongiz, 1960.

29. Pestov, N.Ye., Fiziko-khimicheskiye svoystva poristykh i poroshkoobraznykh produktov (Physico Chemical Properties of Porous and Powdered Products), AN SSSR (Academy of Sciences USSR), 1947.

30. Pozin, M.Ye., Tekhnologiya mineral'nykh soley (Technology of Mineral Salts), Goskhimizdat, 1961.

31. Protsenko, P.I., and Shokina, O.N., "Investigation of a Ternary System Consisting of the Nitrates of Sodium, Potassium, and Barium", ZhNKh, 1959, 4, No. 11, 2554-57.

32. Remy, H., Lehrbuch d. anorg. Chemie, 9 Auft. Band 1 and 2, Leipzig, 1957-1959.

33. Skuratov, S.M., Vorob'yev, A.F., and Privalova, N.M., "On the Problem of the Enthalpy of the Formation of Certain Perchlorates", ZhNKh, 1962, 7, No. 3, 677-679.

34. Smirnov, V.Ya., Pirotekhnicheskiye materialy (Pyrotechnic Materials), Oborongiz, 1939.

35. Spravochnik po plavkosti sistem iz bezvodnykh neorganicheskikh soley (Handbook on the Fusibility of Anhydrous Inorganic Salts), edited by N.K. Voskresenskaya, Vols. 1 and 2, AN SSSR, 1961.

36. Spravochnik po rastvorimosti (Handbook on Solubility) edited by V.B. Kogan, Vol. 1, Books 1 and 2, Goskhimizdat, 1961.

37. Spravochnik khimika (Chemists' Handbook), Second edition, Vols. I-III, Goskhimizdat, Leningrad and Moscow, 1962-1963.

38. Shargorodskiy, S.D., Shor, O.I., "Study of the Thermal Decomposition of the Nitrates and Carbonates of Be, Ca, Sr, and Ba", Ukrainskiy khimicheskiy zhurnal (Ukrainian Chemical Journal), 1954, 20, No. 4, 357.

39. Shidlovskiy, A.A., "Water as an Oxidizer in Pyrotechnic Compositions", DAN SSSR, 1946, 51, No. 2, 127.

40. Shraybman, S.S., Proiizvodstvo bertoletovoy soli i drugikh khloratov (Production of Berthollet's Salt and Other Chlorates), GONTI (State United Scientific and Technical Publishing House), 1939.

41. Fitzpatrick, J.A., Powder Compositions Liberating Hydrogen, American Patent 2.885.277, 1959.

42. Gmelins Handbuch der anorg. Chemie, 8 Aufl. B. 1932-1964 (Publication to be continued). Especially recommended are issues No. 21, Na; No. 22, K; No. 23, Ammonium; No. 27, Mg; No. 29, Sr; No. 30, Ba; and No. 35, Al.

43. Gordon, S., Campbell, C., "Differential Thermal Analysis of Inorganic Compounds: Nitrates and Perchlorates of Metals", Anal. Chem., 1955, 27, No. 7, 1102.

44. Hein, F., Chem. Technik, (Berlin), 1957, 9, 97.

45. Hogan, V.D., Grodon, S., Campbell, C., "Differential Thermal Analysis and Thermogravimetry with Relationship to a System of Potassium Perchlorate - Aluminum - Barium Nitrate", Anal. Chem., 1957, Vol. 29, No. 2, 306.

46. Kirk, R.E., Othmer, D.E., Encyclopedia of Chemical Technology, N.Y., 1947-1956, Vol. 1-15.

47. Leschewsky, K., "Thermal Decomposition of the Nitrate of Sodium and Potassium", Berichte, 1939, 72, 1763-77.

48. Mellor, L.W., A Comprehensive Treatise on Inorganic and Theoretical Chemistry, New Impression, London-N.Y.-Toronto, 1946-1947, Vol. 1-16.

49. Schumacher, J.C., Perchlorates, Their Properties, Manufacture, and Uses, N.Y.-London, 1960.

50. Sneed, M.C., and others, <u>Comprehensive Inorganic Chemistry</u>, N.Y.-London-Toronto, 1956-1963, Vol. 1-8 (Publication to be continued).

51. Ullmanns <u>Encyklopadie d. technischen Chemie</u>, 3 Aufl, Munchen-Berlin, 1951-1964, Bd. 1-14 (Publication to be continued).

52. DD [Note: Reference No. 52 is missing in this translation.]

Chapter 3

53. Abkowitz, S., Burke, D., and Hilz, R., <u>Titanium in Industry</u>, (Russian translatation), Oborongiz, 1957.

54. Eisenkolb, F., <u>Powder Metallurgy</u> (Russian translation from the German), Metallurgizdat, 1959.

55. Belyayev, A.I., <u>Metallurgiya legkikh metallov</u> (Metallurgy of Lightweight Metals), Metallurgizdat, 1954.

56. Gus'kov, V.M., <u>Magniy</u> (Magnesium), article in <u>KKhE</u> [25], 1963, Vol. 2, pp. 1008-12.

57. Deryagin, B.V., "A New Method of Measuring the Specific Surface of Porous Bodies and Powders", <u>Zavodskaya laboratoriya</u> (Plant Laboratory), 1951, 17, No. 3, 324; Deryagin, B.V., and Vlasenko, G.Ya., "Continuous-flow Ultramicroscopic Method of Dispersion Analysis", <u>Kolloidnyy zhurnal</u> (Colloid Journal), 1951, 13, No. 4, 249-55.

58. <u>Issledovaniya pri vysokikh temperaturakh</u> (High Temperature Research), a collection of survey articles edited by V.A. Kirillin, IL, 1962.

59. Krymov, V.V., <u>Tekhnika bezopasnosti pri rabote s magniyevymi splavami</u> (Industrial Safety in Working with Magnesium Alloys), Oborongiz, 1955.

60. Kubashevskiy, O., and Hopkins, B., <u>Oxidation of Metals and Alloys</u> (Russian translation), IL, 1960.

61. Makolkin, I.A., Vernidub, I.I., and others, "Kinetics of the Oxidation of Fine Magnesium Powders at Increased Temperatures", <u>ZhPKh</u> (Journal of Applied Chemistry), 1960, 33, 824-831; Makolkin, I.A., "Kinetics of the Gas Corrosion of Magnesium Alloys", <u>ZhPKh</u>, 1957, 30, 1542.

62. Miller, G.D., <u>Zirconium</u> (Russian translation), IL, 1955; See also <u>Zirconium Metallurgy</u> (Russian translation), edited by G.A. Meerson, IL, 1959.

63. Moroz, L.S., Chechulin, B.B., Polin, I.V., and others, <u>Titan i yego splavy</u> (Titanium and its Alloys), Subpromgiz (State Publishing House for the Shipbuilding Industry), 1960.

64. Ostroushko, Yu.I., and others, <u>Litiy, yego khimiya i tekhnologiya</u> (Lithium, Its Chemistry and Technology), Atomizdat (State Publishing House for Atomic Science and Engineering), 1960.

65. Samsonov, G.V., *et al*, <u>Bor, yego soyedineniya i splevy</u> (Boron, Its Compounds and Alloys), <u>AN SSSR</u>, Kiev, 1960.

66. Samsonov, G.V., and Portnoy, K.I., <u>Splavy na osnove tugoplavkikh soyedeniy</u> (Alloys Based on High-Melting Compounds), Oborongiz, 1961.

67. Sittig, M., <u>Natriy, yego proizvodstvo, svoystva, i premeneniye</u> (Sodium, Its Production, Properties, and Use), Atomizdat, 1961.

68. Strelets, Kh., Tayts, A., and Gulyanitskiy, A., <u>Metallurgiya magniya</u> (Magnesium metallurgy), Metallurgizdat, 1960.

69. Waite, D., and Barks, D., <u>Metallic Beryllium</u>, IL, 1961. <u>Toksikologiya berilliya</u> (Toxicology of Beryllium), collection of articles translated from the English, edited by A.A. Letavet, IL, 1953.

   69A. Filyand, M.A., and Semenova, Ye.I., <u>Svoystva redkikh elementov</u> (Properties of Rare Elements), 1964, Moscow, Metallurgizdat.

70. Herd, D., <u>Introduction to the Chemistry of Hydrides</u> (Russian translation), IL, 1955.

71. Nesmeyanov, An.N., <u>Davleniye para khimicheskikh elementov</u> (Vapor Pressure of Chemical Elements), AN SSSR, 1961.

72. Slavinskiy, M.P., <u>Fiziko-khimicheskiye svoystva elementov</u> (Physico-Chemical Properties of Elements), Metallurgizdat, 1952.

73. Stell, B.D.R., Tables of Vapor Pressures of Specific Explosives (Russian translation), IL, 1949.

74. Tovarov, V., "Measurement of the Specific Surface of Powdered Materials", Zavodskaya laboratoriya, 1948, 14, No. 1, 68-76.

75. Entelis, S.G., and others, "Measurement of the Specific Surface of Porous Bodies and Powders by the Method of Measuring the Escape of a Rarefied Gas", ZhFKh (Journal of Physical Chemistry), 1958, 32, No. 9, 2187.

76. Edwards, Y.D., Wray, R.D., Aluminum Paint and Powder, N.Y., 1955.

77. Hampel, C.A., Rare Metals Handbook N.Y., 1961, second edition.

78. Orr, C., Jr., Delavalle, J.M., Fine Particle Measurement, N.Y., 1959.

79. Pannel, E.V., Magnesium, Its Production and Uses, London, 1949.

80. Pilling, N.B., Bedworth, R.E., "Oxidation of Metals at High Temperatures", J. Inst. Metals, 1923, 29, 529-91.

81. Sartorius, R., "Ascertaining the Possibility of using Silicon in Explosive Mixtures", Mem. poudres, 1952, 34, 205-220.

82. Talley, C.P., "Combustion of Elementary Boron", Aerospace Engng., 1959, 18, No. 6, 37-40, for translation see [225].

Chapter 4

83. Avdeyev, B.A., Tekhnika opredeleniya mekhanicheskikh svoystv metallov (Techniques of Determining the Mechanical Properties of Metals), Metallurgizdat, 1949.

84. Bal'shin, M.Yu., Poro shkovoye metallovedeniye (Powder Metallurgy), Metallurgizdat, 1948.

85. Kiselev, V.S., and Abashkina, A.F., Proizvodstvo lakov, olif. i krasok (Production of Lacquers, Drying Oils, and Paint), Second edition, Khimizdat, 1961.

86. Lazarev, A.I., & Sorokin, M.F., Sinteticheskiye smoly dlya lakov (Synthetic Resins for Lacquers), Khimizdat, 1953.

87. Losev, I.P., & Petrov, G.S., Khimiya iskusstvennykh smol (Chemistry of Artificial Resins), Khimizdat, 1951.

88. Losev, I.P., and Trostyanskaya, Ye.B., Khimiya sinteticheskikh polimerov (Chemistry of Synthetic Polymers), GNTI (State Scientific and Technical Publishing House), 1964.

88A. Paken, A.M., Epoksidnyye soyedineniya i epoksidnyye smoly (Epoxy Compounds and Epoxy Resins), 1962, Goskhimizdat, Leningrad.

89. Petrov, G.S., and Levin, A.N., Termoreakvivnyye smoly i plasticheskiye massy (Termoreactive Resins and Plastics), Goskhimizdat, 1959.

90. Trostyanskaya, Ye.B., Kolachev, B.A., and Sil'vestrovich, S.I., Novyye materialy v tekhnike (New Materials in Engineering), GNTI, 1962.

Chapter 5

91. Demidov, A.N., Vvedeniye v pirotekhniku (Introduction to Pyrotechnics), Voyenizdat (Military Publishing House), 1939.

92. Tsytovich, P.S., Opyt ratsional'noy pirotekhniki (Experience in Rational Pyrotechnics), St. Petersburg, 1694.

Chapter 6

93. Britske, E.V., Kapustinskiy, A.F., and others, Termicheskiye konstanty neorganicheskikh vesckestv (Thermal Constants of Inorganic Substances), Izd. AN SSSR (Publishing House of the Academy of Sciences USSR), Moscow-Leningrad, 1949.

94. Karapet'yants, M.Kh., and Karapet'yants, M.L., "Tables of Certain Thermodynamic Properties of Various Substances", Trudy MKhTI imeni D.I. Mendeleyeva (Transactions of the Moscow Institute of Chemical Technology imeni D.I. Mendeleyev), No. 34, 1961.

95. Kubaschewski, O., Evans, E., Metallurgical Thermochemistry, Second edition, N.Y., 1956.

96. Latimer, V.M., Oxidizing States of Elements and Their Potentials in Aqueous Solutions (Russian translation), IL, 1954.

97. Popov, M.M., Termometriya i kalorimetriya (Thermometry and Calorimetry), Izd. MGU (Publishing House of Moscow State University), 1954.

97A. Skuratov, S.M., Kolesov, V.P., and Vorob'yev, A.F., Termoknimiya (Thermochemistry), Pt I, 1964, Izd. MGU.

98. Rossini, F., Wagman, D., Evans, W., Levin, S., Jaffe, I., Selected Values of Chemical Thermodynamic Properties, Washington, 1952, Circ. 500.

Chapter 8

99. Haydon, A., Spectroscopy and the Theory of Combustion (Russian translation), IL, 1950.

100. Haydon, A., and Wolfguard, H., Flame, Its Structure, Radiation, and Temperature (Russian translation), IL, 1959.

101. Kadyshevich, A.Ye., Izmereniye temperatury plameni (Measurement of Flame Temperature), Metallurgizdat, 1961.

102. Kirillin, V.A., Sheyndlin, A.Ye., Issledovaniya termodinamicheskikh svoystv veshchestv (Investigations of the Thermodynamic Properties of Matter), Energoizdat (Power Engineering Publishing House), Moscow-Leningrad, 1963 (including the determination of high temperatures and calorimetry. A. Sh.).

103. Pokhil, P.F., Mal'tsev, V.M., and Gal'perin, L.M., "A Device for Determining Temperature by the Height of the Flare of a Powder Flame", VhFKh, 1960, 34, No. 5, 1131-33.

104. Sobolev, N.I., "Optical Methods of Measuring the Temperature of Industrial Flames", Trudy FIAN SSSR (Transactions of the Physics Institute, Academy of Sciences, USSR), 1956, 7, 1961.

105. Dean, J.A., Flame Photometry, N.Y., 1960.

106. Doyle, W.L., Conway, J.B., Grosse, A.V., "Combustion of Zirconium in Oxygen", J. Inorg. and Nuclear Chemistry, 1958, 6, No. 2, 138-144.

107. Grosse, A.V., Conway, J.B., "Combustion of Metals In Oxygen", Ind. and Engng. Chem., 1958, 50, No. 4, 668-72.

108. Harrison, P., "Ignition of Titanium and Zirconium", see [147], 7th symposium, pp. 913-918.

109. Venturini, J., "Metallothermic Reactions and Classification of Elements; Calculation of the Temperature of Metallothermic Reactions and Effect of Heating", Metaux et Corrosion, 1953, 28, 293-301, 396-405.

Chapter 9

110. Andreyev, K.K., Termicheskoye razlozheniye i goreniye vzryvchatykh veshchestv (Thermal Decomposition and Combustion of Explosives), Energoizdat, Moscow-Leningrad, 1957.

111. Andreyev, K.K., and Belyayev, A.F., Teoriya vzryvchatykh veshchestv (Theory of Explosives), Oborongiz, 1960.

112. Afanas'yev, G.T., and Bobolev, V.K., "On the Phlegmatization of Explosives", DAN SSSR, 138, No. 1, 186, 1961.

113. Baum, F.A., Stanyukovich, K.P., and Shekhter, B.I., Fizika vzryva (Physics of the Explosion), Fizmatizdat (Publishing House for Physics and Mathematics), 1959.

114. Blinov, I.F., "Effect of the Chemical Nature of Oxidizers on the Sensitivity of Mixtures", Trudy MKhTI imeni D.I. Mendeleyeva, No. 20, 1955, 289-298.

115. Bouden, F., and Ioffe, A., Excitation and Development of an Explosion in Solids and Liquids IL, 1955.

116. Bouden, F., and Ioffe, A., Fast Reactions in solids (Russian translation), IL, 1962.

117. Rempel', G.G., and Likin, V.A., [Note: Russian title is illegible.] (Industrial Safety in working with Explosives), Oborongiz, 1963.

118. Semenov, N.N., O nekotorykh problemakh khimicheskoy kinetike i reaktsionnoy sposobnosti (On Certain Problems of Chemical Kinetics and Reactivity), AN SSSR, 1954.

119. Teoriya vzryvchatykh veshchestv (Theory of Explosives), collection of articles edited by Andreyev, K.K., Belyayev, A.F., Gol'binder, A.I., and Gorst, A.G., Oborongiz, 1963.

120. Freeman, E.S., Gordon, S., "Thermal Ignition of Systems Consisting of: Lithium Nitrate and Magnesium, Sodium Nitrate and Magnesium", J. Phys. Chem., 1956, 60, No. 7, 867-70.

121. Patai, S., Hoffman, E., "Pre-Initial Reactions of Certain Fuels with Solid Oxidizers", J. Applied Chem., London, 1952, 2, 8-11.

122. Tomlinson, W., Audrieth, L., "Unexpected Chemical Explosions", J. Chem. Engr. 1950, 27, No. 11, 606.

Chapter 10

123. Andreyev, K.K., "Dependence of the Rate of Combustion of Bickford Fuze Upon Pressure", DAN SSSR, 1945, 49, No. 6, 437.

124. Andreyev, K.K., "On the Combustion of Explosives at Low Temperatures", ZhFKH, 1946, 20, 484.

125. Bakhman, N.N., and Belyayev, A.F., "Effect of the Dimensions of Particles on the Rate of Combustion of Mixtures Based on $NClO_4$", DAN SSSR, 1960, 133, No. 4, 986.

126. Belyayev, A.F., and Komkova, L.D., "Dependence of the Rate of Combustion of Thermites Upon Pressure", ZhFKh, 1950, 24, No. 11, 1302.

127. Belyayev, A.F., and Maznev, S.F., "Dependence of the Rate of Combustion of Black Powder Upon Pressure", DAN SSSR, 1960, 131, No. 4, 887.

128. Belyayev, A.F., and Lukashenya, G.V., "On the Temperature Coefficient of the Rate of Combustion of Black Powder", ZhFKh, 1962, 36, No. 5, 1050.

129. Belyayev, A.F., and Tsiganov, S.A., "Combustion of Condensed Mixtures with Non-Volatile and Undecomposed Fuels at Increased Pressures", DAN SSSR, 1962, 146, No. 2, 383.

130. Zel'dovich, Ya.V., Goreniye i detonatsiya v gazakh (Combustion and Detonation in Gases), Izd. AN SSSR, 1944.

131. Zel'dovich, Ya.V., "On the Stability of the Combustion Regime of Powder in a Semi-Enclosed Volume", Zhurnal prikladnoy mekhaniki i tekhnicheskoy fiziki (Journal of Applied Mechanics and Engineering Physics), 1963, No. 1, 67-76.

132. Zel'dovich, Ya.V., "Chain Reactions in Hot Flames", Kinetika i kataliz (Kinetics and Catalysis), 1961, 2, No. 3, 305-318.

133. Margolin, A.D., "On the Leading Stage of Combustion", DAN SSSR, 1961, 141, No. 5, 1131.

134. Markshteyn, G., "Combustion of Metals", Raketnaya tekhnika i kosmonavtika (Rocket Engineering and Cosmonautics), 1963, I, No. 3, 3-19.

135. Pokhil, P.F., and Romodanova, L.D., "Dinitrotoluol in Rocket Propellants", DAN SSSR, 1959, 128, No. 1, 133.

136. Semenov, N.N., "Chain Reactions in the Theory of Combustion", Uspekhi khimii (Progress in Chemistry), 1957, 26, No. 3, 273-93.

137. Semenov, N.N., "Theory of Homogeneous Combustion of Homogeneous Gaseous Systems", Izv. AN SSSR, OTN (News of the Academy of Sciences USSSR, Technical Sciences Series), 1953, No. 5, 708.

138. Hill, R., and Cottrell, T., "Investigation of the combustion Processes of Solid Fuels", (Russian translation), article at Fourth Symposium on Problems of Combustion and Detonation Waves, 1958, Moscow, Oborongiz, pp. 246-249.

139. Shidlovskiy, A.A., and Oranzhereyev, S.A., "Investigation of the Combustion Processes of Inorganic Salts, Ammonium Bichromate, and Ammonium Trichromate", ZhPKh, 1953, 26, No. 1, 25.

140. Benture, S., Ish-Shalone, M., Lenji, L., Trocker, M., "Delay of the Combustion of Black Powder by the Use of Foreign Matter", Proc. Roy. Soc., Ser. A., 1955, 280, No. 1180, 33-46.

141. Campbell, C., Weingart, G., "Ignition and Combustion of Black Powder", Trans. Far. Soc., 55, Part 12, 221-228, 1959.

142. Cassel, H.M., Liebman, I., "Combustion of Magnesium Particles", Comb. and Flame, 1962, 6, No. 3, 153-156 and 1963, 7, No. 1, 79-81.

143. Coffin, K.P., "Study of the Combustion of Magnesium Ribbon", see [147], Fifth Symp. 267-276.

144. Fetting, F., "Progress in Investigation of Combustion Processes", Chemieing. Technik, 1961, 33, No. 3, 166-172; Fetting, F., "Flame and Combustion", Separate issue from Fortschritte Verfahrentechnik, 1958-1959, Bd. 4. s. 739-776 (Bibliography of 486 titles).

144A. Friedman, R., Macek, A., "Combustion of Particles of Aluminum in an Atmosphere of Hot Gases", Comb.and Flame, 1962, 6, 9-19.

145. Hershkowitz, F., Schwartz, F., Kaufman, F.V.R., "Combustion of an Uncompacted Mixture of Powders of Potassium Perchlorate and Aluminum", see [147], 8th Symposium, 1962, pp. 720-727.

146. Kirshfeld, Z., "On the Rate of Combustion of Wire Made of Non-Ferrous Metals", Metal, 1960, No. 3, 213-219.

147. I-IX, Symposium on Combustion (Intern.), 1946-1962, U.S.A.-England.

Chapter 11

148. Avanesov, D.S., Praktikum po fiziko-khimicheskim ispytaniyam VV (A Practical Handbook on Physico-Chemical Testing of Explosives), Oborongiz, 1959.

149. Andreyev, K.K., and Gorbunov, V.V., "Investigations on the Transition of the Combustion of Explosives to Explosion", ZhFKh, 1963, 37, No. 9, 1958.

150. Andreyev, K.K., "On the Mechanism of the Origin of a Detonation of Explosives", Izv. AN SSSR, OTN, Energetika i avtomatika (News of the Academy of Sciences USSR, Technical Sciences Division, Power Engineering and Automation Series), 1959, No. 4, 188.

151. Andreyev, K.K., "On the Basic Reasons for the Difference Between Initiating and Secondary Explosives", DAN SSSR, 1962, 1946, No. 2, 413.

152. Godzhello, M.G., Vzryvy promyshlennykh pyley i ikh preduprezhdeniye (Explosions of Industrial Dusts and their Prevention), Izd. MKKh RSFSR (Publishing House of the Ministry of Municipal Services RSFSR), 1952.

152A. Gol'binder, A.I., Laboratornyye raboty po teorii VV (Laboratory Work on the Theory of Explosives), 1963, Rosvuzizdat (Publishing House of the Ministry of Higher Education, Russian Soviet Federated Socialist Republic).

153. Sbornik statey po teorii vzryvchatykh veshchestv (Collection of Articles on the Theory of Explosives), edited by K.K. Andreyev, Oborongiz, 1940.

154. Snitko, K.K., Teoriya vzryvchatykh veshchestv (Theory of Explosives), Second edition parts 1 and 2, 1936, Leningrad, Izd, Artakedemii (Publishing House of the Artillery Academy).

155. Shidlovskiy, A.A., "Explosive Mixtures of Water and Methyl Alcohol with Magnesium and Aluminum", ZhPKh, 1946, 19, No. 4, 371-378.

156. Shilling, N.A., Vzryvchatyye veshchestva i snaryazheniye boyepripasov (Explosives and the Charging of Ammunition) Oborongiz, 1946.

157. Cook, M.A., Science of High Explosives, N.Y., 1958.

158. Grodrinski, J., "Low-Temperature Explosions of Potassium Perchlorate with Fuels", J. Appl. Chem., 1958, 8, No. 8, 528.

159. Langhans, A., "Unexpected Explosions", appendix to Z. Schiess u. Sprengstoffw., 1930.

160. Medard, L., "Explosive Properties of Mixtures of Magnesium or Aluminum with Water or Methanol", Mem. poudres, 1951, 33, 492-503.

161. Schichter, G., "Explosive Properties of Titanium and Zirconium Powders", Neue Hutte, 1958, 3, No. 3, 187.

Chapter 12

162. Andreyev, K.K., Kriger, G.E., and Khotin, V.G., "On for Formation of Fuel Gases in the Reaction of Aluminum with Water and Solutions of Ammonium Nitrate", ZhPKh, 1962, 35, 2569.

163. Akimov, G.V., Osnovy_ucheniya_o_korrozii_i_zashchite_metallov (Foundations of Principles of the Corrosion and Protection of Metals), Metallurgizdat, 1946.

164. Voskresenskiy, N.P., Analiticheskiye_reaktsii_mezhdu_tverdymikhimicheskimi_veshchestvami_i polevoy_Khimicheskiy_analiz (Analytical Reactions Between Solid Chemical substances and Field Chemical Analysis), Geoltekhizdat (Geological Engineering Publishing House), 1963.

165. Mendeleyev, D.I., Osnovy_khimii (Foundations of Chemistry), Goskhimizdat, 1947.

166. Rozman, B.Yu., "On the Rational Selection of Inhibitors of Thermal Decomposition of Ammonium Nitrate", ZhPKh, 1960, 33, 1258.

167. Romanov, V.V., Korroziya_magniya (Corrosion of Magnesium), AN_SSSR, 1961.

167A. Svetlov, B.Ya., and Solntseva, R.M., "On the Chemical Stability of Aluminum in Industrial Explosive Compositions", collection Vzryvnoye_delo (Explosives), 52/9, Gosgortekhizdat (State Mining Engineering Publishing House), 1963, pp. 67-80.

168. Tomashov, N.D., Teoriya_korrozii_i_zashchity_metallov (Theory of Corrosion and Protection of Metals), Izd. AN SSSR, 1959.

169. Shishakov, N.A., and Andrushchenko, N.K., "Oxide Films on Magnesium", ZhFKh, 1956, 30, No. 9, 1966-74.

170. Evans, Yu.R., Korroziya_i_okisleniye_metallov (Corrosion and Oxidation of Metals), Mashgiz (State Publishing House for the Machine-Building Industry), 1962.

171. Danner, C.E., Coldenson, Y., "Analysis of Thermite Incendiary Compositions and Pyrotechnic Mixtures", Anal._Chem., 1947, 19, 627-30.

Chapter 13

172. Ango, M.A., Infrakrasnoye_izlucheniye (Infrared Radition), Energoizdat, 1957.

173. Bel'skiy, M., Bessekerskiy, V., and Donskoy, A., Obshchaya_elektrotekhnika (General Electric Engineering), Gosenergoizdat (State Power Engineering Publishing House), 1951.

174. Bur'yanov, B.P., Magnitoelektricheskiy_ostsillograf (The Magnetoelectric Oscillograph), Gosenergoizdat, 1952.

175. Gershun, A.A., Izbrannyye_trudy_po_fotometrii_i_svetotekhnike (Selected Works on Photometry and Lighting Engineering), Fizmatizdat (Physics and Mathematics Publishing House), 1956.

176. Deribere, M., Prakticheskoye_primeneniye_infrakrasnykh_luchey (Practical Application of Infrared Rays), Gosenergoizdat, Moscow-Leningrad, 1959.

177. Ivanov, Yu.A., and Tyapkin, B.V., Infrakrasnaya_tekhnika_v_voyennom_dele (Infrared Engineering in Military Affairs), "Sovetskoye radio" ("Soviet Radio" Publishing House), 1963.

178. Landsberg, G.S., Optika (Optics), Gostekhizdat (State Engineering Publishing House), 1952.

179. Luk'yanov, S.Yu., Fotoelementy (Photoelements), AN_SSSR 1948.

180. Malyshev, V.I., Infrakrasnoye_izlucheniye (Infrared Radiation), article in Fizicheskiy slovar: (Physics Dictionary), Vol. 2, 1962.

181. Margolin, M.L., and Rumyantsev, N.P., Osnovy_infrakrasnoy_tekhniki (Foundations of Infrared Engineering), Oborongiz, 1957.

182. Meshkov, V.V., Osnovy_svetotekhniki(Foundations of Lighting Engineering), Energoizdat, 1957.

183. Optika_v_voyennom_dele (Optics in Military Affairs), collection of articles edited by S.I. Vavilov, AN SSSR, 1945.

184. Shkurin, G.P., Spravochnik_po_radioezmeritel'noy_i_elektroizmeritelnoy_apparature (Handbook on Radio and Electrical Measuring Apparatus), Voyenizdat, 1956.

185. Tavernier, P., "Study of Illuminating Mixtures", Met. poudres., 1949, 31, 309-426.

Chapter 14

186. Mikhaylov, V.Ya., Fotografiya i aerofotografiya (Photography and Aerial Photography), Geodezizdat (Geodetic Publishing House), 1952.

187. Rozhdestvin, N.P., Aerofotografiya (Aerial Photography), Voyenizdat, 1947.

188. Safronov, L.T., Nochnoye vozdushnoye fotografirovaniye (Night Aerial Photograph), Voyenizdat, 1947.

189. Yutsevich, Yu.K., "Aerial Photography at Night", Vestnik vozdushnogo flota (Air Force Herald), 1956, No. 5, 39-45.

190. Vaid, P.K., "A Flash Bomb", American patent 2.775.938, 1957.

Chapter 15

191. Tret'yakov, G.M., Boyeprapasy artillerii (Artillery Ammunition), Voyenizdat, 1947.

192. Eppig, H., German Tracer Composition, London, 1945.

Chapter 16

193. Gurevich, M.M., Tsvet i yego izmereniye (Color and Its Measurement), Izd. AN SSSR, 1950.

194. Shklover, D.A., and Ioffe, R.S., "The VEI Universal Photoelectric Colorimeter", Izv. AN SSSR, OTN, 1951, NO. 5, pp. 667-681.

195. Eppig, H., The Chemical Composition of German Pyrotechnic Colored Signal Lights, 1945, London.

Chapter 17

196. Belyayev, A.I., and Firsanova, L.A., Odnovalentnyy alyuminiy v metallurgicheskikh protsessakh (Monovalent Aluminum in Metallurgical Processes), Metallurgizdat, 1959.

197. Blinov, V.I., and Khudyakov, S.N., Diffuzionnoye goreniye zhidkostey (Diffusion Combustion of Liquids), AN SSSR, 1961.

198. Van Weser, D.R., Phosphorus and Its Compounds (Russian translation), IL, 1962.

199. Gorlov, A.P., Zazhigatel'nyye sredstva, ikh primeneniye i bor'ba s nimi (Incendiary Devices, Their Use, and Their Control), Izd. Narkomkhoza RSFSR (Publishing House of the People's Commissariat of Economic Affairs RSFSR), Moscow-Leningrad, 1943.

200. Knunyants, I.L., and Fokin, A.V., Pokoreniye nepristupnogo elementa (ftora) (Conquest of an Inaccessible Element (Fluorine)), AN SSSR, 1963.

201. Layner, A.I., Alyuminotermiya (Aluminothermy), article in KKhE [25], Vol. I, page 162.

202. Losev, B.I., Komskiy, M.S., and Troyanskaya, M.A., Tverdyy benzin (Solid Gasoline), Gostoptekhizdat (State Heating Engineering Publishing House), 1955.

203. Ryss, I.G., Khimiya ftora i yego neorganicheskikh soyedineniy (Chemistry of Fluorine and its Inorganic Compounds), Khimizdat, 1956.

204. Tolmachev, I.P., Proizvodstvo alyuminiyevogo poroshka, alyuminiyevoy pudry i termita (Production of Aluminum Powder, Aluminum Dust, and Thermite), Metallurgizdat, Moscow-Leningrad, 1938.

205. Fluorine and Its Compounds (Russian translation), collections of articles edited by G. Simons, Vols. 1 and 2, IL, 1953 and 1956.

206. Khudyakov, G.N., 'A Method of Estimating the Combustibility of a Liquid", Izv. AN SSSR, OTN, No. 4, 579-589, 1948.

207. Shishkov, Yu.D., and Opalovskiy, A.L., "Physico-Chemical Properties of Chlorine Trifluoride", Uspekhi khimii, 1960, 29, No. 6, 760-773.

208. Dautzenberg, W., "Aluminothermy", see [51], Bd. 3, s. 423-436.

209. Hajek, H.V., "Napalm", Explosivstoffe, 1957, 5, No. 6, 121-126.

210. Ritter, H.E., Double-Action Incendiary Mixtures Containing Aluminum, Federal Republic of Germany patent 1.114.419, 1962; Referativnyy zhurnal "Khimiya" (Journal of Abstracts, "Chemistry"), 1963, No. 20, N. 387.
Chapter 18

211. Veytser, Yu.M., and Luchinskiy, G.M., Maskiruyushchiye dymy (Screening Smokes), Khimizdat, 1947; Veytser, Yu.M., and Luchinskiy, G.M., Khimiya i fizika maskiruyushchikh dymov (Chemistry and Physics of Screening Smokes), Oborongiz, Moscow-Leningrad, 1939.

212. Pavlov, V., "The Use of Screening Smokes in the Army of the USA", Voyennyy vestnik (Military Herald), 1950, 1, 55-60.

213. Gordon, S., and Campbell, C., "Pre-Initiation and Ignition Reactions in a $Zn/C_6C_{16}/KClO_4$ System", see [147], 5th symp. pp. 277-284.

Chapter 19

214. Kogan, I.M., Khimiya krasiteley (Chemistry of Dyes), Second edition, 1956, Moscow, Goskhimizdst.

215. Colour Index, 1956, Bradford, Vol. 2, Second Edition, (Handbook).

216. Eppig, H., The Chemical Composition of German Pyrotechnic Smoke Signals, London, 1945.

217. Ninneroff, J., and Wilson, S.W., "A Colorimeter for Determination of the Color Index of Pyrotechnic Smokes', J. Res. Bur. Stand., 1954, 52, No. 4, 195-199.

218. Tuve, P., and Spring, S., "A Pyrotechnic Marker", American patent 2.469.421, 1949.
Chapter 20

219. Abramovich, G.N., Gazovaya dinamika vozdushno-reaktivnykh dvigateley (Gas Dynamics of Air Breathing Jet Engines), Izd. BNT (Publishing House), 1946.

220. Alemasov, V.Ye., Teoriya raketnykh dvigateley (Theory of Rocket Engines), Oborongiz, 1962.

221. Barrer, M., Zhommot, A., Vebek, B., and Vandenkarkkhove, Zh., Haketnyye dvigateli (Rocket Engines), Oborongiz, 1962.

222. Bondaryuk, M.M., and Il'yashchenko, S.M., Pryamotochnyye vozdushno-reaktivnyye dvigateli (Ramjet Air-Breathing Engines), Oborongiz, 1958.

223. Zhidkiye i tverdyye raketnyye topliva (Liquid and Solid Rocket Propellants), collection of articles edited by Yu.Kh. Shaulov, translated from the English, IL, 1959.

224. Zel'dovich, Ya.B., Rivin, M.A., and Frank-Kamenetskiy, D.A., Impul's reaktivnoy sily porokhovykh raket (Impulse of the Reactive Force of Powder Rockets), Oborongiz, 1963.

225. Investigation of Solid Propellant Rocket Engines (Russian translation), edited by M. Summerfield, IL, 1963.

226. Kurov, V.D., and Dolzhanskiy, Yu. M., Osnovy proyektirovaniya porokhovykh raketnykh snaryadov (Foundations of the Designing of Powder Rocket Missiles), Oborongiz, 1961.

227. Maxwell, V., and Young, G., "Large Engines using Rocket Fuel", Voprosy raketnoy tekhniki (Problems of Rocket Engineering), 1962, 12, No. 1 3-20.

228. Paushkin, Ya.M., Khimiya reaktivnykh topliv (Chemistry of Jet Fuels), AN SSSR, 1962.

229. Combustion Processes, collection of articles edited by Lewis, D., Pease, R., and Taylor, H., Fizmatgiz, translation from the English, 1961.

229A. "Solid Poropellant Rocket Engines (Survey)." Voprosy raketnoy tekhniki, 1964, No.6, 32-74.

230. Jet Engines (Russian translation), collection of articles edited by Lancaster, O.K., translation from the English, Voyenizdat, 1962.

231. Serebryakov, M.Ye., Vnutrennyaya ballistika stvol'nykh sistem i porokhovykh raket (Interior Ballistics of Barrel systems and Powder Rockets), Oborongiz, 1962.

232. Sinyarev, G.B., and Dobrovol'skiy, M.V., Zhidkostnyye raketnyye dvigateli (Liquid-Propellant Rocket Engines), Oborongiz, 1957.

233. Termodinamicheskiye svoystva individual'nykh veshchestv (Thermodynamic Properties of Specific Explosives), Handbook in two volumes, edited by V.P. Glyushko, Second edition, AN SSSR, 1962.

234. Feodos'yev, V.I., and Sinyarev, G.B., Vvedeniye v raketnuyu tekhniku (Introduction to Rocket engineering), Second edition, Oborongiz, 1961.

235. Solid-State Chemistry (Russian translation), a collective work of a group of British scientists, edited by V. Garner, IL, 1961.

236. Holmes, G., "Progress in the Field of Production of Solid Propellants", Voprosy raketnoy tekhniki, 1960, No. 6, 20-26; Stuttgart, B., "Basic Tendencies in the Development of Rocket Propellants", ibid., pp. 3-17.

237. Shidlovskiy, A.A., "Thermal Decomposition and Combustion of Ammonium Nitrate with Various Additives at Atmospheric Pressure", Izvestiya VUZov, seriya "Khimiya i khimicheskaya tekhnologiya" (News of the Higher Educational Institutions, "Chemistry and Chemical Technology" Series), 1958, 1, No. 3, 105-110., ibid., "Capability of Inorganic Ammonium Salts for Intra-Molecular Combustion", 3, No. 3, 405-407, 1960; Shidlovskiy, A.A., and Shmagin, L.F., "Thermal Decomposition and Combustion of Ammonium Perchlorate", ibid. 5, No. 4, 529-32, 1962.

238. Shol'moshi, F., and Reves, L., "Thermal Decay of Ammonium Perchlorate in the Presence of Ferric Oxide", Kinetika i kataliz, 1963, No. 1, 88-96.

239. Referativnyy zhurnal, seriya "Aviatsionnyye i raketnyye dvigateli (Journal of Abstracts, "Aviation and Rocket Engines" Series), 1963, No. 1-12.

240. Arden, E.A., Pawling, J., and Smith, W.A.W., "Investigation of the Combustion of Perchlorate", Comb. and Flame, 1962, 6, No. 1, 21-23.

241. Blackman, A.W., and Kuehl, D.K., "Use of Mixtures and Alloys of Non-Ferrous Metals as Additives to Solid Powders", ARS Journal, 1961, 31, No. 9, 1265-72.

242. Brzustowski, T.A., "On the Combustion of Metal Additives in Solid-Propellant Rocket Engines", Canad. Aeron. and Space Journal, 1963, 9, No. 5, 141-149.

243. Dekker, A.O., "Solid Powders", J. Chem. Educ., 1960, No. 11, 597-602.

244. Engel, R., "Limit of Capacity of Classic Solid Propellants", Raketentechnik u. Raumfahrtforschung, 1959, 3, Heft 1, 14-18.

245. Farber, M., "Solid Powders Based on Fluorine Compounds", Astronautics, 1960, No. 8, pp. 34, 40, 42.

246. Friedman, R., Nugent, R.G., Rumbell, R.E., and Scurlock, A., "Combustion of Ammonium Perchlorate", article in [147], 6th Symp. 612-618.

247. Gordon, S., "High-Temperature Chemistry and Rocket Propellants Based on Metals", Jet Propulsion, 28, No. 11, 1956, 769-770.

248. Higgins, H.M., and Gongwer, C.B., "Solid Rocket Propellants", American patent 2.967.097, 1961.

249. Jacobs, P.W.M., and Kureishy, A.R.T., "Thermal Initiation of a System Consisting of Ammonium Perchlorate and Cuprous Oxide", J. Chem. Soc., London, 1962, No. 2, 556-561. Also see articles in [147], 8th Symp. pp. 656-710.

250. Kit, B., Evered, D.S., Rocket Propellant Handbook, N.Y., 1960.

251. Meguire, "The Role of Solid Propellants Increases", Miss. and Rockets, 1959, 5, No. 31, 31-34, 27/VII.

252. Moutet, A., and Moutet, H., "Improvement of Hypergolic Systems Used in Rockets", French patent 76.768, 1961, and 78.084, 1962.

253. Munter, P.Z., "Investigation of the Possibility of Using Solid-Propellant Air-Breathing Ramjet Engines", JAS Paper 1961, No. 129.

254. Olson, W.T., and Setze, P.S., "Certain Problems in the Combustion of High-Efficiency Fuels for Aircraft Engines", see [147], 7th Symp. pp. 883-893.

255. Paleari, C., and Renzanigo, F., "Powder for Rockets", Journal La Rivista dei Combustibili, 1957, 11, No. 12, 841-873.

256. Partel, G., "Selection of Propellant for Jet-Propelled Torpedoes", Revista di Ingegneria, 1962, No. 11, 1199-1206.

257. Ratz, H., and Petters, W., "Rocket Propellants", article in [51], 1963, Bd. 14, 561-574.

258. Slaughter, L., "Development of a Meteorological Rocket with a Solid-Propellant Ramjet Engine", ARS-Preprint, 2343-62.

259. Tavernier, P., and Brisson, J., "Preparation and Properties of Solid Composite Propellants", Mem. poudres, 1960, 42, 305-320.

260. Taylor, J., Solid Propellant and Exothermic Composition, London, 1959.

261. White, W.D., "Lithium and Sodium for Underwater Jet (Reactive) Engines", Astronautics, 1957, No. 4, 38-30.

262. Zaehringer, A.I., Solid Propellant Rockets, 1958, Michigan.

263. Zimney, Ch.M., "An One-Stage ARP Rocket Using Solid Propellants", Jet Propulsion, 1957, 27, No. 3, 274.

264. Zwicky, F., "Reactions Used in Devices Operating Under Water", American patent 2.461.797, 1949.

265. Aviation Week, 1960, 72, No. 14, 78-91; ibid. 1957, 67, No. 15, 123.

266. Miss. and Rockets, 1960, 6, No. 30; ibid., 1959, 5, No. 41.

267. Chem. Week, 1961, 88, No. 2, 61-62.

268. U.S. Dept. Comm., Office Techn. Serv. AD 260, 1961: Chem. Abstr. 1963, 58, No. 1, 401.

Chapter 21

269. Baum, F.A., Trubochnyye porokh i distantsionnyye sostavy (Tubular Powders and Delay Compositions), Oborongiz, 1940.

270. Bubnov, P.F., and Sukhov, I.P., Sredstva initsiirovaniya (Initiating Devices), Oborongiz, 1945.

271. Maxwell, V., Pyrotechnic Whistlers, Article in Fourth Symposium on Problems of Combustion and Detonation Waves, translated from the English into Russian, Oborongiz, 1958.

272. Johnson, L.B., "Inhibiting Compositions", Ind. and Engng. Chem., 1960, 52, No. 10, 868.

273. American patent 2.988.876, 1961.

274. ARS-Preprint, 977-59.

Chapter 22

275. Aksonov, M.Ya., Vernidub, I.I., Gayvoronskiy, I.I., Kartsivadze, A.I., Plaude, M.O., Solov'yev, A.D., and Shishlintsev, V.V., "Obtaining an Ice-Forming Aerosol by Means of Pyrotechnic Compositions", Trudy Tsentr. Aerologich. Observatorii (Transactions of the Central Aerological Observatory), 1962, 44, 63. Aksenov, M.Ya., and others, "Investigation of the Ice-Forming Activity of Silver Iodide, Generated in the Combustion of Pyrotechnic Compositions", Trudy in-ta Geofiziki AN Gruz. SSR (Transactions of the Institute of Geophysics, Academy of Sciences Georgian SSR), 1962, Vol. 20.

276. Aerozoli v sel'skom khozyaystve (Aerosols in Agriculture), Collection of articles edited by A.G. Amelin, Sel'khozgiz (State Agricultural Publishing house), 1956.

277. Bystrov, G.P., Spichechnoye proizvodstvo (Match Production), Lesbumizdat (Lumber and Paper Publishing House), Moscow-Leningrad, 1950.

278. Vernidub, I.I., Zhikharev, A.S., Medaliyev, Kh.Kh., Pravdun, N.S., Sulakvelidze, G.K., anmd Chumakova, G.G., "Investigation of the Ice-Forming Properties of Lead Iodide", Izv. AN SSSR, seriya geofizicheskaya (News of the Academy of Sciences USSR, Geophysics Series), 1962, No. 9, 1286-1293.

279. Gerasimov, B.A., Osnitskaya, Ye.A., and Sidorov, A.I., "Sulfur Smoke Pots", Zashchita rasteniy ot vrediteley i bolezney (Protection of Plants Against Pests and Diseases), 1960, No.10, 34-35.

280. Zhigalov, M.L., "Investigation of Secondary Fragmentation of Ore by Briquette Thermite", Sbornik trudov Mosk. gorn. in-ta (Collection of works of Moscow Mining Institute), 1958, No. 24, 157-158; Kaplunov, R.P., and Zhigalov, M.L., "Secondary Fragmentation (Crushing) of Ore with Briquette Thermite", Bezopasnost' truda v promyshlennosti (Industrial Safety in Industry), 1958, No. 8, 26-28.

281. Zalkover, M., and Gridunov, A., Fasonnoye lit'ye iz termicheskoy stali (Shaped Casting of Thermal Steel), Izd. GUMOS DOR MVD SSSR (Publishing House of the Main Administration of Motor Transport and Highways, Ministry of Internal Affairs USSR), 1950.

282. Kiselev, I.I., "The Thermite Method of Thawing Frozen Ground", Stroitel'nays promyshlennost (Construction Industry), 1957, No. 1, 30.

283. Lapshin, V.V., and Loginov, M.N., Termitnaya svarka rel's na putyakh Moskovskogo tramvaya (Thermite Welding of Rails on the Tracks of the Moscow Trolley), Izd. MKKh RSFSR (Publishing House of the Ministry of Municipal Services RSFSR), 1951.

284. Likhachev, V.A., Pirotekhnika v kino (Pyrotechnics in Motion Pictures), Goskinoizdat (State Motion Picture Publishing House), 1944.

285. Meyson, B.D., Fizika oblakov (Physics of Clouds), Gidrometeoizdat (Hydrometeorological Publishing House), 1961. Mikandrov, V.Ya., Iskusstvennyye vozdeystviya na oblaka i tumany (Artificial Effects on Clouds and Fogs), Gidrometeoizdat, Leningrad, 1959.

286. Nikolayev, V., "New Devices Used in the Working of Frozen Ground", Voyennoinzhenernyy zhurnal (Military Engineering Journal), 1954, No. 1, 26.

287. Sidorov, A.I., "Aerosol Smoke Pots", Zashchita rasteniy ot vrediteley i bolezney, 1960, No.2, 8, 38-39.

288. Slobodkin, D.S., "The Unsuitablility of the Method of Thawing Frozen Ground by Hot Thermite", Gornyy zhurnal (Mining Journal), 1955, No. 10, 58.

289. Spravochnik spichechnika (Handbook for the Match Manufacture), Lestekhizdat (Forest Engineering Publishing House), 1947.

290. Shklovskiy, I.S., and Kurt, V.G., Iskusstvennyye sputniki zemli (Artificial Satellites of the Earth) (ISZ), 1959, No. 3, p. 66, see also the collection On the Threshold of Space (Russian translation), IL, 1960, p. 315.

291. Burchell, T., "The Aluminothermic Process and the Preparation of Commercially Pure Chromium and Manganese", article in the collection The Refining of Nonferrous Metals, London, 1950, pp. 477-505.

292. Flagg, Z.W., and Friedman, R., "Solid Powders as Sources of Cesium Plasma", Raketnaya tekhnika (Rocket Engineering), 1961, No. 1, 122.

293. Marke, D., and Lilly, C., "The Formation of Insecticide Smokes", J. Sci. Food Agr., 1951, 2, 56.

294. Schecter, W.H., Miller, A.A., and Ahers. "Chlorate Candles as a Source of Oxygen", Ind. and Engng. Chem., 1950, 42, No. 11, 2348-58.

Chapter 23

295. Bulavin, I.A., and Silenok, S.G., Mashiny dlya proizvodstva stroitel'nykh materialov (Machines for the Production of Building Materials), Mashgiz, 1959.

296. Vozhnak, Ye., Production of Pyrotechnic Devices, addition to the Polish translation of A.A. Shidlovskiy's book Osnovy pirotekhniki (Foundations of Pyrotechnics), 1957, Warsaw.

297. Vorontsov, I.I., Vspomogatel'nyye protsessy i apparatura anilokrasochnoy promyshlennosti (Auxiliary Processes and Apparatus in the Aniline-Dye Industry), Goskhimizdat, 1949.

298. Lebedev, P.D., Raschet i proyektirovaniye sushil'nykh ustanovok (Calculation and Designing of Drying Plants), Gosenergoizdat, 1963.

299. Planovskiy, A.N., and others, Protsessy i apparaty khimicheskoy tekhnologii (Processes and Apparatus in Chemical Technology), Goskhimizdat, 1962.

300. Smirnov, V.Ya., Vvedeniye v tekhnologiyu pirotekhnicheskikh proizvodstva (Introduction to the Technology of Pyrotechnics Production), Oborongiz, 1939.

Patent Applications

1. Burley, V.V., Nabokov, V.A., and Kazakova, V.I., <u>Sostav dlya izgotovleniya insektitsidnykh dymovykh shishek</u> (Composition for Preparation of Insecticide Smoke Pots), Patent application No. 100910, 1953.

2. Vernidub, I.I., and Shishmintsev, V.V., <u>Pirotekhnicheskiy sostav dlya snaryazheniya protivogradovykh raket i patronov</u> (Pyrotechnic Composition for Charging Hail-Control Rockets and Cartridges), Patent application No. 140630, 1961.

3. Vernidub, I.I., Korolev, A.I., Chumakova, G.G., and Kotova, A.G., <u>Sposob izgotovleniya pirotekhnicheskikh sostavov dlya protivogradovykh raket i patronov</u> (Method of Preparing Pyrotechnic Compositions for Hail-Control Rockets and Cartridges), Patent application No. 146127, 1962.

4. Novikov, V.A., Kazakova, V.I., Ivanova, Z.V., and Malinina, M.S., <u>Insektitsidnyy dymoobrazuyushchiy sostav</u> (Insecticide Smoke-Forming Compositions), Patent application No. 139053, 1961.

5. Novikov, V.A., Kazakova, V.I., Ivanova, Z.V., and Malinina, M.S., <u>Insektitsidnyy dymoobrazuyushchiy sostav</u> (Insecticide Smoke-Forming Compositions), Patent application No. 631302, 1962.

6. Sidorov, A.I., Gerasimov, B.A., Osnitskaya, Ye.A., and Mitrofanov, P.I., <u>Fungitsidnyy sostav dlya izgotovleniya Aerozol'nykh tabletok i shashek</u> (A Fungicide Composition for Preparing Aerosol Tablets and Smoke Pots), Patent application No. 140292, 1961.

7. Sidorov, A.I., and Mitrofanov, P.I., <u>Akaritrsidnaya aerozol'naya smes'</u> (An Acaricide Aerosol Mixture), Patent application No. 147866, 1962.

8. Sidorov, A.I., Mitrofanov, P.I., and Terent'yev, S.N., <u>Akaritsidnaya aerozol'naya smes"</u> (An Acaricide Aerosol Mixture), Patent application No. 148998, 1962.

Index of Subjects

A

ABB 125
Absorptivity 79
Absoslute black body 125
Acrylonitrile 49
Activation energy 91
Additive flame 136
Aerosol 191–95
Air
    pressure 249
    temperature 250
Aluminothermic reaction 175, 228
Aluminum 32, 33, 35, 57, 58, 60, 62, 68, 80, 84, 88, 108, 109, 111, 115, 116, 130, 132, 133, 135, 176, 199
    chloride 63, 247
    flame temperature 80
    fluoride 247
    nitrate 26
    nitride 63, 133
    oxide 32, 35, 126, 176, 247
    sub-oxide 176
    sulfate 179
    sulfide 63, 247
    properties 37
Ammonia 63
Ammonium chlorate 119
    chloride 27, 63, 117, 196
    nitrate 16, 26, 118, 119, 132, 219
    perchlorate 16, 26, 219, 220
Anthracene 63, 196
Antimony 40, 85
    pentoxide 40
    sulfide 40, 41, 88, 90, 92
    trioxide 41
Archimedes screw 44
Arsenic 40
    disulfide 41
    pentoxide 40
    sulfide 88
    trioxide 41
    trisulfide 41
Atomic spectrum 163
Auramine O 204

B

Bakelite 48
Band spectrum 163
Barium 168
    carbonate 72
    chlorate 23, 24, 26, 109, 169
    chloride 72, 168
    fluoride 169
    nitrate 22, 23, 26, 53, 62, 101, 111, 113, 115, 130, 152
    oxide 22, 25, 132, 168
    perchlorate 26, 131, 169
    peroxide 29, 117, 223
    sulfate 132
Benzene 41, 63, 174
Berthelot's principle 17
Beryllium 32, 33, 35, 60, 100, 130
    chloride 247
    floride 247
    nitrate 23
    oxide 32, 35, 126, 247
    sulfide 247
Binder 44–50
    classification 47
Black powder 95, 103, 104
Blackbody 76, 125
Boiling point 35, 41, 177–78, 184, 195, 199
Bomb calorimeter 65
Boron 32, 34, 35, 100, 130, 176
    chloride 247
    fluoride 247
    oxide 32, 35, 176, 177, 247
    sulfide 247
Brightness 125, 128, 129, 133
    temperature 77
Bromine monofluoride 186
    pentafluoride 186
    trifluoride 186
Bulk density 99
Burn rate 95, 98
Burn rate equation 104

C

Calcium 32, 33, 35, 130, 176
    chloride 247
    fluoride 247
    nitrate 23, 26
    oxide 32, 35, 176, 247
    phosphide 184
    resinate 49
    silicide 200
    sulfide 179, 247
Calorimeter, bomb 65
Candle (defined) 124
Carbon 32, 35, 40, 50
Carbon black 109

dioxide 32, 35, 40, 68, 72, 247
disulfide..................................63, 247
monoxide 68, 72
tetrachloride 63, 247
tetrafluoride 247
Catalyst 21, 98, 219
Cellulose 42
Charcoal 50, 84, 88, 101, 109, 113
Chemical bond, covalent.............................40
Chemical stability 26
Chlorate stability 21
Chlorates (production of)...........................27
Chloratite..109
Chlorine..68
monofluoride...............................186
oxidizer..58
trifluoride...................................186
Chlorosulfonic acid..................................195
Chromaticity...................................171, 208
Chromium oxide 126
(III) oxide...............................40, 177
Classification of binders............................47
Classification of pyrotechnics.....................9
Cloud seeding...230
Cobalt...40
oxide..40
Collodion..63
Colophony.....................................49, 50, 85
Color purity....................................171, 208
temperature..........................77, 148
Colored smoke composition..................201–8
Combustion.............................15–17, 95–106
defined 10–11
model 97
products, gaseous 67-70
rate 98–106, 132, 133, 151–53, 161
temperature 68, 71–82, 173, 216
Combustion-to-explosion transition................112
Commercial pyrotechnics 228–33
Compaction coefficient..............................99
Compaction of composition.......................44
Composite rocket engine...................209, 214
Continuous spectrum 77
Copper..72
oxide...................................177, 248
sulfide...248
(I) chloride...................................63
(I) thiocyanate.............................63
(II) carbonate.............................63
(II) chloride..........................63, 248
(II) fluoride..........................60, 248
(II) hydroxide..............................63
Covalent bonds...40
Cryolite..136, 165

D

Decomposition temperature..............22, 91, 98
Delay formula...249
Deliquescence...25
Density...22, 35, 41, 99
bulk...99
fuel...50
Detonation conductor 110
velocity.................................108–9
Dextrin..49
Dicyandiamide - DCDA.......................43, 50
Displacement reaction 26, 117
Drying compositions..............................241
ingredients 235
Oil..50
Dulong & Petit's Law................................72

E

Efflorescence..25
Einstein's law..138
Electromotive series..................................26
"Electron" alloy......................................183
Emissivity...126
Empirical formulas...................................50
Energy, activation....................................91
Erosion combustion................................218
Ethanol...63, 174
Ethyl ether.......................................63, 174
Exothermic reaction..................................14
Explosive properties..........................107–13

F

Ferric chloride.......................................199
Flame additive 136
sensitivity............................83, 86
spectrum....................................77
temperature........................80, 82
temperature defined 81
temperature of elements..............81
Flare composition.............................162–72
formula.................166, 168, 169, 170
Flash duration.................................152, 153
point................................41, 85, 174
Fluorine..185
oxidizer......................................60
Fog...191, 192
Friction sensitivity 83, 89–90
Fuel..15, 30–43
High energy..........................32-38
medium energy..........................40
oil...41
organic................................31, 41
requirements 31

G

Gallic acid 227
Gallium 40
    sesquioxide 40
Gaseous Combustion Products 67–70
Gasoline 41, 174
Gay-Lussac formula 67
Germanium 40
    dioxide 40
Glucose 42
Glycerin 119
GOST 235
Granulation 241
Graphite 35, 85, 88, 109
Gray body 125
Green light formula 64, 70

H

Heat capacity 65, 71, 72
Heat of combustion 18, 34, 41, 130, 132, 133, 176, 177, 205
    table 64
    decomposition 22
    explosion 62
    formation 22, 42, 61–66, 61, 176, 177, 179, 186, 219, 247
    formation, table 32, 40, 63
    fusion 73
    polymerization 49
    reaction 91
    vaporization 73
Hess' law 20, 61, 74
    test 108
Heterogeneous System 15, 96
Hexachlorobenzene 58, 198, 199
Hexachlorocyclohexane 58
Hexachloroethane 18, 58, 63, 198, 199
Hexamethylenetetramine 42
Hexamine 42, 165
Hexogen 63, 69
High energy fuel 32–38
Homogeneous system 16, 97
Hue 171, 208
Humidity, relative 120
Hydrides 33
Hydro jet engine 209, 213
Hydrogen 32, 33, 35, 68, 72
    chloride 63, 68, 247
    fluoride 247
    sulfide 247
Hygroscopicity 26, 27, 116, 120–21, 219
Hygrostat 120

I

Iditol 48, 50, 51, 53, 63, 85, 113
Ignition 95
    composition 222–27
    formula 223–26
    temperature 35, 92, 175
Illumination composition 64, 66, 69, 70, 82, 101, 123–34
    formula 134–37
Impact sensitivity 87–89
Incandescence 80
Incendiary composition 64, 66, 70, 82, 173–90
    formula 178–79
Indigo 204
Inflammation 95
Inorganic fuel 31, 40
Iodine heptafluoride 186
    pentafluoride 186
Iron 25, 40, 58, 72, 118
    disulfide 41
    scale 29
    (III) chloride 63
    (III) oxide 22, 40, 41, 177

K

Kerosene 41, 109
Kirchoff's law 79
Kopp's Law 72

L

Lactose 42, 50, 63, 84, 88, 90
Lambert-Bouguer-Beer formula 194
Lamp type 129
Lead chloride 248
    dioxide 177
    fluoride 60, 248
    nitrate 26, 63
    oxide 248
    sulfide 248
Ligroin 41
Linear burn rate 98
Linseed oil 49
Liquid petroleum 174
Lithium 32, 33, 35
    chloride 247
    fluoride 247
    oxide 32, 35, 247
    perchlorate 219, 220
    sulfide 247
Loading 243
lumen (defined) 124
Luminous efficiency 125, 128, 129, 130, 133
Luminous efficiency factor 128

flux 124
    intensity 133, 140, 155
Lux (defined) 124

M

Maganesium 176
Magnesium 23, 27, 32, 33, 35, 52, 53, 58, 60, 84, 88, 100, 113, 115, 130, 132, 133, 152, 176
    boride 187
    chloride 27, 63, 247
    flame temperature 80
    fluoride 63, 247
    nitrate 26
    nitride 63
    oxide 32, 35, 72, 126, 176, 247
    sulfide 63, 179, 247
    silicide 187
    properties 37
Magnesium-aluminum alloy, properties 38
Manganese 25, 40
    dioxide 22, 29, 177
    nitrate 26
    oxide 25, 40
    sulfide 179
Mass combustion rate 98
Match composition 231–33
    formula 232, 233
Maximum density 99
Medium energy fuel 40
Melting point 22, 23, 35, 41, 176–77, 184, 195, 199, 219
Mercury chloride 248
    fluoride 248
    oxide 248
    sulfide 248
Metal powder 39
Metal powder requirements 38
Metaldehyde 165
Methane 63
Methyl methacrylate 49
Methylene blue 204
Milling 236
Mixing 240
Molecular weight, fuel 50
    spectrum 163
Molybdenum 40
    sesquioxide 40

N

Napalm 181
Naphthalene 42, 50, 63, 88, 198
Nickel 40
    oxide 40
Nitrates (production of) 28

Nitric acid 24
Nitrogen 68, 72
Nitroglycerin 104

O

Octachloronaphthlene 199
Optical pyrometer 77–80
Organic fuel 31, 41
Oxidation-reduction 11
Oxide coating 36
    film 115
    to metal volume ratio, table 36
Oxidizer 14, 18–28
    availability coefficient 56
    hygroscopicity 25
    requirements 27
    properties of 18-20
    properties of, table 22
    types of 19
Oxygen 72
    balance 55, 56
    candle 230

P

Paraffin 42, 50, 88
Paranitroaniline 204
Pentachlorobenzene 199
Petroleum ether 41
    liquid 174
Phosphine 185
Phosphorus 32, 92, 119, 183
    pentasulfide 248
    pentoxide 32, 41, 248
    sesquisulfide 41
    sulfide 184
    trichloride 248
    s trifluoride 248
Photocell 139
Photoflash composition 64, 66, 70, 82, 146–57
    formula 154
    requirements 151
Photometric chamber 143
Pilling and Bedworth number 100
Planck's radiation law 76
Polysulfide rubber 221
Polytetrafluoroethylene (teflon) 18, 60
Polyurethane 221
Polyvinyl chloride 58, 165
Porosity 99
Potassium 40, 186–87
Potassium chlorate 21, 22, 26, 27, 68, 84, 88, 91, 92, 101, 108, 110, 117, 119, 170
    chloride 25, 27, 72
    ferricyanide 90

ferrocyanide 88, 90
nitrate .22, 23, 26, 53, 84, 88, 101, 110, 152, 219
oxide 40
perchlorate 21, 22, 26, 52, 84, 88, 110, 111, 113, 219
permanganate 117, 119, 152
picrate 227
thiocyanate 88, 90
Pressing 46, 241
Pressure, air 250
Propagation 95
Propellant 209–21
formula 216
Properties, aluminum 37
Pyrometer, optical 77–80
Pyrophoric 187
Pyrotechnic composition 11, 96
requirements of 13
Pyrotechnics
classification of 9
defined 9
Pyroxylin 63

R

Ramjet engine 209, 210
Rate constant 100
Rayleigh's formula 193
Realgar 88
Red light composition 64, 70, 74
Relative humidity 120
Resin 48
Rhodamine B 204
Rocket composition 64, 66, 70, 82
Rocket engine 209, 215
Rocket-ramjet engine 209
Russian history 245

S

Screening 238
Screw conveyer 44
Self-ignition temperature 83–85
temperature (defined) 84
Sensitivity 83–93
flame 83, 86
tests 83
factors affecting 91–94
friction 83, 89–90
shock 83
Shellac 49, 50, 63, 85, 101
Shock sensitivity 83
Silicon 32, 34, 100, 130, 176
chloride 248
dioxide 32, 35, 176, 177, 248
disulfide 179

fluoride 248
sulfide 248
tetrachloride 195
Silver fluoride 60
perchlorate 17
Smoke 191, 192
composition 64, 66, 70, 75, 82, 101
dye 203, 204
formula 33, 197–200, 202, 203, 205–8
insecticide formula 229, 230
screen composition 191
Smoke-gas zone 98
Sodium 40, 186–87
carbonate 63
chlorate 23, 26
chloride 25, 27, 72, 248
fluoraluminate 63
fluoride 63, 136, 165, 248
fluosilicate 63, 165
hydrogen carbonate 63
nitrate 21, 22, 26, 101, 116, 131–34, 219
nitrite 63
oxalate 50, 63, 165
oxide 40, 248
perchlorate 22, 24, 26, 131
sulfate 63
sulfide 248
Solubility 26, 219
oxidizer (table) 26
Specific light sum 127, 132, 133, 161
luminous intensity 154
thrust 210, 213, 216
volume 67, 68, 70
Spectral sensitivity 138
Spectrum line and band 163
flame 77
Stability 114–22
chemical 26
chlorate 21
thermal 21
Standard electrode potentials 26
Starch 42, 50, 63
Stearic acid 42, 181
Stearin 42, 50, 181
Stefan-Boltzmann law 76
Stoletov's law 138
Storage period 122
Strength of pressed composition 46
Strontium carbonate 74, 167
chlorate 26, 167
chloride 63, 167
fluoride 167
monochloride 166
Potassium nitrate 22, 23, 26, 57, 152

nitride 63
oxalate 50, 63, 167
oxide 25, 63, 166
sulfate 63, 167
Styrene 49
Sudan I 204
  red 204
Sugar 42, 101, 109
Sulfur 40, 50, 84, 88, 90, 92, 101, 109, 116, 119, 135, 170
  dioxide 40, 68
  trioxide 195

T

Temperature, air 250
  boiling 35, 41, 177-178, 184, 195, 199
  coefficient 103
  decomposition 22, 91, 98
  flame 82
  melting .. 22, 23, 35, 41, 176-177, 184, 195, 199, 219
Tetranitromethane (TNM) 24
Tetryl 69
Thermal head 188
  stability 21
Thermite 14, 23, 117, 175-79, 189
  composition 64, 70, 101
  formula 176-79, 229
  temperature 80
Thermocouple 81
Thorium dioxide 126
  nitrate 152
Threshold illumination 158
Thrust coefficient 210
Tin tetrachloride 195
Titanium 32, 33, 35, 130, 176
  dioxide 32, 176, 247
  flame temperature 80
  oxide 35
  sulfide 247
  tetrachloride 195, 247
  tetrafluoride 247
TNT 109, 113
Total darkness 194

Tracer composition 64, 66, 70, 82, 101, 156–57
  formula 161
Transmissivity 79
Transparency coefficient 158
Trauzl block test 108, 112–13, 200
Trotyl 63, 69
Trouton's formula 73
TU 235
Turpentine 41
Tyndall effect 193

U

Urotropin 42, 50, 63, 165

V

Vinyl acetate 49
Vinyledene chloride 49

W

Water 32, 35, 72, 247
  vapor 68
Wien's displacement law 76
Wood 173
  meal 42, 109

X

Xenon tetrafluoride 60

Y

Yield point 47

Z

Zinc 40, 58, 198
  chloride 63, 197, 248
  fluoride 248
  oxide 40, 199, 248
  sulfide 248
Zirconium 32, 33, 35, 58, 60, 100, 130
  chloride 247
  flame temperature 80
  fluoride 247
  oxide 32, 35, 247
  sulfide 247

Notes

Notes